REACTION ENGINEERING PRINCIPLES

REACTION ENGINEERING PRINCIPLES

Himadri Roy Ghatak

CRC Press
Taylor & Francis Group
Boca Raton London New York

CRC Press is an imprint of the
Taylor & Francis Group, an **informa** business

CRC Press
Taylor & Francis Group
6000 Broken Sound Parkway NW, Suite 300
Boca Raton, FL 33487-2742

Version Date: 20151111

International Standard Book Number-13: 978-1-4987-5856-7 (Hardback)

Library of Congress Cataloging-in-Publication Data

Names: Ghatak, Himadri Roy.
Title: Reaction engineering principles / Himadri Roy Ghatak.
Description: Boca Raton : Taylor & Francis, 2016. | "A CRC title." | Includes bibliographical references and index.
Identifiers: LCCN 2015038787 | ISBN 9781498758567 (alk. paper)
Subjects: LCSH: Chemical engineering--Mathematics. | Chemical reactions.
Classification: LCC TP155.7 .G53 2016 | DDC 660/.2--dc23
LC record available at http://lccn.loc.gov/2015038787

Visit the Taylor & Francis Web site at
http://www.taylorandfrancis.com

and the CRC Press Web site at
http://www.crcpress.com

Contents

Preface

The subject of chemical reaction engineering is undoubtedly at the heart of any chemical engineering educational program may it be undergraduate or postgraduate. The teaching scheme of the majority of the undergraduate programs include two courses of reaction engineering, often suffixed as I and II, and mostly offered in the fifth and sixth semester of an eight-semester module. The postgraduate programs generally have one course of reaction engineering. The subject is fascinating and intimidating at the same time as I have discovered from my experience of teaching the course, and also from students' feedback. Fascinating because it makes it possible for the benefits of chemistry—one of the fundamental pillars of science—to be taken out from the confines of laboratories and delivered to the society at large. It is intimidating because it requires a heavy dose of mathematics to take care of the quantitative nature of the subject. This mathematics, unless suitably explained in the context of the physical phenomenon, clouds rather than enlightens the student.

This book has been written primarily from a student's perspective. The basic presumption is that a student is to be taught rather than lectured. He or she has an inquisitive mind but only the minimal understanding of the basics of scientific principles. Therefore, the phrases like "as we already know" have been consciously avoided as much as possible. This is all the more important with quantitative analyses that require mathematical treatise. Mathematics is central and indispensable to chemical reaction engineering. Nevertheless, every mathematical formula or equation has a rationale behind it that should be clear to the student. It should not appear to him or her as having dropped from the sky. Therefore, the book attempts to provide elaborate detail of the physical phenomenon/model before the ideas are translated into equations and formulae. Another important and, in my opinion, distinct feature of the book is the detailed derivations that it provides to the equations and formulae, including intermediate steps that may look trivial. However, these trivial intermediate steps make mathematics more comprehensible and less intimidating. Again, the focus is on the student.

For the students, I presume that by the time they venture into the subject of chemical reaction engineering, they would have the basic understanding of the following subjects:

- Physical chemistry
- Engineering mathematics
- Material and energy balance
- Chemical engineering thermodynamics
- Mass transfer

As you move through the book, you will come across topics from the above subjects every now and then. Wherever necessary, these topics have been deliberated upon in required detail. This book is divided into nine chapters to gradually build the subject. Chapter 1 is introductory. Chapters 2 through 4 provide the scientific basis for the engineering that would subsequently follow. Among these Chapter 2 deals with stoichiometry. In this chapter as also in the entire book, the emphasis has been made to identify and do the computations in terms of the limiting reactant. I hope this will help the students in problem solving. In the initial sections of Chapter 3, homogeneous reactions have been discussed. Once the students fully understand these concepts and practice their application in problem solving, they can step into the realm of heterogeneous reaction kinetics that is covered in the later part of Chapter 3. Chapter 4 covers the topics of chemical engineering thermodynamics, which find frequent application in dealing with nonisothermal reactor operations, the subject matter for Chapter 7. In between, Chapters 5 and 6 deliberate on the design and performance of ideal isothermal reactors, batch and flow types, respectively. The concepts developed in Chapters 4 through 6 are then utilized in Chapter 7 for the design and performance evaluation of ideal reactors that operate under nonisothermal conditions. Flow reactors that deviate from ideal flow patterns are covered in Chapter 8. Chapter 9 is devoted to some specific reactor types that we frequently encounter in the industry. One of the important inclusions in this chapter is Section 9.1. Reading of Chapters 7 through 9 is advised only after the students are well versed in the application of the ideas and mathematical treatments presented in Chapters 3 through 6.

I have made every possible attempt to make the presentations in the book comprehensible and useful for the students. A sequence of problem solving or application steps, in the form of an algorithm, is provided after each major section and mathematical formulae/equation. These are followed by worked examples in the form of EXAMPLES. Numerous such illustrations are to be found in the entire book. Similarly, pictorial and graphical presentations have been liberally used to facilitate the interpretation of the physical phenomenon/model and mathematical equations/formulae. Exercises at the end of the chapters provide the opportunity for hands-on practice and to sharpen the problem-solving skills of the students.

I am thankful to the publishers for providing an encouraging, interactive, and congenial publishing environment. But for them, the book, in its final form could not have seen the light of day. As with all human endeavors, this book may have shortcomings and scope for improvement. Constructive criticisms and suggestions for betterment are most welcome.

H. R. Ghatak

Author

Himadri Roy Ghatak earned his BE and ME degrees from the erstwhile University of Roorkee (now Indian Institute of Technology Roorkee) and his PhD from Sant Longowal Institute of Engineering and Technology. He worked in the industry from 1991 to 1995 before joining the faculty of Sant Longowal Institute of Engineering and Technology, where he rose to the position of a professor in chemical engineering. He also served as the head of the Department of Chemical Engineering, and associate dean (research and consultancy). He has taught courses in chemical reaction engineering, mass transfer, environmental engineering, and energy technology. Professor Ghatak's research focuses on electrochemical advanced oxidation, hydrogen production, and lignin.

Notations Used in the Book

SYMBOLS

A	Reactant chemical species
A	Atomic weight
AL	Hinterland ratio
Ar	Area
$A_1 \cdot S_v$	A_1 adsorbed on the active site
a	Stoichiometric coefficient of reactant chemical species (in Chapter 2); area per unit volume of catalyst bed (in Chapter 9)
a	Activity of substance
a′	Area per unit mass of catalyst
C	Concentration
C_p	Molar specific heat capacity at constant pressure
D	Diffusivity
E	Residence time distribution function
E(θ)	Residence time distribution as a function of dimensionless time
E	Enhancement factor (in Chapter 3); electrode potential (in Chapter 9)
E_∞	Maximum possible enhancement factor for the instantaneous reaction
E_a	Activation energy
E_{aD}	Activation energy for diffusivity
F	Molar flow rate; fraction of material leaving the reactor from start to a specified time (in Chapter 8)
F	Faraday constant
f	Friction factor
fr	Fraction of material leaving the reactor in differential time
fr*	Limiting solids fraction
f, f'	Function
G	Gibbs free energy
H	Henry's law constant
H	Enthalpy
Ha	Hatta modulus
ΔH_r	Heat of reaction
$(\Delta H_r)_{LR}\big\|_{T_0}$	Heat of reaction per mole limiting reactant at temperature T_0
I	Electrical current
i	Current density
J	Molar flux due to diffusion
K	Equilibrium constant
K_m	Michaelis–Menten constant
k	Specific reaction rate constant based on unit volume

k_{cat}	Specific reaction rate constant based on unit volume of the catalyst
k'	Specific reaction rate constant based on unit area
k_0	Frequency factor
$(k_0)_D$	Frequency factor for diffusivity
k	Mass transfer coefficient
k_v	Volume-based mass transfer coefficient
L	Distance from the reactor inlet along the longitudinal reactor axis
L_R	Length of the reactor
M	Molecular weight
m	Mass
m	Sintering order
N	Moles; number of reactors (in Chapters 6 and 8)
N	Whole number to represent ratio
n	Order of reaction
n	Number of electrons participating in the electrochemical half-cell reaction
n_{per}	Number of perforations
P	Product chemical species
$P_1 \cdot S_v$	P_1 adsorbed on the active site
p	Stoichiometric coefficient of product chemical species
p	Pressure
Q	Heat added to the reacting system
q	Heat added to the reacting system per mole of the limiting reactant converted
q	Electrical charge
$\text{quantity}\vert_{\text{parameter}}$	Value of a quantity at the given value of the parameter
R	Universal gas constant; the radius of a particle
r	Rate of reaction based on unit volume; radius at a general position
r′	Rate of reaction based on unit area
Re	Reynolds number
S	Entropy
Sc	Schmidt number
Sh	Sherwood number
T	Absolute temperature
Th	Thiele modulus
t	Time
U	Internal energy
u	Velocity
V	Volume
V_{max}	Maximum reaction rate in Michaelis–Menten mechanism
v	Volumetric flow rate
v	Volume fraction
W	Mechanical work done on the reacting system

w	Mechanical work done on the reacting system per mole of the limiting reactant converted
X	Conversion
x	Mole fraction; distance
Y	Yield coefficient

GREEK SYMBOLS

α	Parameter used in the equation; the transfer coefficient
Δ	Difference
δ	Film thickness in two-film theory; thickness of the diffusion film
ε	Fractional change in moles per unit volume at complete conversion
ϵ	Bed voidage
η	Contacting efficiency
η	Overpotential
Θ	Sphericity of the particle
θ	Dimensionless time
κ'	Catalyst susceptibility
λ	Dimensionless length
μ	Viscosity, chemical potential
μ_{max}	Maximum specific growth rate
ξ	Effectiveness factor
ρ	Mass or molar density
σ^2	Variance of residence time distribution
τ	Space time
Φ	Selectivity
φ	Dimensionless concentration
χ	Catalyst activity
Ψ	Fractional yield

SUBSCRIPTS

1, 2, 3,.i,.j,..n,..l	Numbers denoting the identity of a chemical species or other parameter
0	Initial condition
(−)	On the immediate left-hand side
(+)	On the immediate right-hand side
1/2	Half-life
I, II	Identity of fluid phases
i	Interface between phases; ith entry of a particular quantity
a	Activity based
act	Actual
an	Anode
ad	Adsorption step

b	Bubble
BFR	Backmix flow reactor
bwd	Backward reaction
BR	Batch reactor
b0	Bubble initial
bm	Bubble maximum
br	Bubble rise
b → c	Bubble to cloud
c	Cloud
cat	Cathode
cell	Electrochemical cell
cen	Center
coke	Coke formed on the catalyst
cr	Central core in fast fluidization
C	Concentration based
CS	Core surface
c → b	Cloud to bubble
c → em	Cloud to emulsion
ES	External surface
d	Dense phase
dth	Microorganism death
des	Desorption step
el	Electrode
em	Emulsion
emg	Emulsion-phase gas
ems	Emulsion-phase solids
eq	Equilibrium
ex	Exchange; reactor exit
em → c	Emulsion to cloud
f	Final condition; formation
FB	Fluidized bed
FBd	Fluidized bed lower dense phase
FBl	Fluidized bed upper lean phase
fl	Fluid phase
fwd	Forward reaction
gen	Generalized
hyp	Hypothetical
in	At reactor inlet
inst	Instantaneous
L	Length
LR	Limiting reactant
l	Lean phase
lim	Limiting
M	Microorganism
mf	Minimum fluidization
max	Maximum, for maximum value

mean	Weighted average
N	Number of reactors
no MT res	No mass transfer resistance
OS	Outer surface
obs	Observed
out	At reactor outlet
ovl	Overall
PB	Packed bed
PFR	Plug flow reactor
p	Particle
p	Pressure based
posn	Catalyst poison
R	Reactor cross section
r	Reaction
real	For a real reactor
rxn	Reaction step
S	Substrate; superficial
sint	Catalyst deactivation by sintering
S_{coke}	Catalyst active sites covered by coke
S_{posn}	Catalyst active sites covered by catalyst poison
S_t	Total active sites on the catalyst surface
S_v	Vacant active site on the catalyst surface
S.E	Substrate–enzyme complex
t	Time; terminal settling
tr	Tracer
w	Wake
wl	Wall region in fast fluidization

SUPERSCRIPT

0	Standard state

1 Introduction

Chemical process industries convert natural raw materials into products that are of higher economic value than the starting material. This is carried out in a sequence of steps involving physical and chemical transformations, the former being known as unit operations and the latter as unit processes. Chemical reactions are thus part and parcel of chemical process industries and often the central idea of such an industrial establishment. It is, therefore, of paramount importance to carry out the reactions in an industry under optimal conditions. The "engineering" approach to the problem at hand requires processing of a diverse set of information to find the optimal solution and forms the basis of the subject "chemical reaction engineering." Figure 1.1 provides the schematic representation of the multidisciplinary nature of the subject. Table 1.1 elaborates some of the important questions to be answered in "chemical reaction engineering" and the information required to arrive at those answers.

Referring to Table 1.1 and Figure 1.1, the subject "chemical reaction engineering" builds on the scientific concepts developed and elaborated in the fields of physical chemistry in combination with the engineering concepts of mechanics of fluids and particles, heat and mass transport processes, and the conservation laws. The following sections briefly introduce the reader to these ideas in relation to chemical reaction engineering before they are elaborately dealt with in the subsequent chapters of the book.

1.1 CONCEPTS FROM PHYSICAL CHEMISTRY

Physical chemistry, a term first coined by Mikhail Lomonosov in the eighteenth century, is the science that understands chemical systems in terms of the principles of physics such as equilibrium, velocity, and energy, to name a few. It provides insight into phenomena such as

- Intermolecular and intramolecular forces
- Rate of transformation of one chemical species into another (chemical kinetics)
- Energy transformations and their viability (thermodynamics and reaction equilibrium)
- Behavior of different phases when brought together (phase equilibrium)
- Interaction of matter with energy that leads to chemical transformation

Of these, the subdisciplines chemical kinetics, thermodynamics, and reaction equilibrium are of particular interest.

1.1.1 Chemical Kinetics

A chemical reaction is the phenomenon that transforms one chemical species into another. A chemical species is identified as a specific grouping of atoms joined

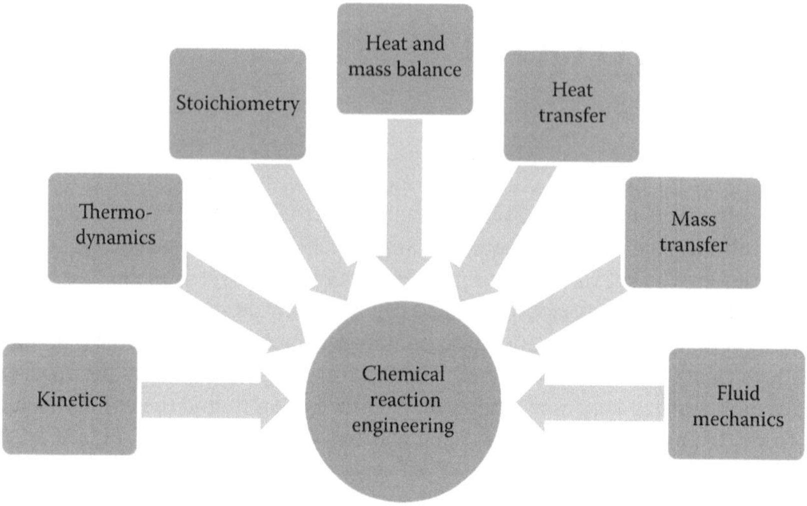

FIGURE 1.1 Multidisciplinary approach to chemical reaction engineering.

together through interatomic forces or bonds, as they are called. Thus, a chemical reaction necessarily involves a rearrangement of atoms, that is, breaking of some of the existing chemical bonds and formation of some new ones. Chemical kinetics deals with the rapidness with which a chemical reaction can be carried out and the factors affecting this. It is common knowledge that not all the reactions proceed with the same rapidness. A solution of sodium hydroxide almost instantly reacts when brought in contact with a solution of hydrochloric acid. On the other hand, a broth of malt brews into alcohol over several days. It is obvious that if the changes are rapid, the desired outcome will be achieved in a short span of time and vice versa. The product quality and the rate of production from a chemical reaction are, therefore, interlinked and influenced by chemical kinetics. It will have a direct bearing on the profitability of the industry. The key factors affecting kinetics are temperature and concentration of reacting species. Another important factor is the presence of catalysts. This knowledge is essential for the engineer to design and operate reacting systems so that the chemical transformations are made to proceed to the desired extent in the least possible time.

1.1.2 Thermodynamics

Breaking of chemical bonds requires energy to be spent while their formation releases energy. As outlined above, a chemical reaction involves breaking of some of the existing chemical bonds and formation of some new ones. If the net energy required for breaking the bonds exceeds that released in the formation of new ones, the process will be energetically unfavorable. The reverse is true if the energy released in the formation of new bonds exceeds that required for bond breaking. In addition, the reacting system may do mechanical work on the surroundings or the surroundings may do work on the reacting system. Therefore, a reaction may proceed with

TABLE 1.1
The Knowledge Base for Chemical Reaction Engineering

Questions to be Answered		Information Needed
Related to design	Which reactor type is to be used?	• Reaction type; homogeneous or heterogeneous • Amount of material to be processed • Temperature to be encountered; isothermal or nonisothermal • Rheological and thermodynamic properties of reacting fluids • Flow regimes in different types of reactors • Economics

	What size reactor is to be used?	• Reactor type
		• Reaction rate, reaction stoichiometry, and reaction equilibrium
		• Amount of reactant(s) to be processed
		• Amount of product required
		• Thermodynamic parameters
		  • Temperature; isothermal or nonisothermal
		  • Pressure; constant or varying
		  • Enthalpy changes
		• Rheological and thermodynamic properties of reacting fluids
		• Posttreatments required

Related to performance	How much product can be obtained?	• Reactor type and size
		• Reaction rate, reaction stoichiometry, and reaction equilibrium
		• Amount of reactant(s) to be processed
		• Thermodynamic parameters
		  • Temperature; isothermal or nonisothermal
		  • Pressure; constant or varying
		  • Enthalpy changes
		• Rheological and thermodynamic properties of reacting fluids
		• Posttreatments required

	What is the composition of the product stream?	• Reactor type and size
		• Reaction rate, reaction stoichiometry, and reaction equilibrium
		• Amount of reactant(s) to be processed
		• Thermodynamic parameters
		  • Temperature; isothermal or nonisothermal
		  • Pressure; constant or varying
		  • Enthalpy changes
		• Rheological and thermodynamic properties of reacting fluids

increase or decrease of enthalpy; that is, it may be exothermic or endothermic. We are aware of the heat released when a lump of coal is burnt or a piece of quick lime is dropped into water. This will impact the temperature prevailing during the reaction, which in turn will influence the kinetics. This is vital information needed by an engineer to design and operate a reacting system.

1.1.3 Reaction Equilibrium

Thermodynamics not only tells us about the energy transformations during a process but also the direction of these energy transformations. Energy transformations, on their own, proceed in the direction of increasing entropy (increasing randomness). Once the entropy has attained its maximum, no more energy transformations are possible. This is the state of equilibrium. The more shifted is the system from its state of equilibrium, the more is its tendency to move toward it. Just as it does for physical systems, this restriction applies equally well for reacting systems. For a reacting system, this is the maximum extent to which the starting chemical species can be transformed into new ones. Irrespective of the efforts made, some untransformed starting chemical species will be left, which cannot be transformed further. This is reaction equilibrium, which puts a theoretical limitation on the maximum extent to which a reaction can be carried out. This sets the constraint for the engineer while designing and operating the reacting system.

1.2 CONCEPTS FROM FLUID AND PARTICLE MECHANICS

Though not always, a reacting system may contain liquids and gases under conditions of flow. For a fluid to flow, the driving force is the pressure drop across the flow path. Maintaining a pressure drop requires some mechanical energy to be spent. The energy requirement depends on several parameters, namely

- The flow regime (laminar or turbulent), most commonly characterized by Reynolds number
- Density of the fluid
- The geometry of the flow path, which includes the area of cross section (constant or varying) and the path length
- Rheological properties of the fluid like viscosity, and deviation from Newtonian behavior

In addition, one may often come across situations where the fluid flows through an assembly of solid particles in the form of a packed or fluidized bed. This is frequently the case with solid-catalyzed reactions, a type of reaction with immense commercial value. Here, the flow path does not conform to any regular shape. Instead, the fluid has to traverse through a number of tortuous, irregular-shaped, narrow channels. This alters the pressure drop requirement, hence the mechanical energy requirement, compared to a flow path sans particles. Besides, for gaseous species, pressure, temperature, and volume are related thereby interlinking the fluid and particle mechanics with kinetics and thermodynamics. All this needs to be given due consideration by the engineer in the design and operation of the reacting system.

1.3 CONCEPTS FROM HEAT AND MASS TRANSPORT PROCESSES

1.3.1 Heat Transport

As outlined in Sections 1.1.1 through 1.1.3, reactions are associated with energy transformations and temperature plays a critical role in deciding the outcome of any chemical reaction. In order to maintain the desired temperature conditions in the reacting system, it may be desired to supply heat into it or to take away heat from it. This is to be done within the timeframe of occurrence of the chemical reaction. Thus, the rate of heat transfer must complement the chemical kinetics. Besides other parameters, the heat-transfer rate depends on the temperature difference across which the heat is to be transferred. In solids, the heat is transferred by conduction and the heat-transfer rate is governed by Fourier's law of heat conduction. Here, the thermal conductivity of the solid and the geometrical configuration are the deciding factors. Even with porous solids, Fourier's law can be applied by using an effective conductivity. In fluids, convection is the dominant mode of heat transfer, which is intimately linked to flow of fluids. Unlike conduction, convection involves more randomness. It is, therefore, less conducive to rigorous mathematical treatment and frequently described with empirical and semiempirical correlations. Factors that affect the rate of convective heat transfer in fluids include

- The flow regime (laminar or turbulent), most commonly characterized by Reynolds number
- Density of the fluid
- The geometry of the flow path, which includes the area of cross section (constant or varying)
- The area across which the heat is transferring
- Rheological properties of the fluid like viscosity, and deviation from Newtonian behavior
- Thermal characteristics of the fluid like specific heat and thermal conductivity

The engineer must be conversant with these ideas and appropriately use them for the design and operation of the reacting system.

1.3.2 Mass Transport

The concept of mass transport comes into play when the reacting system consists of two (or more) phases. In such heterogeneous systems, the site of the reaction does not extend to the entire system but is confined to specific locations. In solid–fluid systems, this may be the interface between the solid and the fluid. When a coal particle burns in air, the reaction takes place at the surface of the coal particle that is exposed to air. In fluid–fluid system, the reaction may be confined to one of the fluid phases, and also within specific zones. Reactive absorption is a good example of such systems. In such cases, the rapidness with which the chemical transformation takes place can be no more and no less than the rapidness with which the reacting chemical species can reach the reaction site. While the former is governed by chemical kinetics, the latter is governed by the rate of mass transport. None of the processes can be understood in isolation from the other. However, there can be instances where the outcome of the entire process is influenced more by (limited by) one of the two processes.

The phenomenon of mass transport and its rate is most conveniently treated in analogy with heat transfer. The driving force for mass transport is concentration difference (more appropriately the difference in chemical potential). Factors that govern the rate of mass transfer include

- The area of interface between the two phases
- The flow regime (laminar or turbulent) in the fluid phases, most commonly characterized by Reynolds number
- Density of the fluid
- Diffusivity of the chemical species
- The geometry of the flow path
- Rheological properties of the fluid like viscosity, and deviation from Newtonian behavior
- Equilibrium behavior between the phases

It is imperative on the engineer to appropriately apply these principles when designing and operating the heterogeneous-reacting system.

1.4 CONSERVATION LAWS

Irrespective of the predominant phenomena in operation and their governing principles, a reacting system, at all times, must conform to the Law of Conservation of Mass and the Law of Conservation of Energy. For a closed-reacting system, the Law of Conservation of Mass can be expressed as

Total initial mass = Total final mass

and for a flow system on the basis of unit time

Mass entering the system = Mass going out of the system + Mass accumulating within the system

The Law of Conservation of Energy for a closed system under adiabatic conditions takes the form

Energy released/absorbed in reaction = Increase/decrease in enthalpy of reacting system

and for a flow system under adiabatic conditions on the basis of unit time

Enthalpy of the entering material + Energy released/absorbed in reaction = Enthalpy of the leaving material + Increase/decrease in enthalpy of reacting system

2 Stoichiometry

As noted in Chapter 1, a chemical reaction is the phenomenon that transforms one chemical species into another. The reaction will start with one or more types of chemical species. These are known as the reactants. As the reaction proceeds, the reactants will deplete in mass. In their place, one or more types of new chemical species will be formed. These are known as products. The mass of the products go on increasing with the progress of the reaction. Stoichiometry is the discipline of science that allows us to calculate the mass of any chemical species (reactant or product) in the reacting system in relation to the other species present (IUPAC Gold Book, 2014). It is founded in two fundamental laws of chemistry, namely the Law of Definite Proportions and the Law of Multiple Proportions, along with the Law of Conservation of Mass.

2.1 LAW OF DEFINITE PROPORTIONS

First put forward by the French chemist Joseph Proust at the end of the eighteenth century (Proust, 1799), the law states, "a chemical compound always contains exactly the same proportion of elements by mass." Suppose a compound Z is made up of "n" elements, $E_1, E_2, E_3, \ldots E_n$. A sample of this compound with a mass m_{1_Z} kg has $m_{1_{E1}}$ kg of the element E_1, $m_{1_{E2}}$ kg of the element E_2, $m_{1_{E3}}$ kg of the element E_3, and so on. Another sample of this compound with a mass m_{2_Z} kg has $m_{2_{E1}}$ kg of the element E_1, $m_{2_{E2}}$ kg of the element E_2, $m_{2_{E3}}$ kg of the element E_3, and so on. Then, from the Law of Definite Proportions we have

$$\frac{m_{1_{E1}}}{m_{1_Z}} = \frac{m_{2_{E1}}}{m_{2_Z}}, \frac{m_{1_{E2}}}{m_{1_Z}} = \frac{m_{2_{E2}}}{m_{2_Z}}, \frac{m_{1_{E3}}}{m_{1_Z}} = \frac{m_{2_{E3}}}{m_{2_Z}} \ldots \frac{m_{1_{En}}}{m_{1_Z}} = \frac{m_{2_{En}}}{m_{2_Z}} \tag{2.1}$$

and this will be true for any other sample of the compound Z. If we have other samples of the compound Z with masses $m_{3_Z}, m_{4_Z}, m_{5_Z}$, etc., with masses of element E_1 being $m_{3_{E1}}, m_{4_{E1}}, m_{5_{E1}}$, etc., then for the element E_1 we will have

$$\frac{m_{1_{E1}}}{m_{1_Z}} = \frac{m_{2_{E1}}}{m_{2_Z}} = \frac{m_{3_{E1}}}{m_{3_Z}} = \frac{m_{4_{E1}}}{m_{4_Z}} = \frac{m_{5_{E1}}}{m_{5_Z}} \text{ and so on} \tag{2.2}$$

Similar relationships will also hold good for the elements $E_2, E_3, \ldots E_n$. Nonstoichiometric compounds are exceptions to this law.

EXAMPLE 2.1

77 g ferrous sulfide contains 49 g iron and 28 g sulfur. In an experiment, 3.92 g iron was dissolved in hydrochloric acid and excess of hydrogen sulfide was passed

through the solution. What is the amount of ferrous sulfide produced and how much sulfur does it contain?

SOLUTION

Mass of sample 1 of ferrous sulfide $m_{1_{FeS}} = 77$ g
Mass of iron in sample 1 of ferrous sulfide $m_{1_{Fe}} = 49$ g
Mass of iron in sample 2 of ferrous sulfide $m_{2_{Fe}} = 3.92$ g
Mass of sulfur in sample 1 of ferrous sulfide $m_{1_S} = 28$ g

Let the mass of sample 2 of ferrous sulfide produced in the experiment be $m_{2_{FeS}}$ and the mass of sulfur in it be m_{2S}. From Equation 2.1 for the first element, iron

$$\frac{m_{1_{Fe}}}{m_{1_{FeS}}} = \frac{m_{2_{Fe}}}{m_{2_{FeS}}}$$

$$\Rightarrow \frac{49}{77} = \frac{3.92}{m_{2_{FeS}}}, \quad \Rightarrow m_{2_{FeS}} = 6.16 \text{ g}$$

Again from Equation 2.1 for the second element, sulfur

$$\frac{m_{1_S}}{m_{1_{FeS}}} = \frac{m_{2_S}}{M_2}$$

$$\Rightarrow \frac{28}{77} = \frac{m_{2_S}}{6.16}, \quad \Rightarrow m_{2_S} = 2.24 \text{ g}$$

The law can also be expressed in terms of the moles of elements E_1, E_2, etc., and the moles of the compound Z. From Equation 2.1 for element E_1,

$$\frac{m_{1_{E1}}}{m_{1_Z}} = \frac{m_{2_{E1}}}{m_{2_Z}}$$

Dividing the numerator on both sides by the atomic weight of element E_1, we have

$$\frac{(m_{1_{E1}})/A_{E1}}{m_{1_Z}} = \frac{(m_{2_{E1}})/A_{E1}}{m_{2_Z}} \Rightarrow \frac{N_{1_{E1}}}{m_{1_Z}} = \frac{N_{2_{E1}}}{m_{2_Z}} \tag{2.3}$$

Further, dividing the denominator on both sides by the molecular weight of Z, we have

$$\frac{N_{1_{E1}}}{(m_{1_Z})/M_Z} = \frac{N_{2_{E1}}}{(m_{2_Z})/M_Z} \Rightarrow \frac{N_{1_{E1}}}{N_{1_Z}} = \frac{N_{2_{E1}}}{N_{2_Z}} \tag{2.4}$$

2.2 LAW OF MULTIPLE PROPORTIONS

Framed by the famous British chemist John Dalton in the early nineteenth century (Dalton, 1808), this law states, "If two elements form more than one compound between them, then the ratios of the masses of the second element which combine with a fixed mass of the first element will be ratios of small whole numbers." Suppose elements E_1 and E_2 form n different compounds, Z_1, Z_2, Z_3,...Z_n, between them. For m_{E1} kg of the element E_1 compound Z_1 has $m_{1_{E2}}$ kg of element E_2, compound Z_2 has $m_{2_{E2}}$ kg of element E_2, compound Z_3 has $m_{3_{E2}}$ kg of element E_2, and so on. Then, from the Law of Multiple Proportions we have

$$m_{1_{E2}} : m_{2_{E2}} : m_{3_{E2}} : \ldots : m_{n_{E2}} :: \mathbf{N}_1 : \mathbf{N}_2 : \mathbf{N}_3 : \ldots : \mathbf{N}_n \quad (2.5)$$

where $N_1 : N_2 : N_3 : \ldots : N_n$ are small whole numbers. In addition, these numbers are constant for a group of compounds comprising the elements E_1 and E_2. Nonstoichiometric compounds do not follow this law as well. Further, for some high molecular weight compounds, the whole numbers in the ratios of Equation 2.5 may not be small. One such example is the higher members of the homologous series of saturated hydrocarbons. Nevertheless, they are always whole numbers.

In terms of moles, Equation 2.5 can be written as

$$N_{1_{E2}} : N_{2_{E2}} : N_{3_{E2}} : \ldots : N_{n_{E2}} :: \mathbf{N}_1 : \mathbf{N}_2 : \mathbf{N}_3 : \ldots : \mathbf{N}_n \quad (2.6)$$

EXAMPLE 2.2

Nitrous oxide has 57.1 g oxygen per 100 g nitrogen while nitrogen dioxide has 228 g oxygen per 100 g nitrogen. 2356.5 g nitrous oxide is oxidized to nitrogen dioxide. How much oxygen will be required?

Solution

Here element E_1 is nitrogen (N_2) and element E_2 is oxygen (O_2). The compound Z_1 is nitrous oxide (N_2O) and the compound Z_2 is nitrogen dioxide (NO_2).

Let the amount of oxygen in 2356.5 g nitrous oxide be $(m_{O_2})_{N_2O}$.

For the compound nitrous oxide (Z_1), by the Law of Definite Proportions using Equation 2.2

$$\frac{57.1}{(100+57.1)} = \frac{(m_{O_2})_{N_2O}}{2356.5}, \Rightarrow (m_{O_2})_{N_2O} = 856.5 \text{ g}$$

Let the amount of oxygen present in the nitrogen dioxide obtained by the oxidation of 2356.5 g nitrous oxide $(m_{O_2})_{NO_2}$. From Equation 2.5

$$\frac{(m_{O_2})_{NO_2}}{(m_{O_2})_{N_2O}} = \frac{\mathbf{N}_1}{\mathbf{N}_2}$$

$$\Rightarrow \frac{(m_{O_2})_{NO_2}}{856.5} = \frac{228/100}{57.1/100} \Rightarrow (m_{O_2})_{NO_2} = 3420 \text{ g}$$

$$\text{Amount of oxygen required} = (m_{O_2})_{NO_2} - (m_{O_2})_{N_2O} = 3420 - 856.5 = 2563.5 \text{ g}$$

2.3 REACTION STOICHIOMETRY

Consider a chemical reaction in which n reactants, $A_1, A_2, \ldots A_n$ undergo chemical transformation to make l products, $P_1, P_2, \ldots P_l$. We further consider the situation when all the reactants are completely transformed into products and none of the chemical species $A_1, A_2, \ldots A_n$, are left. For this reaction, the "Law of Conservation of Mass" requires

$$\sum_{i=1}^{n} m_{A_i} = \sum_{j=1}^{l} m_{P_j} \tag{2.7}$$

Equation 2.7 does not impose any limitation on the individual masses $m_{A_1}, m_{A_2}, \ldots m_{A_n}$ or $m_{P_1}, m_{P_2}, \ldots m_{P_l}$, as long as the sum of the masses on the two sides of the equation are equal. Then, can the reactants react among themselves in any proportion of mass (or moles) to form the products in any proportion of mass (or moles)? The answer is "No" (Brown et al., 2012). The reactants always react among themselves in a fixed proportion of mass (or moles). Similarly, the products formed are always in a fixed proportion of mass (or moles). This leads to the concept of balanced chemical equation. The reaction in question can be expressed as a balanced chemical equation in the form

$$\sum_{i=1}^{n} a_i A_i \rightarrow \sum_{j=1}^{l} p_j P_j \tag{2.8}$$

What is expressed by the chemical Equation 2.8? It tells us that in the said reaction a_1 moles of the reactant A_1 react with a_2 moles of the reactant A_2, a_3 moles of the reactant A_3, and so on to form p_1 moles of the product P_1, p_2 moles of the product P_2, and so on. For this reaction, the ratios $a_1/a_2, a_1/a_3, \ldots a_1/a_n$, and $a_1/p_1, a_1/p_2, \ldots a_1/p_l$ are all fixed. If they are not fixed, the "Law of Definite Proportions" and the "Law of Multiple Proportions" will be violated. The quantities $a_1, a_2, \ldots a_n$, and $p_1, p_2, \ldots p_l$ are known as the stoichiometric coefficients of the respective chemical species $A_1, A_2, \ldots A_n$, and $P_1, P_2, \ldots P_l$. Therefore, it follows that in the balanced chemical equation of a given reaction, the stoichiometric coefficients are in a fixed ratio.

Again consider the chemical reaction represented by the balanced chemical Equation 2.8. Writing Equation 2.7 in terms of moles of respective chemical species reacting or forming, we have

$$\sum_{i=1}^{n} N_{A_i} M_{A_i} = \sum_{j=1}^{l} N_{P_j} M_{P_j} \tag{2.9}$$

From the discussion of the previous paragraph, it follows that

$$\frac{N_{A_2}}{N_{A_1}} = \frac{a_2}{a_1}, \frac{N_{A_3}}{N_{A_1}} = \frac{a_3}{a_1}, \ldots \frac{N_{A_n}}{N_{A_1}} = \frac{a_n}{a_1} \tag{2.10}$$

and

$$\frac{N_{P_1}}{N_{A_1}} = \frac{p_1}{a_1}, \frac{N_{P_2}}{N_{A_1}} = \frac{p_2}{a_1}, \ldots \frac{N_{P_1}}{N_{A_1}} = \frac{p_1}{a_1} \tag{2.11}$$

In terms of the masses of the respective chemical species

$$\frac{m_{A_2}}{m_{A_1}} \times \frac{M_{A_1}}{M_{A_2}} = \frac{a_2}{a_1}, \frac{m_{A_3}}{m_{A_1}} \times \frac{M_{A_1}}{M_{A_3}} = \frac{a_3}{a_1}, \ldots \frac{m_{A_n}}{m_{A_1}} \times \frac{M_{A_1}}{M_{A_n}} = \frac{a_n}{a_1} \tag{2.12}$$

and

$$\frac{m_{P_1}}{m_{A_1}} \times \frac{M_{A_1}}{M_{P_1}} = \frac{p_1}{a_1}, \frac{m_{P_2}}{m_{A_1}} \times \frac{M_{A_1}}{M_{P_2}} = \frac{p_2}{a_1}, \ldots \frac{m_{P_n}}{m_{A_1}} \times \frac{M_{A_1}}{M_{P_n}} = \frac{p_n}{a_1} \tag{2.13}$$

Equation 2.9 then takes the form

$$\sum_{i=1}^{n} N_{A_1}\left(\frac{a_i}{a_1}\right) M_{A_i} = \sum_{j=1}^{1} N_{A_1}\left(\frac{p_j}{a_1}\right) M_{P_j} \tag{2.14}$$

It is obvious that in writing Equations 2.10 through 2.14, the species A_1 has been taken as the reference substance. Similar equations can be written by taking any of the chemical species involved in the reaction as the reference substance.

EXAMPLE 2.3

Combustion of ethyne can be represented by the balanced chemical equation

$$2C_2H_2 + 5O_2 \rightarrow 4CO_2 + 2H_2O$$

How much oxygen is required for the combustion of 130 kg of ethyne and how much carbon dioxide is produced in the process?

SOLUTION

We take C_2H_2 as the reference substance. Here,

$$a_{C_2H_2} = 2, a_{O_2} = 5, p_{CO_2} = 4, \text{ and } m_{C_2H_2} = 130 \text{ kg}$$

Using Equation 2.12 for the reactant O_2

$$\frac{m_{O_2}}{m_{C_2H_2}} \times \frac{M_{C_2H_2}}{M_{O_2}} = \frac{a_{O_2}}{a_{C_2H_2}}$$

$$\Rightarrow \quad \frac{m_{O_2}}{130} \times \frac{26}{32} = \frac{5}{2} \Rightarrow \quad m_{O_2} = 400 \text{ kg}$$

Using Equation 2.13 for the product CO_2

$$\frac{m_{CO_2}}{m_{C_2H_2}} \times \frac{M_{C_2H_2}}{M_{CO_2}} = \frac{a_{CO_2}}{a_{C_2H_2}}$$

$$=> \quad \frac{m_{CO_2}}{130} \times \frac{26}{44} = \frac{4}{2} => \quad m_{CO_2} = 440\ \text{kg}$$

EXAMPLE 2.4

Industrial urea production involves the three reactions

$$3H_2 + N_2 \rightarrow 2NH_3$$

$$2NH_3 + CO_2 \rightarrow H_2NCOONH_4$$

and

$$H_2NCOONH_4 \rightarrow (NH_2)_2CO + H_2O$$

How many kilomoles of urea will be formed per hour if 1000 kmoles of hydrogen is consumed per hour?

Solution

First, we combine the three balanced chemical equations to obtain the equation representing the overall reaction

$$3H_2 + N_2 \rightarrow 2NH_3$$

$$2NH_3 + CO_2 \rightarrow H_2NCOONH_4$$

$$H_2NCOONH_4 \rightarrow (NH_2)_2CO + H_2O$$

Adding the three equations, we get

$$3H_2 + N_{2-} + CO_2 \rightarrow (NH_2)_2CO + H_2O$$

We take H_2 as the reference substance. Here,

$$a_{H_2} = 3, p_{urea} = 1, N_{H_2} = 1000\ \text{kmoles}$$

Using Equation 2.11 for the product urea

$$\frac{N_{urea}}{N_{H_2}} = \frac{p_{urea}}{a_{H_2}}$$

$$\Rightarrow \quad \frac{N_{urea}}{1000} = \frac{1}{3} \Rightarrow \quad N_{urea} = 333.33\ \text{kmoles}$$

2.3.1 Limiting Reactant

In the above discussion, a condition was imposed that all the reactants are completely transformed into products and none of the chemical species A_1, A_2, ... A_n, are left. If this is not the case and some quantity of each reactant is left, the Law of Conservation of Mass will be modified as

$$\left(\sum_{i=1}^{n} m_{A_i}\right)_0 = \sum_{j=1}^{1} m_{P_j} + \left(\sum_{i=1}^{n} m_{A_i}\right)_f \tag{2.15}$$

and in terms of moles

$$\left(\sum_{i=1}^{n} N_{A_i} M_{A_i}\right)_0 = \sum_{j=1}^{1} N_{P_j} M_{P_j} + \left(\sum_{i=1}^{n} N_{A_i} M_{A_i}\right)_f \tag{2.16}$$

If, initially, the moles of the reactants $(N_{A_1})_0, (N_{A_2})_0, \ldots (N_{A_n})_0$, are all conforming to the ratios of the respective stoichiometric coefficients as required by Equation 2.10 the moles of the reactants left $(N_{A_1})_f, (N_{A_2})_f, \ldots (N_{A_n})_f$, will also conform to the same ratios. However, if the initial moles of the reactants do not conform to the ratios provided by Equation 2.10 and the reaction is allowed to proceed for infinite time (practically for a very long time so that no further change in the composition of the reacting system is observed), only one of the reactants will be absent in the final reaction mixture. In other words, only one of the reactants will be fully consumed (Brown et al., 2012). This reactant is known as the limiting reactant, A_{LR}. In terms of the initial moles of the limiting reactant, Equation 2.10 will now be modified as

$$\frac{(N_{A_2})_0}{(N_{A_{LR}})_0} > \frac{a_2}{a_1}, \frac{(N_{A_3})_0}{(N_{A_{LR}})_0} > \frac{a_3}{a_1}, \ldots \frac{(N_{A_n})_0}{(N_{A_{LR}})_0} > \frac{a_n}{a_1} \tag{2.17}$$

Equation 2.11 will be modified as

$$\frac{(N_{P_1})_f}{(N_{A_{LR}})_0} = \frac{p_1}{a_{LR}}, \frac{(N_{P_2})_f}{(N_{A_{LR}})_0} = \frac{p_2}{a_{LR}}, \ldots \frac{(N_{P_l})_f}{(N_{A_{LR}})_0} = \frac{p_l}{a_{LR}} \tag{2.18}$$

EXAMPLE 2.5

For the reaction of Example 2.3, 208 kg ethyne is combusted in the presence of 700 kg oxygen. Which is the limiting reactant?

Solution

$$a_{C_2H_2} = 2, a_{O_2} = 5, (m_{C_2H_2})_0 = 208\text{ kg}, (m_{O_2})_0 = 700\text{ kg}$$

$$(N_{C_2H_2})_0 = \frac{208}{26} = 8 \text{ kmoles}$$

$$(N_{O_2})_0 = \frac{700}{32} = 21.88 \text{ kmoles}$$

First, assume oxygen as the limiting reactant. Using Equation 2.17, we check

$$\frac{(N_{C_2H_2})_0}{(N_{O_2})_0} = \frac{8}{21.88} = 0.366$$

and

$$\frac{a_{C_2H_2}}{a_{O_2}} = \frac{2}{5} = 0.4$$

So, $\frac{(N_{C_2H_2})_0}{(N_{O_2})_0} < \frac{a_{C_2H_2}}{a_{O_2}}$ and oxygen is not the limiting reactant.

$$\frac{(N_{O_2})_0}{(N_{C_2H_2})_0} = \frac{21.88}{8} = 2.735$$

and

$$\frac{a_{O_2}}{a_{C_2H_2}} = \frac{5}{2} = 2.5$$

So,

$$\frac{(N_{O_2})_0}{(N_{C_2H_2})_0} > \frac{a_{O_2}}{a_{C_2H_2}}$$

and ethyne is the limiting reactant.

2.3.2 Conversion

When the reaction is not complete, and some quantity of each reactant is left including the limiting reactant, then the following conditions will prevail:

$$\left.\begin{aligned} \frac{(N_{A_1})_0 - (N_{A_1})_f}{(N_{A_{LR}})_0 - (N_{A_{LR}})_f} &= \frac{a_1}{a_{LR}}, \\ \frac{(N_{A_2})_0 - (N_{A_2})_f}{(N_{A_{LR}})_0 - (N_{A_{LR}})_f} &= \frac{a_2}{a_{LR}}, \\ &\vdots \\ \frac{(N_{A_n})_0 - (N_{A_n})_f}{(N_{A_{LR}})_0 - (N_{A_{LR}})_f} &= \frac{a_n}{a_{LR}} \end{aligned}\right\} \quad (2.19)$$

And for the products

$$\left.\begin{aligned} \frac{N_{P_1}}{(N_{A_{LR}})_0-(N_{A_{LR}})_f} &= \frac{p_1}{a_{LR}}, \\ \frac{N_{P_2}}{(N_{A_{LR}})_0-(N_{A_{LR}})_f} &= \frac{p_2}{a_{LR}}, \\ &\vdots \\ \frac{N_{P_l}}{(N_{A_{LR}})_0-(N_{A_{LR}})_f} &= \frac{p_l}{a_{LR}} \end{aligned}\right\} \tag{2.20}$$

For a particular reactant, A_n, we define the quantity conversion, X_{A_n} as

$$X_{A_n} = \frac{\text{Number of moles of } A_n \text{ consumed}}{\text{Initial number of moles of } A_n} \tag{2.21}$$

$$\Rightarrow \quad X_{A_n} = \frac{(N_{A_n})_0-(N_{A_n})_f}{(N_{A_n})_0}$$

$$\Rightarrow (N_{A_n})_f = (N_{A_n})_0(1-X_{A_n}) \tag{2.22}$$

For the limiting reactant

$$X_{A_{LR}} = \frac{(N_{A_{LR}})_0-(N_{A_{LR}})_f}{(N_{A_{LR}})_0} \tag{2.23}$$

It is immediately apparent from Equations 2.21 and 2.22 that the minimum possible value for conversion is "zero" at the start of the reaction when $(N_{A_n})_0 = (N_{A_n})_f$. For the limiting reactant, the maximum possible value is "one" at the end of the reaction when it is entirely consumed and $(N_{A_{LR}})_f$ = zero. For all other reactants, the maximum possible conversion is less than one as $(N_{A_n})_f$ is not zero even at the end of the reaction.

In many instances, the composition of a reacting system is described in terms of the molar concentration (moles per unit volume) of the different chemical species, C_{A_1}, C_{A_2}, etc. If the volume of the reacting system does not change with the progress of the reaction—practically the case with liquids and solids—we can divide the numerator and the denominator in Equation 2.21 by the volume of the reacting system "V." We then have

$$X_{A_n} = \frac{\dfrac{(N_{A_n})_0}{V} - \dfrac{(N_{A_n})_f}{V}}{\dfrac{(N_{A_n})_0}{V}} = \frac{(C_{A_n})_0-(C_{A_n})_f}{(C_{A_n})_0} \tag{2.24}$$

and

$$(C_{A_n})_f = (C_{A_n})_0(1-X_{A_n}) \tag{2.25}$$

EXAMPLE 2.6

For the problem of Example 2.5, how much oxygen will be consumed for 60% conversion of ethyne?

SOLUTION

$$a_{C_2H_2} = 2,\ a_{O_2} = 5,\ (m_{C_2H_2})_0 = 208\ \text{kg},\ (m_{O_2})_0 = 700\ \text{kg},\ X_{C_2H_2} = 0.6$$

$$(N_{C_2H_2})_0 = \frac{208}{26} = 8\ \text{kmoles}$$

From Equation 2.22

$$(N_{C_2H_2})_f = (N_{C_2H_2})_0(1 - X_{C_2H_2}) = 8 \times 0.4 = 3.2\ \text{kmoles}$$

From Equation 2.19

$$\frac{(N_{O_2})_0 - (N_{O_2})_f}{(N_{C_2H_2})_0 - (N_{C_2H_2})_f} = \frac{a_{O_2}}{a_{C_2H_2}}$$

$$=> \quad \frac{(N_{O_2})_0 - (N_{O_2})_f}{8 - 3.2} = \frac{5}{2}$$

$$=> \quad (N_{O_2})_0 - (N_{O_2})_f = 12\ \text{kmoles} = 12 \times 32 = 384\ \text{kg}$$

2.3.3 Gas Laws and Reaction Stoichiometry

Unlike in liquids and solids, where the volume is practically unchanged during the course of a reaction, in reacting systems consisting of gases the volume is related to the total number of moles, total pressure, and temperature of the system by the equation of state. As a first—and often reasonable—approximation, we assume the gases to behave as ideal gas and apply the ideal gas law (Clapeyron, 1834) to the system. Consider the reaction represented by the balanced chemical Equation 2.8. For the ith reactant, A_i,

$$p_{A_i} V = N_{A_i} RT$$

$$=> \quad p_{A_i} = \frac{N_{A_i}}{V} RT$$

$$=> \quad C_{A_i} = \frac{p_{A_i}}{RT} \tag{2.26}$$

and, partial pressure of the ith reactant,

$$p_{A_i} = x_{A_i} p = \frac{N_{A_i}}{\sum_{i=1}^{n} N_{A_i} + \sum_{j=1}^{1} N_{P_j} + N_{\text{inerts}}} p \tag{2.27}$$

When all the reactants are present in the ratio of their stoichiometric coefficients, we have

$$p_{A_i} = \frac{N_{A_i}}{\sum_{i=1}^{n} N_{A_{LR}}\left(\frac{a_i}{a_{LR}}\right) + \sum_{j=1}^{1} N_{A_{LR}}\left(\frac{p_j}{a_{LR}}\right) + N_{inerts}}\, p \tag{2.28}$$

Also, in terms of total pressure

$$pV = \left[\sum_{i=1}^{n} N_{A_i} + \sum_{j=1}^{1} N_{P_j} + N_{inerts}\right] RT \tag{2.29}$$

$$\Rightarrow V = \left[\sum_{i=1}^{n} N_{A_i} + \sum_{j=1}^{1} N_{P_j} + N_{inerts}\right] RT/p \tag{2.30}$$

2.3.3.1 Systems at Constant Temperature and Pressure

Again, consider the reaction represented by the balanced chemical Equation 2.8. From Equation 2.19,

$$\frac{(N_{A_i})_0 - (N_{A_i})_f}{(N_{A_{LR}})_0 - (N_{A_{LR}})_f} = \frac{a_i}{a_{LR}}$$

$$\Rightarrow \quad (N_{A_i})_f = (N_{A_i})_0 - \frac{a_i}{a_{LR}}[(N_{A_{LR}})_0 - (N_{A_{LR}})_f]$$

$$\Rightarrow \quad \sum_{i=1}^{n}(N_{A_i})_f = \sum_{i=1}^{n}(N_{A_i})_0 - [(N_{A_{LR}})_0 - (N_{A_{LR}})_f]\sum_{i=1}^{n}\frac{a_i}{a_{LR}} \tag{2.31}$$

And, in combination with Equation 2.23

$$\sum_{i=1}^{n}(N_{A_i})_f = \sum_{i=1}^{n}(N_{A_i})_0 - (N_{A_{LR}})_0 X_{A_{LR}} \sum_{i=1}^{n}\frac{a_i}{a_{LR}} \tag{2.32}$$

Similarly, from Equation 2.20, in combination with Equation 2.23,

$$\sum_{j=1}^{1}(N_{P_j})_f = (N_{A_{LR}})_0 X_{A_{LR}} \sum_{j=1}^{1}\frac{p_j}{a_{LR}} \tag{2.33}$$

Adding Equations 2.32 and 2.33,

$$\sum_{i=1}^{n}(N_{A_i})_f + \sum_{j=1}^{l}(N_{P_j})_f = \sum_{i=1}^{n}(N_{A_i})_0 + (N_{A_{LR}})_0 X_{A_{LR}}\left[\sum_{j=1}^{l}\frac{p_j}{a_{LR}} - \sum_{i=1}^{n}\frac{a_i}{a_{LR}}\right] \quad (2.34)$$

If, $\Sigma_{i=1}^{n} a_i = \Sigma_{j=1}^{l} p_j$, Equation 2.34 reduces to

$$\sum_{i=1}^{n}(N_{A_i})_f + \sum_{j=1}^{l}(N_{P_j})_f = \sum_{i=1}^{n}(N_{A_i})_0 \quad (2.35)$$

and, combining with Equation 2.30,

$$V_f = V_0 \quad (2.36)$$

Thus, if, $\Sigma_{i=1}^{n} a_i = \Sigma_{j=1}^{l} p_j$, the volume of the reacting system does not change with the progress of the reaction.

If, $\Sigma_{i=1}^{n} a_i > \Sigma_{j=1}^{l} p_j$, the second term on the right-hand side (RHS) of Equation 2.34 is negative. So,

$$\sum_{i=1}^{n}(N_{A_i})_f + \sum_{j=1}^{l}(N_{P_j})_f < \sum_{i=1}^{n}(N_{A_i})_0 \quad (2.37)$$

and

$$V_f < V_0 \quad (2.38)$$

Thus, if, $\Sigma_{i=1}^{n} a_i > \Sigma_{j=1}^{l} p_j$, the volume of the reacting system decreases with the progress of the reaction. Similarly, when $\Sigma_{i=1}^{n} a_i < \Sigma_{j=1}^{l} p_j$, the second term on the RHS of Equation 2.34 is positive. So,

$$\sum_{i=1}^{n}(N_{A_i})_f + \sum_{j=1}^{l}(N_{P_j})_f > \sum_{i=1}^{n}(N_{A_i})_0 \quad (2.39)$$

and

$$V_f > V_0 \quad (2.40)$$

Thus, if, $\Sigma_{i=1}^{n} a_i < \Sigma_{j=1}^{l} p_j$, the volume of the reacting system increases with the progress of the reaction.

For the limiting reactant, from Equation 2.23

$$X_{A_{LR}} = \frac{(N_{A_{LR}})_0 - (N_{A_{LR}})_f}{(N_{A_{LR}})_0} = \frac{(C_{A_{LR}})_0 V_0 - (C_{A_{LR}})_f V_f}{(C_{A_{LR}})_0 V_0}$$

$$\Rightarrow \quad (C_{A_{LR}})_f V_f = (C_{A_{LR}})_0 V_0 (1 - X_{A_{LR}})$$

$$\Rightarrow \quad (C_{A_{LR}})_f = \frac{(C_{A_{LR}})_0 (1 - X_{A_{LR}})}{V_f / V_0} \tag{2.41}$$

From Equation 2.30, together with Equation 2.34

$$\frac{V_f}{V_0} = \frac{\sum_{i=1}^{n} (N_{A_i})_0 + (N_{A_{LR}})_0 X_{A_{LR}} \left[\sum_{j=1}^{1} \frac{p_j}{a_{LR}} - \sum_{i=1}^{n} \frac{a_i}{a_{LR}} \right] + N_{inerts}}{\sum_{i=1}^{n} (N_{A_i})_0 + N_{inerts}}$$

$$\Rightarrow \quad \frac{V_f}{V_0} = 1 + \frac{(N_{A_{LR}})_0 X_{A_{LR}} \left[\sum_{j=1}^{1} \frac{p_j}{a_{LR}} - \sum_{i=1}^{n} \frac{a_i}{a_{LR}} \right]}{\sum_{i=1}^{n} (N_{A_i})_0 + N_{inerts}} \tag{2.42}$$

$$\Rightarrow \frac{V_f}{V_0} = 1 + \varepsilon_{LR} X_{A_{LR}} \tag{2.43}$$

Here, we define a new quantity, fractional change in volume at complete conversion, ε_{LR}. From Equations 2.43 and 2.42, when $X_{A_{LR}} = 1$,

$$\varepsilon_{LR} = \frac{V_f - V_0}{V_0} = \frac{(N_{A_{LR}})_0 \left[\sum_{j=1}^{1} \frac{p_j}{a_{LR}} - \sum_{i=1}^{n} \frac{a_i}{a_{LR}} \right]}{\sum_{i=1}^{n} (N_{A_i})_0 + N_{inerts}} \tag{2.44}$$

Therefore, the quantity ϵ_{LR} is defined as the ratio of change in moles of the reacting system at complete conversion to the initial moles. Equation 2.41 now becomes

$$(C_{A_{LR}})_f = \frac{(C_{A_{LR}})_0 (1 - X_{A_{LR}})}{1 + \varepsilon_{LR} X_{A_{LR}}} \tag{2.45}$$

2.3.3.2 Systems at Constant Temperature and Volume

From Equation 2.29

$$p = \left[\sum_{i=1}^{n} N_{A_i} + \sum_{j=1}^{1} N_{P_j} + N_{inerts} \right] RT/V \tag{2.46}$$

From the previous discussion in Section 2.3.1.1, if $\sum_{i=1}^{n} a_i = \sum_{j=1}^{l} p_j$,

$$p_f = p_0 \tag{2.47}$$

The pressure of the reacting system does not change with the progress of the reaction. If $\sum_{i=1}^{n} a_i > \sum_{j=1}^{l} p_j$,

$$p_f < p_0 \tag{2.48}$$

The pressure of the reacting system decreases with the progress of the reaction. If $\sum_{i=1}^{n} a_i < \sum_{j=1}^{l} p_j$,

$$p_f > p_0 \tag{2.49}$$

The pressure of the reacting system increases with the progress of the reaction. For the constant volume system $V_f/V_0 = 1$ and Equation 2.25 is applicable. However, in such systems, at $X_{A_{LR}} = 1$

$$(\varepsilon_{LR})_p = \frac{p_f - p_0}{p_0} = \frac{(N_{A_{LR}})_0\left[\sum_{j=1}^{l}\frac{p_j}{a_{LR}} - \sum_{i=1}^{n}\frac{a_i}{a_{LR}}\right]}{\sum_{i=1}^{n}(N_{A_i})_0 + N_{inerts}} \tag{2.50}$$

and, at any conversion,

$$p_f = p_0[1 + (\varepsilon_{LR})_p X_{A_{LR}}] \tag{2.51}$$

EXAMPLE 2.7

Reaction of Example 2.3 is carried out at 1 atm pressure and 950 K. In the beginning, the ethyne and oxygen are present at 1:3 molar ratio. Find the concentration of ethyne at 50% conversion.

SOLUTION

$$a_{C_2H_2} = 2,\ a_{O_2} = 5,\ p_{CO_2} = 4,\ p_{H_2O} = 2,\ \frac{(N_{C_2H_2})_0}{(N_{O_2})_0} = \frac{1}{3},\ p = 1\,\text{atm},\ T = 950\,\text{K},\ X_{C_2H_2} = 0.5$$

Therefore,

$$\frac{(N_{O_2})_0}{(N_{C_2H_2})_0} > \frac{a_{O_2}}{a_{C_2H_2}}$$

and ethyne is the limiting reactant.

From Equation 2.27

$$(p_{C_2H_2})_0 = x_{C_2H_2} p = \frac{(N_{C_2H_2})_0}{(N_{C_2H_2})_0 + (N_{O_2})_0} p = \frac{1}{1+3} p = 0.25 \text{ atm}$$

From Equation 2.26

$$(C_{C_2H_2})_0 = \frac{(p_{C_2H_2})_0}{RT} = \frac{0.25}{0.082 \times 950} = 0.0032 \text{ mol/L}$$

From Equation 2.44

$$\varepsilon_{C_2H_2} = \frac{(N_{C_2H_2})_0 \left[\sum_{j=1}^{l} \frac{p_j}{a_{C_2H_2}} - \sum_{i=1}^{n} \frac{a_i}{a_{C_2H_2}} \right]}{\sum_{i=1}^{n} (N_{A_i})_0} = \frac{1 \times [(2+1) - (1+2.5)]}{1+3} = -0.125$$

From Equation 2.45

$$(C_{C_2H_2})_f = \frac{(C_{C_2H_2})_0 (1 - X_{C_2H_2})}{1 + \varepsilon_{C_2H_2} X_{C_2H_2}} = \frac{0.0032 \times (1 - 0.5)}{1 - 0.125 \times 0.5} = 0.0017 \text{ mol/L}$$

EXERCISES

2.1 Boron trichloride is reduced by hydrogen according to the reaction

$$2BCl_3 + 3H_2 \rightarrow 2B + 6HCl$$

In carrying out this reaction in the laboratory, a certain amount of BCl_3 is reacted with excess of hydrogen and the resulting HCl is absorbed in 1 L of water. 10 mL of this solution was titrated with 0.2 N NaOH and 35 mL of the NaOH solution was consumed. What was the initial amount of BCl_3 taken? Take the atomic weight of boron as 11.

2.2 A certain amount of a hydrocarbon on complete combustion produces 22 g CO_2 and 8 g H_2O. What is the mass percentage of carbon and hydrogen in the hydrocarbon?

2.3 The following reactions are involved in the industrial production of sulfuric acid:

$$S_8(s) + 8O_2 \rightarrow 8SO_2$$

$$2SO_2 + O_2 \rightarrow 2SO_3$$

$$SO_3 + H_2O(l) \rightarrow H_2SO_4(l)$$

A factory produces 98 metric tons of sulfuric acid per day. What is the daily requirement of sulfur in the industry?

2.4 In Exercise 2.3, what is the daily requirement of sulfur in the industry if the first reaction gives 95% conversion of sulfur, and the second reaction gives 85% conversion of SO_2? Assume the third reaction to be 100% complete.

2.5 In Exercise 2.4, what is the total daily requirement of oxygen if it is taken 15% in excess than the stoichiometric amount required?

2.6 In Exercises 2.3 and 2.4, what is the daily water requirement if the final product is 98 wt% solution of sulfuric acid?

2.7 A limekiln produces quick lime from limestone according to the calcination reaction

$$CaCO_3 \rightarrow CaO + CO_2$$

The resulting quick lime is slaked in excess of water to produce calcium hydroxide through the following reaction:

$$CaO + H_2O \rightarrow Ca(OH)_2$$

The industry gets limestone with 90% purity, the first reaction has a conversion of 90%, and the second reaction goes to 60% completion. What is the purity of calcium hydroxide produced?

2.8 In Exercise 2.7, the industry calcines 100 metric tons of limestone every day. What is the amount of quick lime produced and its purity?

2.9 In Exercise 2.8, what is the annual emission of carbon dioxide assuming that the industry operates for 330 days in a year?

2.10 The reactant A produces B and C according to the reaction

$$A \rightarrow 2B \rightarrow C$$

We start with a beaker containing 10 L of 0.2 M solution of A. After some time it is found that the concentration of B in the beaker is 0.2 M. In order to find the conversion of A, what additional information is required?

2.11 Hydrolysis of starch produces glucose by the reaction

$$C_{6n}H_{(10n+2)}O_{(5n+1)} + (n-1)H_2O \rightarrow nC_6H_{12}O_6$$

A starch sample with degree of polymerization (n) 200 is hydrolyzed to glucose. An industry hydrolyzes 8104.5 kg of starch per batch to produce a 2 molal glucose syrup. How much water is required per batch? Assume complete conversion.

2.12 The reaction of Example 2.3 is carried out in a reaction bomb maintained at constant temperature. In the beginning, ethyne and oxygen are present at 1:2 molar ratio, and the total pressure is 1 atm. What will be the pressure inside the reaction bomb when half the ethyne is consumed? What will be the partial pressure of oxygen at this stage?

2.13 Solve Exercise 2.12 if the reaction is carried out in the presence of air instead of pure oxygen.

2.14 In Exercise 2.13, the reaction is carried out in a variable volume bomb, with an initial volume of 10 L, working at a constant pressure of 1 atm. What will be the volume of the bomb when half the ethyne is consumed?

REFERENCES

Brown, T.L., LeMay Jr., H.E., Bursten, B.E., Murphy, C.J., and Woodward, P.M., *Chemistry the Central Science*, Prentice-Hall, New York, 76–104, 2012.

Clapeyron, P.E., Puissance Motrice De La Chaleur. *J. de l'Ecole Polytechn.*, 14, 153, 1834.

Compendium of Chemical Terminology Gold Book, Version 2.3.3, IUPAC, 1459, 2014. Available online; http://goldbook.iupac.org/PDF/goldbook.pdf.

Dalton, J., *A New System of Chemical Philosophy Part I*, 212–216, Manchester, 1808. Available online; https://ia700402.us.archive.org/31/items/newsystemofchemi01daltuoft/newsystemofchemi01daltuoft.pdf.

Proust, J.L., Researches on copper. *Ann. Chim.*, 32, 26–54, 1799.

3 Chemical Kinetics

In a chemical reaction, reactants decrease in quantity while the quantity of products increases. Stoichiometry provides us the means to calculate the change in quantities (masses and/or moles) of all the different chemical species (reactants and products) provided the change in quantity of any one of them is known. It does not tell us about the time required for this change in quantity to take place. Chemical kinetics deals with this aspect of chemical reactions. Chemical reactions involve breaking of existing bonds and formation of new bonds. The time scale on which the breaking and making of bonds take place decides the time required for the accomplishment of a chemical reaction. When all the reactants are in the same phase, all the reactant molecules are accessible to one another. Here, sluggishness or otherwise the breaking and making of bonds alone decides the kinetics. In reacting systems consisting of more than one phase, phase barriers separate different reactant molecules from each other. For the chemical reaction to take place, close proximity of reactant molecules is a prerequisite. Reactant molecules, therefore, need to be transported across the phase boundaries so as to be in the vicinity of other reactant molecules. Besides the ease of breaking and formation of bonds, the speed of the physical processes that transport the molecules across the phase boundaries also influence the overall rate of the chemical reaction.

3.1 THE RATE OF A HOMOGENEOUS CHEMICAL REACTION

Homogeneous reactions are those where the reacting system consists of a single phase. Consider a reacting system where the chemical reaction represented by the balanced chemical Equation 2.8, Chapter 2, is taking place. Consider a part of this reacting system that has a volume, V, at time t. Let the moles of the ith reactant, A_i, within this part of the reacting system at time, t, be $\left(N_{A_i}\right)_t$ and that at time $t + \Delta t$ be $\left(N_{A_i}\right)_{t+\Delta t}$ Therefore, the change in moles of A_i in time interval Δt

$$\Delta N_{A_i} = (N_{A_i})_{t+\Delta t} - (N_{A_i})_t \tag{3.1}$$

Then, the change per unit volume of the reacting system will be

$$\Delta\left(\frac{N_{A_i}}{V}\right) = \left(\frac{N_{A_i}}{V}\right)_{t+\Delta t} - \left(\frac{N_{A_i}}{V}\right)_t \tag{3.2}$$

$$\Rightarrow \quad \Delta(C_{A_i}) = (C_{A_i})_{t+\Delta t} - (C_{A_i})_t \tag{3.3}$$

Dividing Equation 3.3 by the time interval Δt,

$$\frac{\Delta(C_{A_i})}{\Delta t} = \frac{(C_{A_i})_{t+\Delta t} - (C_{A_i})_t}{\Delta t} \tag{3.4}$$

In the limit $\Delta t \rightarrow 0$, Equation 3.4 reduces to

$$\frac{d(C_{A_i})}{dt} = \frac{(C_{A_i})_{t+dt} - (C_{A_i})_t}{dt} \tag{3.5}$$

Since A_i is a reactant, its concentration decreases during the course of the reaction. From Equation 3.5, therefore, the quantity $d(C_{A_i})/dt$ is negative. In terms of the ith reactant, A_i, we define the rate of the chemical reaction as "the rate of decrease of concentration of A_i with time." Therefore,

$$-r_{A_i} = -\frac{d(C_{A_i})}{dt} = \frac{(C_{A_i})_t - (C_{A_i})_{t+dt}}{dt} \tag{3.6}$$

It may be noted that the mention of the term *reaction rate* does not make any sense in itself unless a reactant is referred to. The reaction rate can be expressed in terms of any of the reactants, and most conveniently, in terms of the *limiting reactant*, A_{LR}. We then have

$$-r_{A_{LR}} = -\frac{d(C_{A_{LR}})}{dt} = \frac{(C_{A_{LR}})_t - (C_{A_{LR}})_{t+dt}}{dt} \tag{3.7}$$

Using Equation 2.19, Chapter 2, we can write the rates for each of the reactants in terms of the limiting reactant as

$$\frac{-r_{A_i}}{-r_{A_{LR}}} = \frac{(C_{A_i})_t - (C_{A_i})_{t+dt}}{(C_{A_{LR}})_t - (C_{A_{LR}})_{t+dt}} = \frac{(N_{A_i}/V)_t - (N_{A_i}/V)_{t+dt}}{(N_{A_{LR}}/V)_t - (N_{A_{LR}}/V)_{t+dt}}$$

$$\Rightarrow \quad \frac{-r_{A_i}}{-r_{A_{LR}}} = \frac{(N_{A_i})_t - (N_{A_i})_{t+dt}}{(N_{A_{LR}})_t - (N_{A_{LR}})_{t+dt}} = \frac{a_i}{a_{LR}} \tag{3.8}$$

Similarly, for the product P_i, the rate of chemical reaction is "the rate of increase of concentration of P_i with time."

From Equation 2.45, Chapter 2,

$$C_{A_{LR}} = \frac{(C_{A_{LR}})_0(1 - X_{A_{LR}})}{1 + \varepsilon_{LR} X_{A_{LR}}}$$

$$\Rightarrow \quad \frac{d(C_{A_{LR}})}{dt} = (C_{A_{LR}})_0 \left[\frac{(1+\varepsilon_{LR}X_{A_{LR}})(-d(X_{A_{LR}})/dt) - (1 - X_{A_{LR}})\varepsilon_{LR}(d(X_{A_{LR}})/dt)}{(1+\ \varepsilon_{LR}X_{A_{LR}})^2} \right]$$

$$\Rightarrow \quad \frac{d(C_{A_{LR}})}{dt} = (C_{A_{LR}})_0 \left[\frac{-(1+\varepsilon_{LR})(d(X_{A_{LR}})/dt)}{(1+\varepsilon_{LR}X_{A_{LR}})^2} \right] \tag{3.9}$$

For the constant volume system, Equation 3.9 reduces to

$$\frac{d(C_{A_{LR}})}{dt} = -(C_{A_{LR}})_0 \frac{d(X_{A_{LR}})}{dt} \tag{3.10}$$

Combining with Equation 3.7,

$$-r_{A_{LR}} = \left(C_{A_{LR}}\right)_0 \left[\frac{(1+\varepsilon_{LR})}{\left(1+\varepsilon_{LR}X_{A_{LR}}\right)^2}\right] \frac{d\left(X_{A_{LR}}\right)}{dt} \tag{3.11}$$

For the constant volume system, Equation 3.11 reduces to

$$-r_{A_{LR}} = \left(C_{A_{LR}}\right)_0 \frac{d\left(X_{A_{LR}}\right)}{dt} \tag{3.12}$$

Equations 3.7 and 3.11 can also be graphically interpreted as depicted in Figure 3.1. For a given chemical reaction we can plot the concentration of the limiting reactant as a function of time. Negative of the slope of the tangent to this plot at any time, t, is equal to the reaction rate, $-r_{A_{LR}}$, at that time. Similarly, for Equation 3.11 we can plot the conversion of the limiting reactant as a function of time. Here, the slope of the tangent to this plot at any time, t, multiplied by the quantity $\left(C_{A_{LR}}\right)_0\left[(1+\varepsilon_{LR})/(1+\varepsilon_{LR}X_{A_{LR}})^2\right]$ is equal to the reaction rate, $-r_{A_{LR}}$, at that time.

EXAMPLE 3.1

In a liquid-phase chemical reaction, the concentration of the limiting reactant can be expressed as a function of time as

$$C_{A_{LR}} = \frac{100}{10t+1}$$

where t is expressed in minutes, and the concentration is expressed in mol/L. Find the rate of the reaction after 5 min from the start of the reaction.

Solution

Given

$$C_{A_{LR}} = \frac{100}{10t+1}$$

$$\Rightarrow \quad \frac{d(C_{A_{LR}})}{dt} = \frac{-10\times 100}{(10t+1)^2}$$

$$\Rightarrow \quad -r_{A_{LR}} = \frac{1000}{(10t+1)^2}$$

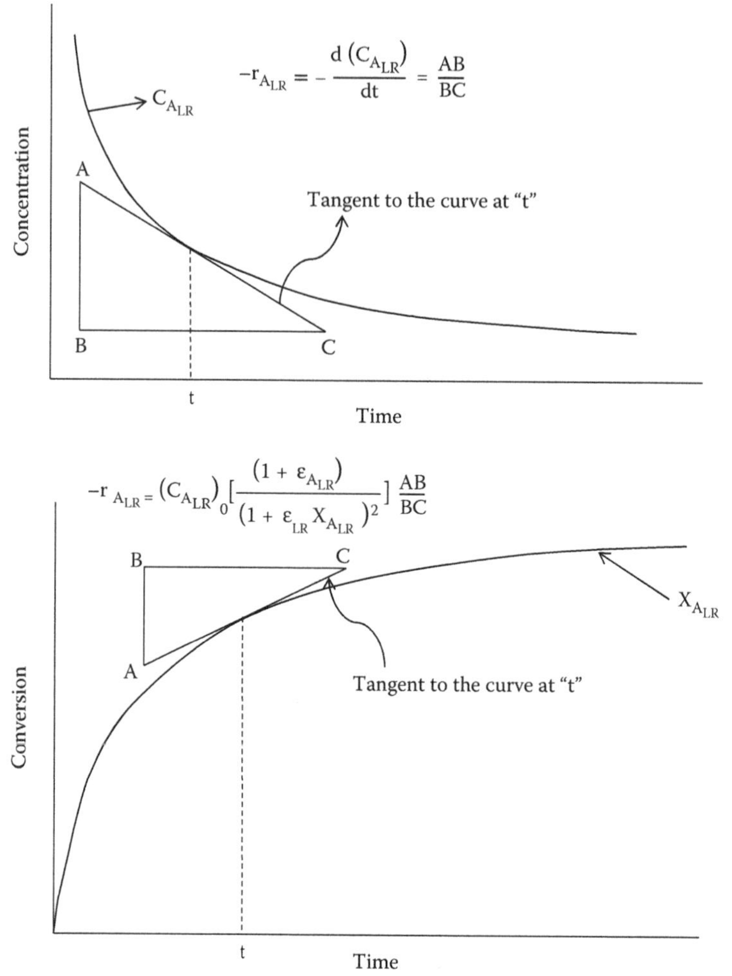

FIGURE 3.1 Graphical interpretation of reaction rate.

At t = 5 min

$$-r_{A_{LR}} = \frac{1000}{(10 \times 5 + 1)^2} = 0.38 \text{ mol/(L} \cdot \text{min)}$$

EXAMPLE 3.2

In a liquid-phase chemical reaction, the concentration of the limiting reactant varies with time as follows:

Concentration (mol/L)	100	67	45	32	24	20	18	17	16
Time (min)	0	2	4	6	8	10	12	14	16

Find the rate of the reaction at 4 min.

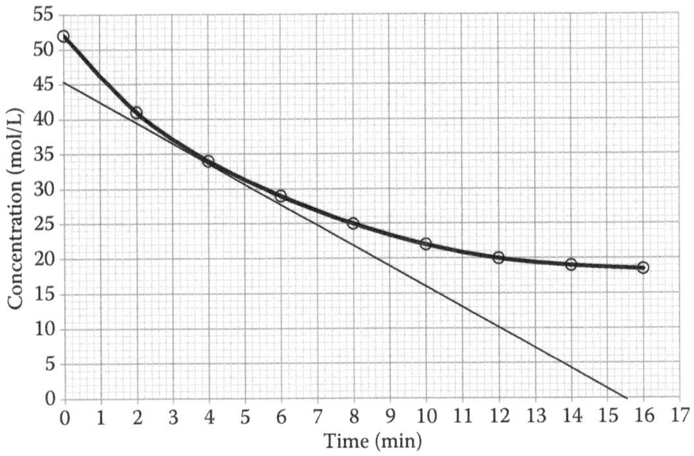

FIGURE EXE3.2 Concentration time plot.

Solution

Plotting the concentration time as explained in Figure 3.1 we get the plot Figure EXE3.2. To this plot a tangent AB is drawn at t = 4 min. The tangent intersects with the concentration axis and the time axis as shown in Figure EXE3.2. Therefore,

$$-r_{A_{LR}} = \frac{45}{15.4} = 2.92 \text{ mol/(L} \cdot \text{min)}$$

3.2 FACTORS AFFECTING THE RATE OF A CHEMICAL REACTION

Suppose we monitor the progress of the chemical reaction, represented by the balanced chemical Equation 2.8, Chapter 2, in terms of the reactant, A_i, commonly the limiting reactant, A_{LR}. We consider unit volume of the reacting system. Within this unit volume, the number of molecules (hence, the moles) of A_{LR} continuously decrease over time as they transform into products. During a time interval between t and t + dt, what number of molecules of A_{LR} transform into products? Given the submicroscopic size of the individual molecules, from a macroscopic standpoint we are observing a very large number of molecules of A_{LR} within the unit volume. For such a large congregation of molecules, the concepts of statistical probability will be applicable with fairly reasonable accuracy. Within the unit volume of the reacting system, at any moment, the number of molecules of A_{LR} transforming into products will depend on the number of molecules of A_{LR} present at that moment and the probability of each molecule of A_{LR} transforming into products. However, the number of molecules of A_{LR} transforming into products will not depend only on the number of molecules of A_{LR} present. The reaction cannot take place if the molecules of any of the reactants are absent, howsoever large the number of molecules of A_{LR} present may be. For the reaction to take place, it is mandatory that molecules of all the reactants are present. Moreover, the number of molecules of each of the reactants present will influence the rate of the chemical reaction.

3.2.1 Concentration Dependence of Reaction Rate

The number of molecules of A_{LR} present within the unit volume of the reacting system at any moment is proportional to the concentration of A_{LR} at that moment. Therefore, one of the factors affecting the rate of a chemical reaction is the instantaneous concentration of the reactant A_{LR}. And by the same reasoning, the rate of a chemical reaction at any moment is dependent on the instantaneous concentrations of all the reactants. In general form, this dependence can be represented by a mathematical function as

$$-r_{A_{LR}} = f(C_{A_1}, C_{A_2}, \ldots C_{A_{LR}}, \ldots C_{A_n}) \tag{3.13}$$

From Equation 2.19, Chapter 2,

$$N_{A_1} = (N_{A_1})_0 - \frac{a_1}{a_{LR}}\{(N_{A_{LR}})_0 - N_{A_{LR}}\}$$

$$\Rightarrow \quad C_{A_1}V = (C_{A_1})_0 V_0 - \frac{a_1}{a_{LR}}\{(C_{A_{LR}})_0 V_0 - C_{A_{LR}}V\}$$

$$\Rightarrow \quad C_{A_1} = (C_{A_1})_0 \frac{V_0}{V} - \frac{a_1}{a_{LR}}\left\{(C_{A_{LR}})_0 \frac{V_0}{V} - C_{A_{LR}}\right\} \tag{3.14}$$

From Equation 2.43, Chapter 2,

$$\frac{V_0}{V} = \frac{1}{1 + \varepsilon_{LR} X_{A_{LR}}} \tag{3.15}$$

and from Equation 2.45, Chapter 2,

$$X_{A_{LR}} = \frac{(C_{A_{LR}})_0 - C_{A_{LR}}}{(C_{A_{LR}})_0 + \varepsilon_{LR} C_{A_{LR}}} \tag{3.16}$$

Combining Equations 3.14 through 3.16, we have

$$C_{A_1} = (C_{A_1})_0 \left[\frac{(C_{A_{LR}})_0 + \varepsilon_{LR} C_{A_{LR}}}{(C_{A_{LR}})_0 (1 + \varepsilon_{LR})}\right] - \frac{a_1}{a_{LR}}\left\{(C_{A_{LR}})_0 \left[\frac{(C_{A_{LR}})_0 + \varepsilon_{LR} C_{A_{LR}}}{(C_{A_{LR}})_0 (1 + \varepsilon_{LR})}\right] - C_{A_{LR}}\right\} \tag{3.17}$$

Similar equations can be written for all the other reactants. On the RHS of Equation 3.17, all quantities except $C_{A_{LR}}$ are constants for a given reacting system. It is, therefore, apparent that for a given reacting system the concentrations $C_{A_1}, C_{A_2}, \ldots C_{A_n}$ are unique functions of $C_{A_{LR}}$ as represented by Equation 3.17. If we replace $C_{A_1}, C_{A_2}, \ldots C_{A_n}$ in Equation 3.13 by these unique functions of $C_{A_{LR}}$, Equation 3.13 reduces to a single variable function of $C_{A_{LR}}$ represented by

$$-r_{A_{LR}} = f'\left(C_{A_{LR}}\right) \tag{3.18}$$

It now remains to be seen that what form the function "*f*" in Equation 3.13 or the function "*f'*" in Equation 3.18 can take. On the basis of the form of these functions, reactions can be classified into different types.

3.2.1.1 Elementary Reactions

The first attempt to provide a mathematical expression for the concentration dependence of reaction rate was made by the Swedish chemists Guldberg and Waage. Based on their experimental observations, they proposed the *law of mass action* (Lund, 1965). The law states that for a chemical reaction, for example, the one represented by the balanced chemical Equation 2.8, Chapter 2, the rate at any instant in terms of the limiting reactant, $-r_{A_{LR}}$, is proportional to the concentration of each reactant raised to the power of its respective stoichiometric coefficient. Mathematically,

$$-r_{A_{LR}} \propto C_{A_1}^{a_1} C_{A_2}^{a_2} \ldots C_{A_{LR}}^{a_{LR}} \ldots C_{A_n}^{a_n}$$

$$\Rightarrow -r_{A_{LR}} = k \prod_{i=1}^{n} C_{A_i}^{a_i} \tag{3.19}$$

where k is a constant of proportionality known as the *specific reaction rate constant.* Since its advent many reactions have been found not to follow this law.

The reactions that rigorously conform to this law under all circumstances are known as *elementary reactions* (IUPAC Gold Book, 2014). We can see that the reaction rate for such reactions faithfully corresponds to the balanced chemical equation as provided by the reaction stoichiometry. To our observation, such a reaction is a "single event" at the molecular level. That is, during the course of the reaction we cannot observe any other chemical species but the reactants A_1, A_2, … A_n, and the products P_1, P_2, … P_l. An example of an elementary reaction is

$$2HI \rightarrow H_2 + I_2 \tag{3.20}$$

With the rate expressed as

$$-r_{HI} = kC_{HI}^2$$

Another example of an elementary reaction is

$$NO + O_3 \rightarrow NO_2 + O_2 \tag{3.21}$$

With the rate expressed as

$$-r_{NO} = kC_{NO}C_{O_3}$$

It is believed that the reactants A_1, A_2, ... A_n transit through states with one or more chemical bonds at excited energy states before the products P_1, P_2, ... P_n finally appear. These are known as transition states or transition complexes and, by definition, cannot be isolated. The sum of the stoichiometric coefficients of the reactants is known as the *molecularity* (IUPAC Gold Book, 2014) of the elementary reaction. For the representative reaction under consideration, the molecularity is $\Sigma_{i=1}^{n} a_i$. Since A_1, A_2, ... A_n are all positive whole numbers, the molecularity is always a positive whole number. Essentially, molecularity is the minimum number of molecules required for the elementary reaction to take place.

3.2.1.2 Nonelementary Reactions

Elementary reactions are exceptions rather than the rule. For the vast majority of reactions, the transformation of reactants into products is "not a single event" at the molecular level. In such reactions, though the overall outcome is the transformation of reactants into products, chemical species other than the reactants A_1, A_2, ... A_n, and the products P_1, P_2, ... P_l, are observed to be formed during the course of the reaction. However, the concentration of these chemical species is so less that the overall outcome still remains the transformation of reactants A_1, A_2, ... A_n, into products P_1, P_2, ... P_l. Such reactions are known as nonelementary reactions. The reaction rate for such reactions may or may not follow the law of mass action and it is not mandatory for the reaction rate to correspond to the balanced chemical equation as provided by the reaction stoichiometry. Nonelementary reactions can be explained as combinations of elementary reactions. Chemical species other than the original reactants and the final products are known as the *reaction intermediates*. Depending on the nature of the particular reacting system, these intermediates can be free radicals (moieties containing one or more unpaired electrons), ions (moieties containing one or more units of positive or negative electrical charge), or molecules with very short life spans. When a nonelementary reaction is represented as the combination of its constituent elementary reactions, such a scheme is commonly known as the *reaction mechanism* for the nonelementary reaction. Some illustrative examples of nonelementary reactions and their mechanisms are presented in Table 3.1.

As can be seen from Table 3.1, the reaction rate for the reaction

$$C_2H_6 \rightarrow C_2H_4 + H_2 \tag{3.22}$$

follows the law of mass action. But, for the reactions

$$2N_2O_5 \rightarrow 4NO_2 + O_2 \tag{3.23}$$

and

$$H_2 + Br_2 \rightarrow 2HBr \tag{3.24}$$

the reaction rates do not follow the law of mass action. Thus, a nonelementary reaction may or may not follow the law of mass action or, conversely, merely following the law of mass action does not make a reaction elementary.

TABLE 3.1
Examples of Nonelementary Reactions and Their Mechanism

Nonelementary Reaction	Constituent Elementary Reactions	Reaction Intermediates	Observed Reaction Rate
$C_2H_6 \rightarrow C_2H_4 + H_2$	$C_2H_6 \rightarrow 2CH_3\bullet$ $CH_3\bullet + C_2H_6 \rightarrow CH_4 + C_2H_5\bullet$ $CH_4 \rightarrow CH_3\bullet + H\bullet$ $C_2H_5\bullet \rightarrow C_2H_4 + H\bullet$ $H\bullet + C_2H_6 \rightarrow H_2 + C_2H_5\bullet$ $C_2H_5\bullet + H\bullet \rightarrow C_2H_6$ $2\,H\bullet \rightarrow H_2$	Free radicals: $CH_3\bullet$, $H\bullet$, $C_2H_5\bullet$ Molecule: CH_4	$-r_{C_2H_6} = kC_{C_2H_6}$
$2N_2O_5 \rightarrow 4NO_2 + O_2$	$N_2O_5 \rightarrow NO_3 + NO_2$ $NO_3 + NO_2 \rightarrow N_2O_5$ $NO_3 + NO_2 \rightarrow NO_2 + NO + O_2$ $NO + NO_3 \rightarrow 2NO_2$	Molecules: NO_3, NO	$-r_{N_2O_5} = kC_{N_2O_5}$
$H_2 + Br_2 \rightarrow 2HBr$	$Br_2 \rightarrow 2Br\bullet$ $Br\bullet + H_2 \rightarrow HBr + H\bullet$ $H\bullet + Br_2 \rightarrow HBr + Br\bullet$	Free radicals: $Br\bullet$, $H\bullet$	$-r_{H_2} = \dfrac{k_1 C_{H_2} C_{Br_2}^{3/2}}{(k_2 C_{Br_2} + C_{HBr})}$
$NO_2 + CO \rightarrow NO + CO_2$	$2NO_2 \rightarrow NO_3 + NO$ $NO_3 + CO \rightarrow NO_2 + CO_2$	Molecules: NO_3	$-r_{NO_2} = kC_{NO_2}^2$

3.2.1.3 Order of Reaction

The reaction rates for both reactions 3.22 and 3.23 are of the general form of Equation 3.19. In both the rate equations shown in Table 3.1, the reaction rate is proportional to the reactant concentration. There are many nonelementary reactions where the rate equation is of the form shown in Equation 3.19. For an elementary reaction the powers to which the individual concentration terms are raised are equal to the corresponding stoichiometric coefficients. For nonelementary reactions these powers are not necessarily equal to the corresponding stoichiometric coefficients—for example, in reaction 3.23. They can even be proper or improper fractions. The power to which the concentration term of a particular reactant is raised in the rate equation is known as the *order of the reaction* (Atkins and Paula, 2006) with respect to that reactant. The sum of all these powers in the rate equation is known as the overall order of the reaction. For reactions 3.20 and 3.21 the reaction order is two. It is important to note that the molecularity of these reactions is also two. This is a distinct feature of elementary reaction. For reactions 3.22 and 3.23 the reaction order is one. The order of a reaction is purely an experimentally determined number which need not be endorsed by a theoretical justification. It merely represents the best fit of the experimental observation of the reaction kinetics for a particular reaction. What about reaction 3.24? For such reactions, the term "order of reaction" has no meaning. Based on the discussions in Sections 3.2.1.1 through 3.2.1.3, a comparison of elementary and nonelementary reactions is presented in Table 3.2.

TABLE 3.2
Salient Features of Elementary and Nonelementary Reactions

Elementary Reactions	Nonelementary Reactions
Rigorously conform to the law of mass action.	May or may not follow the law of mass action.
Perfect match between reaction rate and reaction stoichiometry.	Reaction rate and reaction stoichiometry may or may not match.
Occur as a single event at the molecular level.	Occur as multiple events at the molecular level.
Transition states or transition complexes are involved.	Reaction intermediates are involved.
	Combinations of elementary reactions are involved.
Concept of molecularity applicable.	Concept of reaction order may or may not be applicable.
Molecularity is equal to the overall reaction order.	
Molecularity and overall order of reaction always equal.	Reaction order may be whole number, proper fractions, or improper fractions.

3.2.1.4 Reaction Mechanism and Observed Rate Equation

Is there a correspondence between the reaction mechanism and the observed reaction rate of a nonelementary reaction? For a reaction mechanism to be acceptable it must satisfactorily explain the observed reaction rate that is experimentally verified. In doing so, we make use of an approximation known as the *pseudo-steady-state approximation.* It states that—the concentrations of the reaction intermediates during the course of a nonelementary reaction remain unchanged so that their net rate of formation is zero. Let us check this with the help of an example.

EXAMPLE 3.3

For reaction 3.23, show that the reaction mechanism given in Table 3.1 satisfactorily explains the observed reaction rate.

Solution

From Table 3.1, the constituent elementary reactions are

$$N_2O_5 \rightarrow NO_3 + NO_2 \tag{3.25}$$

$$NO_3 + NO_2 \rightarrow N_2O_5 \tag{3.26}$$

$$NO_3 + NO_2 \rightarrow NO_2 + NO + O_2 \tag{3.27}$$

$$NO + NO_3 \rightarrow 2NO_2 \tag{3.28}$$

From reaction 3.25, the rate of disappearance of N_2O_5 is

$$-r_{N_2O_5} = k_1\, C_{N_2O_5} \tag{3.3.1}$$

From reaction 3.26, the rate of formation of N_2O_5 is

$$r_{N_2O_5} = k_2\ C_{NO_3}C_{NO_2} \tag{3.3.2}$$

Therefore, the net rate of disappearance of N_2O_5 is

$$\left(-r_{N_2O_5}\right)_{net} = k_1C_{N_2O_5} - k_2C_{NO_3}C_{NO_2} \tag{3.3.3}$$

For reaction 3.23, NO_3 and NO are intermediates. From the pseudo-steady-state approximation, their net rate of formation is zero. Reactions 3.27 and 3.28 involve the intermediate NO. Therefore, we have

$$(r_{NO})_{net} = k_3\ C_{NO_3}C_{NO_2} - k_4\ C_{NO_3}C_{NO} = 0$$

$$\Rightarrow \quad C_{NO} = \frac{k_3\ C_{NO_3}C_{NO_2}}{k_4\ C_{NO_3}} = \frac{k_3\ C_{NO_2}}{k_4} \tag{3.3.4}$$

Similarly, for reactions 3.25 through 3.28, we have

$$\left(r_{NO_3}\right)_{net} = k_1\ C_{N_2O_5} - k_2\ C_{NO_3}C_{NO_2} - k_3\ C_{NO_3}C_{NO_2} - k_4\ C_{NO_3}C_{NO} = 0$$

$$\Rightarrow \quad k_1\ C_{N_2O_5} - k_2\ C_{NO_3}C_{NO_2} - k_3\ C_{NO_3}C_{NO_2} - k_4\ C_{NO_3}\frac{k_3\ C_{NO_2}}{k_4} = 0$$

$$\Rightarrow \quad (k_2 + 2k_3)\ C_{NO_3}C_{NO_2} = k_1\ C_{N_2O_5}$$

$$\Rightarrow \quad C_{NO_3}C_{NO_2} = \frac{k_1\ C_{N_2O_5}}{(k_2 + 2k_3)} \tag{3.3.5}$$

Substituting the value of $C_{NO_3}C_{NO_2}$ from Equation 3.3.5 into Equation 3.3.3

$$\left(-r_{N_2O_5}\right)_{net} = k_1\ C_{N_2O_5} - k_2\frac{k_1\ C_{N_2O_5}}{(k_2 + 2k_3)}$$

$$\Rightarrow \quad \left(-r_{N_2O_5}\right)_{net} = \left[k_1 - \frac{k_1k_2}{(k_2 + 2k_3)}\right] C_{N_2O_5}$$

$$\Rightarrow \quad \left(-r_{N_2O_5}\right)_{net} = k'\ C_{N_2O_5} \tag{3.3.6}$$

Therefore, the net rate of disappearance of N_2O_5 is first order in $C_{N_2O_5}$, which is in accordance with the observed reaction rate.

3.2.2 Temperature Dependence of Reaction Rate for Homogeneous Reactions

Referring to Section 3.2, the reaction rate depends on the probability of the reactant molecules converting into products. In an elementary reaction, the reactant molecules must pass through transition states with one or more chemical bonds at excited energy states. Consequently, the molecules with higher energy will have a higher probability of acquiring the transition state and converting into products. Since the average energy of molecules depends on temperature, it will be natural to expect the reaction rate to also depend on temperature. In Equation 3.19, the specific reaction rate constant "k" is constant only if the temperature of the reacting system does not change. The specific reaction rate constant has different values at different temperatures. The higher the temperature, the higher is the specific reaction rate constant. The dependence of a specific reaction rate constant with temperature is best represented by *Arrhenius' equation* (Arrhenius, 1889; Logan, 1982). It is an empirical equation that fits the experimental observation excellently over a wide range of temperature for almost all reactions. Mathematically, by this equation the specific reaction rate constant for a reaction is expressed as

$$k = k_0 \exp\left(-\frac{E_a}{RT}\right) \tag{3.29}$$

$$\Rightarrow \quad \ln(k) = \ln(k_0) - \frac{E_a}{RT} \tag{3.30}$$

k_0 is known as the preexponential factor or the *frequency factor* and E_a is termed the *activation energy* of the reaction. At two different temperatures T_1 and T_2, the specific reaction rate constants are related as

$$\frac{k_{T_1}}{k_{T_2}} = \exp\left\{\frac{E_a}{R}\left(\frac{1}{T_2} - \frac{1}{T_1}\right)\right\} \tag{3.31}$$

$$\Rightarrow \quad \ln\left(\frac{k_{T_1}}{k_{T_2}}\right) = \frac{E_a}{R}\left(\frac{1}{T_2} - \frac{1}{T_1}\right) \tag{3.32}$$

Equation 3.30 is helpful in finding the activation energy and frequency factor for a reaction if the specific reaction rate constants at different temperatures are known. This equation provides a linear relationship between ln(k) and 1/T with a slope of $-E_a/R$ and an intercept of $\ln(k_0)$ on the ln(k) axis.

EXAMPLE 3.4

For a reaction the experiment suggests that the specific reaction rate constant, k, varies with temperature as follows:

T (K)	323	353	373	393	413	433
k (min^{-1})	4.48	7.17	9.39	12.18	16.44	21.76

Determine the activation energy and frequency factor for the reaction.

Solution

Graphical method:

From the given data we complete Table EXE3.4 by computing 1/T and ln(k).

TABLE EXE3.4

T	k	1/T	ln(k)
323	4.481689	0.003096	1.5
353	7.170676	0.002833	1.97
373	9.393331	0.002681	2.24
393	12.18249	0.002545	2.5
413	16.44465	0.002421	2.8
433	21.7584	0.002309	3.08

Next, we plot the calculated values of ln(k) against the corresponding calculated values of 1/T and draw a straight line passing through all these points as shown in Figure EXE3.4.

From the slope of the straight line

$$\frac{E_a}{R} = \frac{(3.2-1.5)}{(0.00307-0.0022)} = 1954$$

$$\Rightarrow E_a = 1954 \times 8.314 = 16245.6 \text{ J/mole}$$

For the determination of frequency factor, we must produce the straight line to the point where it intersects with the ln(k) axis. Obviously, this will be difficult on a graph sheet of limited dimension. Alternatively, we can write the equation of the plotted straight line as

$$\frac{\{\ln(k)-3.2\}}{\{(1/T)-0.0022\}} = -1954$$

$$\Rightarrow \quad \ln(k) = -1954\left(\frac{1}{T}\right) + 7.4988$$

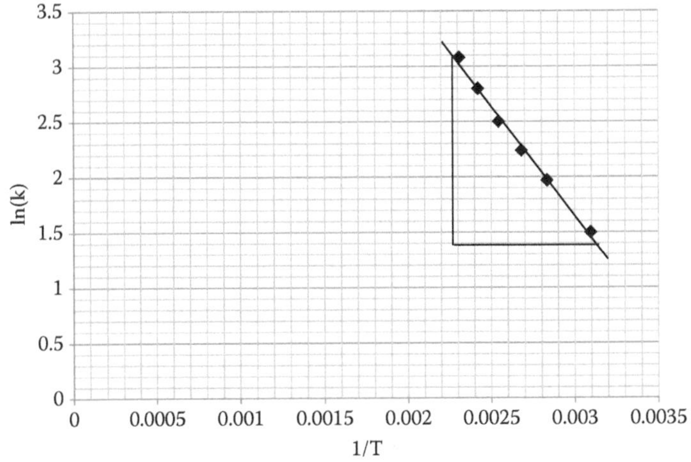

FIGURE EXE3.4 Plot between ln(k) and 1/T.

So,

$$\ln(k_0) = 7.4988$$

$$\Rightarrow \quad k_0 = 1805.9 \text{ min}^{-1}$$

Alternatively, the values for ln(k) and 1/T can be fitted into a straight line using a suitable numerical technique like the least-square method.

Arrhenius' equation will be rigorously applicable only for elementary reactions. For nonelementary reactions, like reactions 3.23 and 3.24, each constituent elementary reaction has its own specific reaction rate constant and activation energy. To study the temperature dependence of such reactions, each constituent elementary reaction has to be given due consideration.

EXAMPLE 3.5

For reaction 3.23, deduce an expression relating the reaction rate with temperature.

Solution

From Equation 3.3.6, the reaction rate for this reaction can be expressed as

$$\left(-r_{N_2O_5}\right)_{net} = k'C_{N_2O_5}$$

where

$$k' = k_1 - \frac{k_1k_2}{(k_2 + 2k_3)}$$

$$\Rightarrow \quad k' = \frac{2k_1k_3}{(k_2 + 2k_3)} \tag{3.5.1}$$

Let the activation energies for the first three constituent elementary reactions for reaction 3.23 be E_{a1}, E_{a2}, and E_{a3}, respectively. Then,

$$k_1 = k_{01} \exp\left(-\frac{E_{a1}}{RT}\right) \tag{3.5.2}$$

$$k_2 = k_{02} \exp\left(-\frac{E_{a2}}{RT}\right) \tag{3.5.3}$$

$$k_3 = k_{03} \exp\left(-\frac{E_{a3}}{RT}\right) \tag{3.5.4}$$

Substituting the values from Equations 3.5.2 through 3.5.4 into Equation 3.5.1

$$k' = \frac{2k_{01}k_{03} \exp\left\{-(E_{a1} + E_{a3})/RT\right\}}{\{k_{02} \exp(-E_{a2}/RT) + 2k_{03} \exp(-E_{a3}/RT)\}} \tag{3.5.5}$$

For the same concentration of N_2O_5 if the reaction rates at temperatures T_1 and T_2 are $\left(-r_{N_2O_5}\right)_{T1}$ and $\left(-r_{N_2O_5}\right)_{T2}$, respectively, then

$$\frac{\left(-r_{N_2O_5}\right)_{T1}}{\left(-r_{N_2O_5}\right)_{T2}} = \frac{2k_{01}k_{03}\exp\{-(E_{a1}+E_{a3})/RT_1\}}{\{k_{02}\exp(-E_{a2}/RT_1)+2k_{03}\exp(-E_{a3}/RT_1)\}}$$

$$\times\frac{\{k_{02}\exp(-E_{a2}/RT_2)+2k_{03}\exp(-E_{a3}/RT_2)\}}{2k_{01}k_{03}\exp\{-E_{a1}+E_{a3}/RT_2\}}$$

$$\Rightarrow \frac{\left(-r_{N_2O_5}\right)_{T1}}{\left(-r_{N_2O_5}\right)_{T2}} = \exp\left\{\frac{(E_{a1}+E_{a3})}{R}\left(\frac{1}{T_2}-\frac{1}{T_1}\right)\right\}\times\frac{\{k_{02}\exp(-E_{a2}/RT_2)+2k_{03}\exp(-E_{a3}/RT_2)\}}{\{k_{02}\exp(-E_{a2}/RT_1)+2k_{03}\exp(-E_{a3}/RT_1)\}}$$

$$\Rightarrow \frac{\left(-r_{N_2O_5}\right)_{T1}}{\left(-r_{N_2O_5}\right)_{T2}} = \exp\left\{\frac{(E_{a1}+E_{a3}-E_{a2})}{R}\left(\frac{1}{T_2}-\frac{1}{T_1}\right)\right\}\times\frac{\{1+2k_{03}/k_{02}\exp((E_{a2}-E_{a3})/RT_2)\}}{\{1+2k_{03}/k_{02}\exp((E_{a2}-E_{a3})/RT_1)\}} \tag{3.5.6}$$

3.2.3 Multiple Reactions

There are reacting systems where a single stoichiometric equation and a single rate equation are not sufficient to describe the changes observed to be occurring. Since more than one stoichiometric equation and rate equation are needed to completely describe the system, these reactions are known as *multiple reactions*. First, we must clearly distinguish between nonelementary reactions and multiple reactions. It may be argued that a nonelementary reaction is actually a combination of more than one elementary reaction. For each of these elementary reactions, we can write a separate reaction rate. Why then are they not called multiple reactions? The difference lies in the time scale in which the reaction intermediates exist in these constituent elementary reactions and their abundance. Their presence can be felt but they cannot be isolated from the reacting system as stable products. Contrary to this, multiple reactions are combinations of single reactions, which can stand out on their own. Each of these single reactions leads to stable products that can be isolated from the reacting system. Multiple reactions can be *series reactions, parallel reactions*, or mixed schemes of series and parallel reactions.

3.2.3.1 Series Reactions

Simplest of the series reactions will have two sequential steps and can be schematically represented as

$$\sum_{i=1}^{n} a_i A_i \rightarrow \sum_{j=1}^{l} p_j P_j \rightarrow \sum_{r=1}^{r} q_r Q_r \tag{3.33}$$

where

$$-r_{A_i} = -\frac{d\left(C_{A_i}\right)}{dt} = f_1'(C_{A_i}) \tag{3.34}$$

$$-r_{P_j} = -\frac{d\left(C_{P_j}\right)}{dt} = f_2'\left(C_{P_j}\right) - f_1'(C_{A_i}) \tag{3.35}$$

$$r_{Q_r} = \frac{d\left(C_{Q_r}\right)}{dt} = f_2'\left(C_{P_j}\right) \tag{3.36}$$

$$N_{P_j} - N_{P_{j0}} = \frac{p_j}{a_i}(N_{A_0} - N_A) - \frac{p_j}{q_r}(N_Q - N_{Q_0})$$

$$\Rightarrow \quad \frac{N_{A_0}}{a_i} + \frac{N_{P_0}}{p_j} + \frac{N_{Q_0}}{q_r} = \frac{N_A}{a_i} + \frac{N_P}{p_j} + \frac{N_Q}{q_r} \tag{3.37}$$

Similar treatment can be extended for a series reaction having more than two sequential steps.

3.2.3.2 Parallel Reactions

The simplest of the parallel reactions, having two competitive reaction paths, can be schematically represented as

$$\sum_{i=1}^{n} a_i A_i \begin{array}{l} \nearrow \sum_{j=1}^{1} p_j P_j \\ \searrow \sum_{r=1}^{r} q_r Q_r \end{array} \tag{3.38}$$

$$r_{P_j} = \frac{d\left(C_{P_j}\right)}{dt} = f_3'\left(C_{A_i}\right) \tag{3.39}$$

$$r_{Q_r} = \frac{d\left(C_{Q_r}\right)}{dt} = f_4'\left(C_{A_i}\right) \tag{3.40}$$

$$-r_{A_i} = -\frac{d\left(C_{A_i}\right)}{dt} = f_3'(C_{A_i}) + f_4'\left(C_{A_i}\right) \tag{3.41}$$

$$\frac{a_i}{p_j}\left(N_{P_j} - N_{P_{j0}}\right) + \frac{a_i}{q_r}\left(N_{Q_r} - N_{Q_{r0}}\right) = \left(N_{A_{i0}} - N_{A_i}\right)$$

$$\Rightarrow \quad \frac{N_{A_{i0}}}{a_i} + \frac{N_{P_{j0}}}{p_j} + \frac{N_{Q_{r0}}}{q_r} = \frac{N_{A_i}}{a_i} + \frac{N_{P_j}}{p_j} + \frac{N_{Q_r}}{q_r} \tag{3.42}$$

Similar treatment can be extended for a parallel reaction having more than two competitive reaction paths.

3.2.3.3 Mixed Series Parallel Reactions

These are combinations of series and parallel reactions, for example,

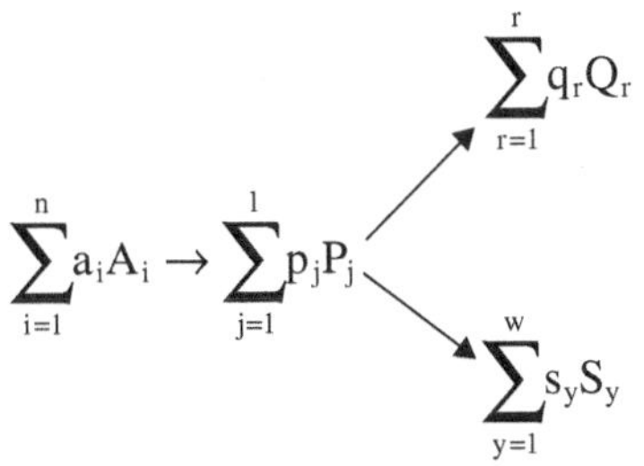

(3.43)

Reaction 3.43 can be resolved into two competitive series reactions

$$\sum_{i=1}^{n} a_i A_i \rightarrow \sum_{j=1}^{1} p_j P_j \rightarrow \sum_{r=1}^{r} q_r Q_r \tag{3.44}$$

and

$$\sum_{i=1}^{n} a_i A_i \rightarrow \sum_{j=1}^{1} p_j P_j \rightarrow \sum_{y=1}^{w} s_y S_r \tag{3.45}$$

for which

$$-r_{A_i} = -\frac{d\left(C_{A_i}\right)}{dt} = f_5'(C_{A_i}) \tag{3.46}$$

$$r_{Q_r} = \frac{d\left(C_{Q_r}\right)}{dt} = f_6'\left(C_{P_j}\right) \tag{3.47}$$

$$r_{S_y} = \frac{d\left(C_{S_y}\right)}{dt} = f_7'\left(C_{P_j}\right) \tag{3.48}$$

$$-r_{P_j} = -\frac{d\left(C_{P_j}\right)}{dt} = f_6'(C_{P_j}) + f_7'\left(C_{P_j}\right) - f_5'\left(C_{A_i}\right) \tag{3.49}$$

$$\frac{p_j}{a_i}\left(N_{A_{i0}} - N_{A_i}\right) - \frac{p_j}{s_y}\left(N_{S_y} - N_{S_{y0}}\right) - \frac{p_j}{q_r}\left(N_{Q_r} - N_{Q_{r0}}\right) = N_{P_j} - N_{P_{j0}}$$

$$\Rightarrow \quad \frac{N_{A_{i0}}}{a_i} + \frac{N_{P_{j0}}}{p_j} + \frac{N_{Q_{r0}}}{q_r} + \frac{N_{S_{y0}}}{s_y} = \frac{N_{A_i}}{a_i} + \frac{N_{P_j}}{p_j} + \frac{N_{Q_r}}{q_r} + \frac{N_{S_y}}{s_y} \tag{3.50}$$

EXAMPLE 3.6

For the multiple reaction

$$2C_2H_4 + O_2 \rightarrow 2C_2H_4O \tag{3.6.1}$$

$$2C_2H_4O + 5O_2 \rightarrow 4CO_2 + 4H_2O \tag{3.6.2}$$

$$C_2H_4 + 3O_2 \rightarrow 2CO_2 + 2H_2O \tag{3.6.3}$$

identify the constituent series and parallel reactions and the required rate equations.

Solution

Constituent series reaction is

$$2C_2H_4 \xrightarrow{O_2} 2C_2H_4O \xrightarrow{5O_2} CO_2 + H_2O \tag{3.6.4}$$

Constituent parallel reactions are

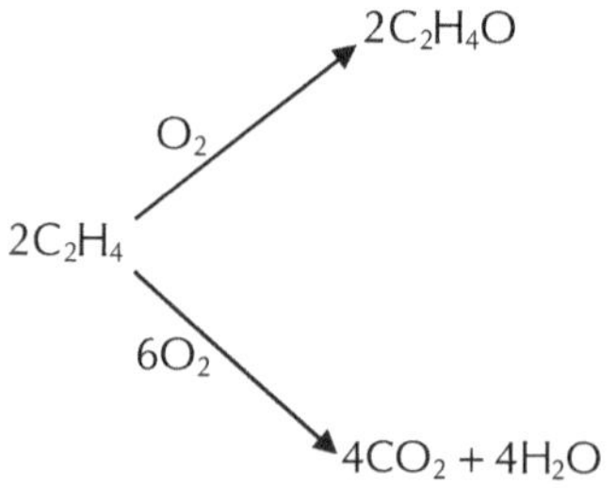

(3.6.5)

Required rate equations are

$$-r_{C_2H_4} = -\frac{d\left(C_{C_2H_4}\right)}{dt} = f_{1(C_2H_4)'}\left(C_{C_2H_4}\right) + f_{2(C_2H_4)'}\left(C_{C_2H_4}\right) \tag{3.6.6}$$

$$-r_{C_2H_4O} = -\frac{d\left(C_{C_2H_4O}\right)}{dt} = f_{C_2H_4O'}\left(C_{C_2H_4O}\right) - f_{1C_2H_4'}\left(C_{C_2H_4}\right) \tag{3.6.7}$$

$$r_{CO_2} = \frac{d\left(C_{CO_2}\right)}{dt} = f_{C_2H_4O'}\left(C_{C_2H_4O}\right) + f_{2(C_2H_4)'}\left(C_{C_2H_4}\right) \tag{3.6.8}$$

3.2.4 Homogeneous Catalytic Reactions

Certain reactions undergo an appreciable change in the reaction rate upon the addition of a substance other than the reactants. This phenomenon is known as *catalysis* and the substance other than the reactants that changes the reaction rate is called a *catalyst*. In principle, the phenomenon of catalysis comprises both the increase and decrease of the reaction rate. However, practically we are most interested in cases where the addition of the catalyst increases the reaction rate because of the commercial importance of such cases. Homogeneous catalytic reactions are those where the reactants and the catalyst are in the same phase. Some of the essential features of catalysts and catalytic reactions are

- In most cases, the catalysts are required at a very less amount.
- Catalysts change the reaction kinetics but not the reaction stoichiometry.

- In reversible elementary reactions, the catalyst does not change the equilibrium.

The catalyst is believed to function by lowering the activation energy of the reaction. It interacts with the reactant moiety to form a transition intermediate other than the one that is formed in the absence of the catalyst. Less energy is required for the formation of this catalyst-assisted transition intermediate compared to the noncatalytic one. Thus, in essence the presence of catalyst provides an energetically less demanding alternative pathway to the reactant to transform into product. This in turn manifests in the increase in the reaction rate. The catalyst-assisted transition intermediate transforms into the product and the free catalyst that can reinteract with another reactant moiety.

3.3 KINETICS OF HETEROGENEOUS REACTIONS

Heterogeneous reactions are those where the reacting system consists of more than one phase. A general classification of heterogeneous reactions is presented in Figure 3.2.

In noncatalytic heterogeneous reactions, the reacting species are separated by a phase boundary. The reaction can occur at this boundary as in the case of many fluid–solid noncatalytic reactions as well as some instances of fluid–fluid noncatalytic reactions. In such case, the reactant within the fluid phase must move from the bulk of the fluid phase to the interface. The reaction can also occur within the interior of one of the phases as observed in many fluid–fluid noncatalytic reactions. In such cases, not only the reactant within one of the fluid phases must move from the bulk to the interface but also cross over the interface into the second fluid phase and move within that phase. In heterogeneous catalytic reactions, the chemical transformation takes place on the surface of a solid catalyst. These reaction sites mostly lie in the

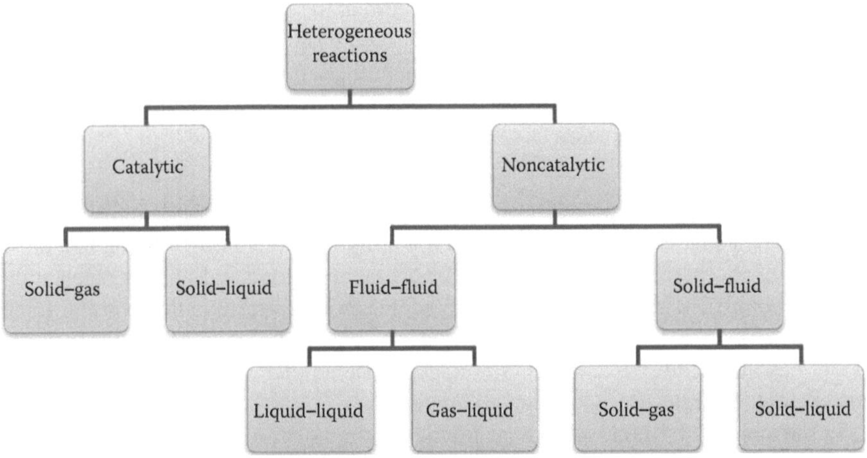

FIGURE 3.2 Classification of heterogeneous reactions.

interior of the catalyst having a porous structure. The reactant within the fluid phase must move from the bulk of the fluid phase all the way to these reaction sites. Thus, for heterogeneous reactions, the physical movement of reacting moieties (and also the products sometimes), besides the actual chemical transformation, is an integral part of the overall process. Essentially, we are looking at problems of combined mass transfer and chemical reaction. We will first discuss the noncatalytic heterogeneous reactions and subsequently the heterogeneous catalytic reactions.

3.3.1 Fluid–Solid Noncatalytic Reactions

These are the reactions where some of the reactants are in the fluid (gas or liquid) phase while some are in the solid phase. The reactant from the fluid migrates to the surface of the solid where the reaction takes place. Depending on the type of solid and the products generated, these reactions can be of various types as presented in Table 3.3. For these reactions, the solid can be characterized as nonporous or porous.

In Table 3.3, the first three possible cases pertain to the nonporous solids. All these possible cases have one common feature. The reaction occurs at the outer surface of the unreacted solid, which decreases in size with the progress of the reaction. The last three cases pertain to the porous solids. Here, the reaction at all times occurs throughout the bulk of the solid. However, the availability and hence the concentration of the reactant decreases as we move deeper inside the solid. Even in the case of porous solids, if the reaction is very fast compared to the migration of reactant from the fluid to the reaction site, the zone of reaction will be practically limited to a thin layer on the outer surface of the unreacted solid. In such cases, even the porous solid can be approximated as a nonporous solid. Based on this discussion, we can perceive two physical models and their corresponding mathematical treatments for the solid–fluid noncatalytic reactions. These are

1. Shrinking core model. The reaction occurs at the outer surface of the unreacted solid, which decreases in size with the progress of reaction.
2. Progressive conversion model. The reaction at all times occurs throughout the bulk of the solid. However, the availability and hence the concentration of the reactant decreases as we move deeper inside the solid.

Researchers have proposed several variations to these models. It must be clear to the student that each model is based on a simplified picture of the phenomenon in terms of the perceived geometrical arrangement of the solid and its orientation in space coupled with the movement of reactant moieties relative to the solid. Each model has its set of assumptions that may not be rigorously true. The success of the model lies in its ability to conform to the experimental data. In this context, there is no substitute for real-time experimental data.

In the majority of fluid–solid noncatalytic reactions, the experimental data fit well into the mathematical analyses of the *shrinking core model*. On this account, this model has wide acceptance and is presented here.

TABLE 3.3
Possible Types of Fluid–Solid Noncatalytic Reactions

Type of Solid	Type of Product	Example	Schematic Representation of the Process
Nonporous	Fluid	C(graphite) + O_2(air) → CO_2(g)	1. Migration of reactant from the bulk of fluid to the surface of solid. 2. Chemical reaction. 3. Migration of product from the surface of solid to the bulk of fluid.
Nonporous	Solid; forming an outer layer over the inner unreacted solid	Fe(s) + O_2(air) → Fe_2O_3(s)	1. Migration of reactant from the bulk of fluid to the outer surface of solid product layer. 2. Migration of reactant through the solid product layer to the surface of unreacted solid. 3. Chemical reaction.
Nonporous	Solid; detaching and separating from the remaining unreacted solid	Zn(s) + I_2(solution) → 2ZnI	1. Migration of reactant from the bulk of fluid to the surface of solid. 2. Chemical reaction.
Porous	Fluid	C(coal) + O_2(air) → CO_2(g)	1. Migration of reactant from the bulk of fluid to the external surface of solid. 2. Migration of reactant from the external surface of solid to the internal surface of the solid through the pores inside the solid. 3. Chemical reaction. 4. Migration of product from the internal surface of solid to the external surface of solid through the pores inside the solid. 5. Migration of product from the external surface of solid to the bulk of fluid.
Porous	Solid	CaO(s) + H_2O(l) → $Ca(OH)_2$(s)	1. Migration of reactant from the bulk of fluid to the external surface of solid. 2. Migration of reactant from the external surface of solid to the internal surface of the solid through the pores inside the solid. 3. Chemical reaction.
Porous	Solid + fluid	Na_2CO_3(solution) + $CaOH_2$(s) → 2NaOH(solution) + $CaCO_3$(s)	1. Migration of reactant from the bulk of fluid to the external surface of solid. 2. Migration of reactant from the external surface of solid to the internal surface of the solid through the pores inside the solid. 3. Chemical reaction. 4. Migration of fluid product from the internal surface of solid to the external surface of solid through the pores inside the solid. 5. Migration of fluid product from the external surface of solid to the bulk of fluid.

3.3.1.1 Shrinking Core Model

The mathematical model for the simplest case of a single spherical solid particle, where the reaction is first order in the gaseous reactant, was first developed by Yagi and Kunii (1961). Consider a fluid–solid noncatalytic reaction

$$A_1(fl) + a_2A_2(s) \rightarrow p_1P_1(fl) + p_2P_2(s) \quad (3.51)$$

In a general sense, the overall process consists of the following steps taking place sequentially (Table 3.3):

1. Migration of A_1 from the bulk of fluid to the outer surface of the layer of P_2
2. Migration of reactant A_1 through the layer of P_2 to the surface of unreacted A_2
3. Chemical reaction between A_1 and A_2
4. Migration of P_1 through the layer of P_2 to the outer surface of the layer of P_2
5. Migration of P_1 from the outer surface of the layer of P_2 to the bulk of fluid

When no solid product is formed (case 1 in Table 3.3) or when the solid product detaches and separates from the remaining unreacted solid (cases 3 and 5 in Table 3.3), Steps 2 and 4 will be absent. As a further simplification, if the reaction is irreversible, the presence of P_1 will not influence the rate of the reaction and that of the overall process. The steps influencing the rate of the overall process then reduce to

1. Migration of A_1 from the bulk of fluid to the outer surface of the layer of P_2
2. Migration of reactant A_1 through the layer of P_2 to the surface of unreacted A_2
3. Chemical reaction between A_1 and A_2

with Step 2 either present or absent.

The situation is pictorially represented in Figure 3.3. We analyze the two cases, one where a layer of solid product is formed on the surface of the unreacted core, and the other when no solid product is formed or when the solid product detaches and separates from the remaining unreacted solid.

3.3.1.1.1 Solid Product Is Formed on the Surface of the Unreacted Core

At steady state

$$\left(-\frac{dN_{A_1}}{dt}\right)_{\text{Step 1}} = \left(-\frac{dN_{A_1}}{dt}\right)_{\text{Step 2}} = \left(-\frac{dN_{A_1}}{dt}\right)_{\text{Step 3}} = -\frac{dN_{A_1}}{dt} \quad (3.52)$$

For Step 1,

$$\left(-\frac{dN_{A_1}}{dt}\right)_{\text{Step 1}} = k_{fl} \times 4\pi R^2\left[\left(C_{A_1}\right)_{fl} - \left(C_{A_1}\right)_{OS}\right]$$

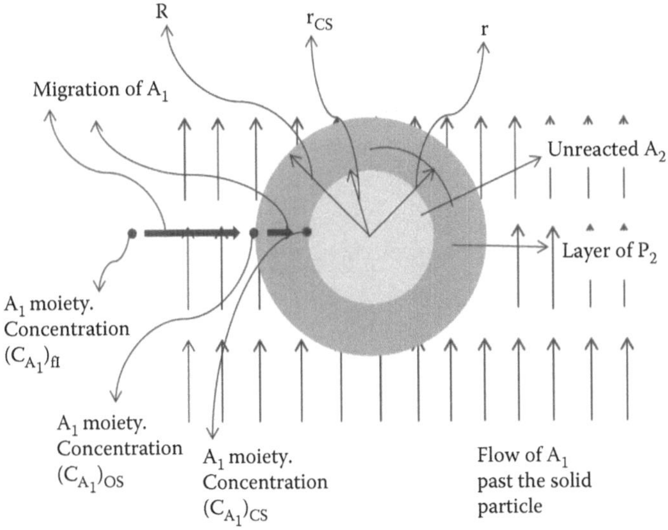

FIGURE 3.3 Physical interpretation of shrinking core model.

$$\frac{1}{k_{fl} \times 4\pi R^2}\left(-\frac{dN_{A_1}}{dt}\right)_{\text{Step 1}} = (C_{A_1})_{fl} - (C_{A_1})_{OS} \tag{3.53}$$

Using the equality of Equation 3.52,

$$\frac{1}{k_{fl} \times 4\pi R^2}\left(-\frac{dN_{A_1}}{dt}\right) = (C_{A_1})_{fl} - (C_{A_1})_{OS} \tag{3.54}$$

For Step 2,

$$\left(-\frac{dN_{A_1}}{dt}\right)_{\text{Step 2}} = \left(-\frac{dN_{A_1}}{dt}\right) = 4\pi r^2 D \frac{dC_{A_1}}{dr} \tag{3.55}$$

Since $(-dN_{A_1}/dt)$ is independent of location, we separate the variables C_{A_1} and r in Equation 3.56 and integrate as

$$\Rightarrow \frac{1}{4\pi D}\left(-\frac{dN_{A_1}}{dt}\right)\int_{r_{CS}}^{R}\frac{dr}{r^2} = \int_{(C_{A_1})_{CS}}^{(C_{A_1})_{OS}} dC_{A_1}$$

$$\Rightarrow \frac{1}{4\pi D}\left(-\frac{dN_{A_1}}{dt}\right)\left(\frac{1}{r_{CS}} - \frac{1}{R}\right) = (C_{A_1})_{OS} - (C_{A_1})_{CS} \tag{3.56}$$

For Step 3, in general terms the rate for the reaction represented by the chemical Equation 3.51 will be proportional to $(C_{A_1})_{CS}C_{A_2}^{a2}$. However, since the reaction is occurring only at the surface of the unreacted core, which is progressively receding inward, the concentration of A_2 within the unreacted core remains unchanged. Therefore, the concentration term of A_2 is essentially constant and can be merged with the rate constant. We then have

$$\left(-\frac{dN_{A_1}}{dt}\right)_{\text{Step 3}} = \left(-\frac{dN_{A_1}}{dt}\right) = 4\pi r_{CS}^2 k'(C_{A_1})_{CS} \tag{3.57}$$

$$\frac{1}{k' \times 4\pi r_{CS}^2}\left(-\frac{dN_{A_1}}{dt}\right) = (C_{A_1})_{CS} \tag{3.58}$$

Adding Equations 3.54, 3.56, and 3.58, we have

$$\Rightarrow \left[\frac{1}{k_{fl} \times 4\pi R^2} + \frac{1}{4\pi D}\left(\frac{1}{r_{CS}} - \frac{1}{R}\right) + \frac{1}{k' \times 4\pi r_{CS}^2}\right]\left(-\frac{dN_{A_1}}{dt}\right) = (C_{A_1})_{fl}$$

$$\Rightarrow \frac{1}{4\pi R^2}\left(-\frac{dN_{A_1}}{dt}\right) = \frac{(C_{A_1})_{fl}}{1/k_{fl} + R(R - r_{CS})/Dr_{CS} + R^2/k' r_{CS}^2} \tag{3.59}$$

$$\Rightarrow (-r_{A_1})' = \frac{(C_{A_1})_{fl}}{1/k_{fl} + R(R - r_{CS})/Dr_{CS} + R^2/k' r_{CS}^2} \tag{3.60}$$

It is interesting to note here that in Equations 3.57, 3.59, and 3.60, the rate is expressed on the basis of area instead of volume. It is customary of most heterogeneous reactions.

A detailed discussion on the estimation of *mass transfer coefficients* is beyond the scope of this book. For that purpose, the reader is referred to texts dealing exclusively with the subject of mass transfer and transport phenomena. In brief, mass transfer coefficient, k_{fl} to be used with Equations 3.59 and 3.60, is determined by the use of empirical correlations. A number of such correlations are available for different conditions. For fluid moving past a single spherical particle, Froessling (1938) provided the following correlation:

$$Sh = 2 + 0.6\, Re^{1/2} Sc^{1/3} \tag{3.61}$$

$$\Rightarrow \frac{k_{fl} d_p}{D_{A_1}} = 2 + 0.6\left(\frac{u\rho d_p}{\mu}\right)^{1/2}\left(\frac{\mu}{D_{A_1}\rho}\right)^{1/3} \tag{3.62}$$

Fluid passing through a packed bed of particles with Re > 80, Ranz (1952) has suggested the following correlation:

$$\mathrm{Sh} = 2 + 1.8\,\mathrm{Re}^{1/2}\mathrm{Sc}^{1/3} \tag{3.63}$$

From the balanced chemical Equation 3.51,

$$\left(-r_{A_2}\right)' = \frac{1}{4\pi R^2}\left(-\frac{dN_{A_2}}{dt}\right) = a_2\left(-r_{A_1}\right)' = \frac{a_2\left(C_{A_1}\right)_{fl}}{1/k_{fl} + R(R - r_{CS})/Dr_{CS} + R^2/k'r_{CS}^2} \tag{3.64}$$

$$\Rightarrow \left(-\frac{dN_{A_2}}{dt}\right) = \frac{a_2\left(C_{A_1}\right)_{fl}}{1/k_{fl}\times 4\pi R^2 + (1/4\pi D)\left(1/r_{CS} - 1/R\right) + 1/k'\times 4\pi r_{CS}^2} \tag{3.65}$$

$$\left(-\frac{d\left(4/3\pi r_{CS}^3\rho_{A_2}\right)}{dt}\right) = \frac{a_2\left(C_{A_1}\right)_{fl}}{1/k_{fl}\times 4\pi R^2 + (1/4\pi D)(1/r_{CS} - 1/R) + 1/k'\times 4\pi r_{CS}^2}$$

$$\Rightarrow -\frac{4}{3}\pi\rho_{A_2}\frac{d(r_{CS}^3)}{dt} = \frac{a_2\left(C_{A_1}\right)_{fl}}{1/k_{fl}\times 4\pi R^2 + (1/4\pi D)(1/r_{CS} - 1/R) + 1/k'\times 4\pi r_{CS}^2}$$

$$\Rightarrow -4\pi\rho_{A_2}r_{CS}^2\frac{d(r_{CS})}{dt} = \frac{a_2\left(C_{A_1}\right)_{fl}}{1/k_{fl}\times 4\pi R^2 + (1/4\pi D)(1/r_{CS} - 1/R) + 1/k'\times 4\pi r_{CS}^2}$$

$$\Rightarrow -\int_R^{r_{CS}}\left[\frac{\rho_{A_2}}{k_{fl}R^2}r_{CS}^2 + \frac{\rho_{A_2}}{D}r_{CS} - \frac{\rho_{A_2}}{DR}r_{CS}^2 + \frac{\rho_{A_2}}{k'}\right]d(r_{CS}) = a_2\left(C_{A_1}\right)_{fl}\int_0^t dt$$

$$\Rightarrow \left[\frac{\rho_{A_2}}{3k_{fl}R^2}(R^3 - r_{CS}^3) + \frac{\rho_{A_2}}{2D}(R^2 - r_{CS}^2) - \frac{\rho_{A_2}}{3DR}(R^3 - r_{CS}^3) + \frac{\rho_{A_2}}{k/}(R - r_{CS})\right] = a_2\left(C_{A_1}\right)_{fl}t$$

$$\Rightarrow t = \frac{\rho_{A_2}}{a_2\left(C_{A_1}\right)_{fl}}\left[\frac{1}{3k_{fl}R^2}(R^3 - r_{CS}^3) + \frac{1}{2D}(R^2 - r_{CS}^2) - \frac{1}{3DR}(R^3 - r_{CS}^3) + \frac{1}{k'}(R - r_{CS})\right] \tag{3.66}$$

This is the time required in which the solid particle with an initial radius of R will be left with an unreacted core of radius r_{CS}. From the definition of conversion

$$X_{A_2} = \frac{\left(N_{A_2}\right)_0 - N_{A_2}}{\left(N_{A_2}\right)_0} = 1 - \frac{4/3\pi r_{CS}^3\rho_{A_2}}{4/3\pi R^3\rho_{A_2}} = 1 - \frac{r_{CS}^3}{R^3}$$

$$\Rightarrow r_{CS} = R(1 - X_{A_2})^{1/3} \tag{3.67}$$

Using this value of r_{CS} in Equation 3.63,

$$t = \frac{\rho_{A_2}}{a_2\left(C_{A_1}\right)_{fl}}\left\{\frac{RX_{A_2}}{3k_{fl}} + \frac{R^2}{2D}\left[\left(1-\left(1-X_{A_2}\right)^{2/3}\right)\right] - \frac{R^2X_{A_2}}{3D} + \frac{R}{k'}\left[1-\left(1-X_{A_2}\right)^{1/3}\right]\right\} \quad (3.68)$$

At complete conversion, no unreacted core will be left in the solid particle and r_{CS} will be zero. Therefore, we have, from Equation 3.63

$$\Rightarrow \quad t_{total} = \frac{\rho_{A_2}}{a_2\left(C_{A_1}\right)_{fl}}\left[\frac{R}{3k_{fl}} + \frac{R^2}{6D} + \frac{R}{k'}\right] \quad (3.69)$$

3.3.1.1.2 *The Limiting Cases*

Looking at Equation 3.61, the rate of the overall process is dependent on the three resistances, the first due to the fluid covering the particle, the second due to the layer of P_2 around the unreacted core of A_2, and the third due to the chemical reaction proper. These resistances are, respectively, $1/k_{fl}$, $R(R - r_{CS})/Dr_{CS}$, $R^2 / k' r_{CS}^2$. Depending on the relative values of these resistances, three different limiting situations may arise.

1. When $1/k_{fl}$, the resistance due to the fluid covering the particle, is much larger compared to the other two resistances, this resistance controls the rate of the overall process. Here, the step "migration of A_1 from the bulk of fluid to the outer surface of the layer of P_2" acts as the rate-limiting step. For this limiting case

$$\left(-r_{A_2}\right)' = a_2\left(-r_{A_1}\right)' = \frac{a_2\left(C_{A_1}\right)_{fl}}{1/k_{fl}} \quad (3.70)$$

$$t = \frac{\rho_{A_2}(R^3 - r_{CS}^3)}{3k_{fl}a_2\left(C_{A_1}\right)_{fl} R^2} = \frac{\rho_{A_2} RX_{A_2}}{3k_{fl}a_2\left(C_{A_1}\right)_{fl}} \quad (3.71)$$

In the Stokes regime, $k_{fl} = \left(D_{A_1}\right)_{fl}/R$, and substituting this value in Equation 3.69,

$$t = \frac{\rho_{A_2} R(R^3 - r_{CS}^3)}{3D_{fl}a_2\left(C_{A_1}\right)_{fl} R^2} = \frac{\rho_{A_2} R^2 X_{A_2}}{3\left(D_{A_1}\right)_{fl} a_2\left(C_{A_1}\right)_{fl}} \quad (3.72)$$

$$t_{total} = \frac{\rho_{A_2} R}{3k_{fl}a_2\left(C_{A_1}\right)_{fl}} \quad (3.73)$$

From Equations 3.69 and 3.70,

$$\frac{t}{t_{total}} = \frac{(R^3 - r_{CS}^3)}{R^3} = X_{A_2} \tag{3.74}$$

2. When $R(R - r_{CS})/Dr_{CS}$, the resistance due to the layer of P_2 around the unreacted core of A_2, is much larger compared to the other two resistances, this resistance controls the rate of the overall process. Here, the step "migration of reactant A_1 through the layer of P_2 to the surface of unreacted A_2" acts as the rate-limiting step. For this limiting case

$$\left(-r_{A_2}\right)' = a_2\left(-r_{A_1}\right)' = \frac{a_2\left(C_{A_1}\right)_{fl}}{R(R - r_{CS})/Dr_{CS}} \tag{3.75}$$

$$t = \frac{\rho_{A_2}}{a_2\left(C_{A_1}\right)_{fl}}\left[\frac{1}{2D}(R^2 - r_{CS}^2) - \frac{1}{3DR}(R^3 - r_{CS}^3)\right] \tag{3.76}$$

$$t = \frac{\rho_{A_2}}{a_2\left(C_{A_1}\right)_{fl}}\left\{\frac{R^2}{2D}\left[1-\left(1-X_{A_2}\right)^{2/3}\right] - \frac{R^2X_{A_2}}{3D}\right\} \tag{3.77}$$

$$t_{total} = \frac{\rho_{A_2}R^2}{6a_2\left(C_{A_1}\right)_{fl}D} \tag{3.78}$$

$$\frac{t}{t_{total}} = 3 - 3(1 - X_{A_2})^{2/3} - 2X_{A_2} \tag{3.79}$$

3. When $R^2 / k'r_{CS}^2$, the resistance due to the chemical reaction proper, is much larger compared to the other two resistances, this resistance controls the rate of the overall process. Here, the step "chemical reaction between A_1 and A_2" acts as the rate-limiting step. For this limiting case

$$\left(-r_{A_2}\right)' = a_2\left(-r_{A_1}\right)' = \frac{a_2\left(C_{A_1}\right)_{fl}}{R^2 / k'r_{CS}^2} \tag{3.80}$$

$$t = \frac{\rho_{A_2}}{a_2\left(C_{A_1}\right)_{fl}}\frac{(R - r_{CS})}{k'} \tag{3.81}$$

$$t = \frac{\rho_{A_2}R}{a_2\left(C_{A_1}\right)_{fl}k'}\left[1-\left(1-X_{A_2}\right)^{1/3}\right] \tag{3.82}$$

$$t_{total} = \frac{\rho_{A_2} R}{a_2 \left(C_{A_1}\right)_{fl} k'} \tag{3.83}$$

$$\frac{t}{t_{total}} = 1 - \left(1 - X_{A_2}\right)^{1/3} \tag{3.84}$$

3.3.1.1.3 Effect of Parameters and the Limiting Resistances

For other parameters remaining the same, Equation 3.71 or 3.72, 3.77, and 3.82 relate the time of reaction with the conversion of the solid reactant A_2. The same is graphically presented in Figure 3.4a. If the process is limited by mass transfer in the fluid

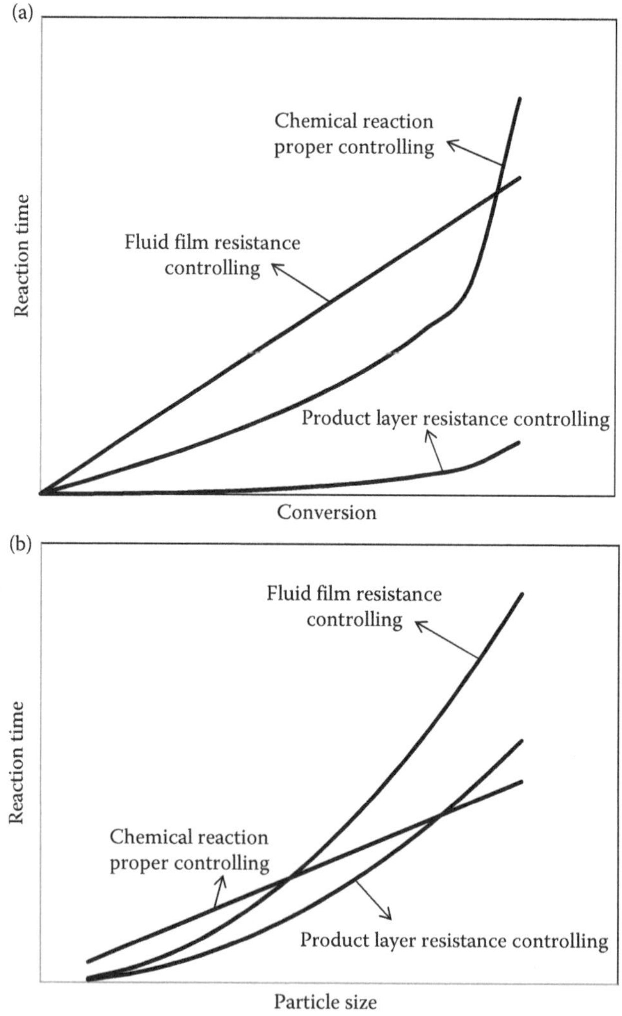

FIGURE 3.4 Controlling resistances in shrinking core model.

phase, the time of reaction linearly varies with the conversion attained. Such a linear dependence is, therefore, a good indication of the controlling resistance being the fluid film. From the same equations, if the other parameters remain the same, the time required for any given conversion varies with the initial radius of the solid particle, R. This variation is presented in Figure 3.4b. If the process is limited by chemical reaction proper, the time of reaction for any given conversion linearly varies with the initial radius of the solid particle. Such a linear dependence serves as a good indication of the controlling resistance being the chemical reaction proper. As a rule of thumb, for solid–fluid noncatalytic reactions where a layer of solid product is formed, we can say that

- If the time of reaction linearly increases with the increase in conversion for a given particle size, but nonlinearly increases with the increase in particle size for a given conversion, the process is limited by mass transfer in the fluid phase.
- If the time of reaction nonlinearly increases with the increase in conversion for a given particle size, but linearly increases with the increase in particle size for a given conversion, the process is limited by chemical reaction proper.
- If the time of reaction nonlinearly increases with the increase in conversion for a given particle size, as well as with the increase in particle size for a given conversion, the process is limited by mass transfer in the product layer formed.

A qualitative discussion on the dependence of k_{fl} on different parameters can be informative. An increase in the fluid velocity relative to the solid particle results in increase in the mass transfer coefficient, k_{fl}. Similarly, an increase in the mass diffusivity of A_1 in the fluid, $\left(D_{A_1}\right)_{fl}$ enhances the mass transfer coefficient, k_{fl}. Conversely, increase in the viscosity of the fluid lowers the mass transfer coefficient. Among these the fluid velocity is the most convenient operating parameter that can be controlled and often acts as the tool of choice for the designer. Mass diffusivity and viscosity are temperature dependent. While mass diffusivity increases with temperature for both liquids and gases, the viscosity increases with temperature for gases but decreases for liquids. The effect of temperature is most pronounced in the case of specific reaction rate constant, k′. For heterogeneous non-catalytic reactions limited by chemical reaction proper, an increase in operating temperature can increase the overall rate of the process.

EXAMPLE 3.7

For the noncatalytic gas–solid reaction

$$A_1(g) + A_2(s) \rightarrow P(s)$$

derive an expression for t/t_{total} when A_2 is in the form of a solid cylindrical particle of radius R and length L with $L \gg R$ and the process is controlled by mass transfer of A_1 in the gas.

Solution

Since $L \gg R$, and $a_2 = 1$

$$(-r_{A_2})' = \frac{1}{2\pi RL}\left(-\frac{dN_{A_2}}{dt}\right) = (-r_{A_1})' = \frac{(C_{A_1})_{fl}}{1/k_{fl}}$$

$$\Rightarrow \quad \frac{\rho_{A_2}}{2R}\left(-\frac{d(r_{CS}^2)}{dt}\right) = \frac{(C_{A_1})_{fl}}{1/k_{fl}}$$

$$\Rightarrow \quad -\frac{\rho_{A_2}}{R} r_{CS}\frac{dr_{CS}}{dt} = \frac{(C_{A_1})_{fl}}{1/k_{fl}}$$

$$\Rightarrow \quad -\frac{\rho_{A_2}}{R}\int_R^{r_{CS}} r_{CS} d(r_{CS}) = (C_{A_1})_{fl} k_{fl}\int_0^t dt$$

$$\Rightarrow \quad \frac{\rho_{A_2}\left(R^2 - r_{CS}^2\right)}{2R} = (C_{A_1})_{fl} k_{fl} t$$

$$\Rightarrow \quad t = \frac{\rho_{A_2}\left(R^2 - r_{CS}^2\right)}{2R(C_{A_1})_{fl} k_{fl}} \tag{3.7.1}$$

$$X_{A_2} = \frac{(N_{A_2})_0 - N_{A_2}}{(N_{A_2})_0} = 1 - \frac{\pi r_{CS}^2 L\rho_{A_2}}{\pi R^2 L\rho_{A_2}} = 1 - \frac{r_{CS}^2}{R^2}$$

$$\Rightarrow \quad r_{CS} = R\left(1 - X_{A_2}\right)^{1/2} \tag{3.7.2}$$

Substituting this value in Equation 3.7.1,

$$t = \frac{\rho_{A_2} R}{2(C_{A_1})_{fl} k_{fl}} X_{A_2} \tag{3.7.3}$$

At complete conversion,

$$t_{total} = \frac{\rho_{A_2} R}{2(C_{A_1})_{fl} k_{fl}} \tag{3.7.4}$$

Hence,

$$\frac{t}{t_{total}} = X_{A_2} \tag{3.7.5}$$

3.3.1.1.4 No Solid Product Layer Is Formed on the Surface of the Unreacted Core

When no solid product is formed or when the solid product detaches and separates from the remaining unreacted solid, the overall process consists of the following steps:

1. Migration of A_1 from the bulk of fluid to the outer surface of the unreacted core of A_2.
2. Chemical reaction between A_1 and A_2.

At steady state,

$$\left(-\frac{dN_{A_1}}{dt}\right)_{Step\ 1} = \left(-\frac{dN_{A_1}}{dt}\right)_{Step\ 2} = -\frac{dN_{A_1}}{dt} \tag{3.85}$$

For Step 1,

$$\left(-\frac{dN_{A_1}}{dt}\right)_{Step\ 1} = -\frac{dN_{A_1}}{dt} = k_{fl} \times 4\pi r_{CS}^2\left[\left(C_{A_1}\right)_{fl} - \left(C_{A_1}\right)_{CS}\right] \tag{3.86}$$

$$\Rightarrow \quad \frac{1}{k_{fl} \times 4\pi r_{CS}^2}\left(-\frac{dN_{A_1}}{dt}\right)_{Step\ 1} = \left(C_{A_1}\right)_{fl} - \left(C_{A_1}\right)_{CS} \tag{3.87}$$

For Step 2,

$$\left(-\frac{dN_{A_1}}{dt}\right)_{Step\ 2} = \left(-\frac{dN_{A_1}}{dt}\right) = 4\pi r_{CS}^2 k'\left(C_{A_1}\right)_{CS} \tag{3.88}$$

$$\Rightarrow \quad \frac{1}{k' \times 4\pi r_{CS}^2}\left(-\frac{dN_{A_1}}{dt}\right) = \left(C_{A_1}\right)_{CS} \tag{3.89}$$

Adding Equations 3.87 and 3.89,

$$\left[\frac{1}{k_{fl} \times 4\pi r_{CS}^2} + \frac{1}{k' \times 4\pi r_{CS}^2}\right]\left(-\frac{dN_{A_1}}{dt}\right) = \left(C_{A_1}\right)_{fl}$$

$$\Rightarrow \quad \frac{1}{4\pi R^2}\left(-\frac{dN_{A_1}}{dt}\right) = \frac{\left(C_{A_1}\right)_{fl}}{R^2/k_{fl}r_{CS}^2 + R^2/k'r_{CS}^2} \tag{3.90}$$

$$\Rightarrow \quad (-r_{A_1})' = \frac{\left(C_{A_1}\right)_{fl}}{R^2/k_{fl}r_{CS}^2 + R^2/k'r_{CS}^2} \tag{3.91}$$

Now,

$$(-r_{A_2})' = \frac{1}{4\pi R^2}\left(-\frac{dN_{A_2}}{dt}\right) = a_2(-r_{A_1})' = \frac{a_2\left(C_{A_1}\right)_{fl}}{R^2/k_{fl}r_{CS}^2 + R^2/k'r_{CS}^2} \tag{3.92}$$

$$\left(-\frac{dN_{A_2}}{dt}\right) = \frac{a_2\left(C_{A_1}\right)_{fl}}{1/k_{fl} \times 4\pi r_{CS}^2 + 1/k' \times 4\pi r_{CS}^2} \tag{3.93}$$

$$\Rightarrow \quad \left(-\frac{d(4/3\pi r_{CS}^3 \rho_{A_2})}{dt}\right) = \frac{a_2\left(C_{A_1}\right)_{fl}}{1/k_{fl} \times 4\pi r_{CS}^2 + 1/k' \times 4\pi r_{CS}^2}$$

$$\Rightarrow \quad -\frac{4}{3}\pi\rho_{A_2}\frac{d(r_{CS}^3)}{dt} = \frac{a_2\left(C_{A_1}\right)_{fl}}{1/k_{fl} \times 4\pi r_{CS}^2 + 1/k' \times 4\pi r_{CS}^2}$$

$$\Rightarrow \quad -4\pi\rho_{A_2} r_{CS}^2 \frac{d(r_{CS})}{dt} = \frac{a_2\left(C_{A_1}\right)_{fl}}{1/k_{fl} \times 4\pi r_{CS}^2 + 1/k' \times 4\pi r_{CS}^2}$$

$$=> \quad -\int_{R}^{r_{CS}} [\rho_{A_2}\left[\frac{1}{k_{fl}} + \frac{1}{k'}\right] d(r_{CS}) = a_2\left(C_{A_1}\right)_{fl} \int_{0}^{t} dt \tag{3.94}$$

If the fluid-phase mass transfer coefficient, k_{fl} is independent of the particle size, Equation 3.94 can be integrated to have

$$\rho_{A_2}\left[\frac{1}{k_{fl}} + \frac{1}{k'}\right](R - r_{CS})] = a_2\left(C_{A_1}\right)_{fl} t$$

$$\Rightarrow \quad t = \frac{\rho_{A_2}}{a_2\left(C_{A_1}\right)_{fl}}\left[\frac{1}{k_{fl}} + \frac{1}{k'}\right](R - r_{CS})] \tag{3.95}$$

And, in combination with Equation 3.67, we have

$$t = \frac{\rho_{A_2} R}{a_2\left(C_{A_1}\right)_{fl}}\left[\frac{1}{k_{fl}} + \frac{1}{k'}\right]\left[1 - \left(1 - X_{A_2}\right)^{1/3}\right]\} \tag{3.96}$$

And, at complete conversion

$$t_{total} = \frac{\rho_{A_2} R}{a_2\left(C_{A_1}\right)_{fl}}\left[\frac{1}{k_{fl}} + \frac{1}{k'}\right] \tag{3.97}$$

However, experience shows that k_{fl}, besides the fluid properties and the flow regime, is dependent on the particle size. We discuss here the particular case of Stokes regime where

$$k_{fl} = \frac{D_{fl}}{r_{CS}} \tag{3.98}$$

Substituting this value in Equation 3.94,

$$-\int_{R}^{r_{CS}} [\rho_{A_2}\left[\frac{r_{CS}}{D_{fl}} + \frac{1}{k'}\right] d(r_{CS}) = a_2\left(C_{A_1}\right)_{fl} \int_0^t dt$$

$$\Rightarrow \quad \rho_{A_2}\left[\frac{(R^2 - r_{CS}^2)}{2D_{fl}} + \frac{(R - r_{CS})}{k'}\right] = a_2\left(C_{A_1}\right)_{fl} t \tag{3.99}$$

$$=> \quad t = \frac{\rho_{A_2}}{a_2\left(C_{A_1}\right)_{fl}}\left[\frac{(R^2 - r_{CS}^2)}{2D_{fl}} + \frac{(R - r_{CS})}{k'}\right] \tag{3.100}$$

In terms of conversion,

$$t = \frac{\rho_{A_2}}{a_2\left(C_{A_1}\right)_{fl}}\left[\frac{R^2\left[1-(1-X_{A_2})^{2/3}\right]}{2D_{fl}} + \frac{R\left[1-(1-X_{A_2})^{1/3}\right]}{k'}\right] \tag{3.101}$$

And,

$$t_{total} = \frac{\rho_{A_2}}{a_2\left(C_{A_1}\right)_{fl}}\left[\frac{R^2}{2D_{fl}} + \frac{R}{k'}\right] \tag{3.102}$$

3.3.1.1.4.1 The Limiting Cases

i. When $k_{fl} \ll k'$ or $R^2/k_{fl}r_{CS}^2 \gg R^2/k'r_{CS}^2$, the resistance due to the fluid covering the particle controls the rate of the overall process. Here, the step "migration of A_1 from the bulk of fluid to the outer surface of the unreacted core of A_2" acts as the rate-limiting step. For this limiting case

$$(-r_{A_2})' = a_2(-r_{A_1})' = \frac{a_2\left(C_{A_1}\right)_{fl}}{R^2 / k_{fl}r_{CS}^2} \tag{3.103}$$

If, k_{fl} is independent of the particle size,

$$t = \frac{\rho_{A_2}}{a_2\left(C_{A_1}\right)_{fl} k_{fl}}(R - r_{CS})] \tag{3.104}$$

$$t = \frac{\rho_{A_2} R}{a_2 \left(C_{A_1}\right)_{fl} k_{fl}} \left[1 - \left(1 - X_{A_2}\right)^{1/3}\right]\} \quad (3.105)$$

$$t_{total} = \frac{\rho_{A_2} R}{a_2 \left(C_{A_1}\right)_{fl} k_{fl}} \quad (3.106)$$

$$\frac{t}{t_{total}} = 1 - \left(1 - X_{A_2}\right)^{1/3} \quad (3.107)$$

In the Stokes regime,

$$t = \frac{\rho_{A_2} (R^2 - r_{CS}^2)}{2a_2 \left(C_{A_1}\right)_{fl} D_{fl}} \quad (3.108)$$

$$t = \frac{\rho_{A_2} R^2}{2a_2 \left(C_{A_1}\right)_{fl} D_{fl}} \left[\left(1 - \left(1 - X_{A_2}\right)^{2/3}\right)\right] \quad (3.109)$$

$$t_{total} = \frac{\rho_{A_2} R^2}{2a_2 \left(C_{A_1}\right)_{fl} D_{fl}} \quad (3.110)$$

$$\frac{t}{t_{total}} = 1 - \left(1 - X_{A_2}\right)^{2/3} \quad (3.111)$$

ii. When $k' \ll k_{fl}$ or $R^2/k'r_{CS}^2 \gg R^2/k_{fl}r_{CS}^2$, the resistance due to the chemical reaction proper controls the rate of the overall process. Here, the step "chemical reaction between A_1 and A_2" acts as the rate-limiting step. For this limiting case

$$\left(-r_{A_2}\right)' = a_2 \left(-r_{A_1}\right)' = \frac{a_2 \left(C_{A_1}\right)_{fl}}{R^2 / k' r_{CS}^2} \quad (3.112)$$

$$t = \frac{\rho_{A_2} (R - r_{CS})}{a_2 \left(C_{A_1}\right)_{fl} k'} \quad (3.113)$$

$$t = \frac{\rho_{A_2} R \left[1 - \left(1 - X_{A_2}\right)^{1/3}\right]}{a_2 \left(C_{A_1}\right)_{fl} k'} \quad (3.114)$$

$$t_{total} = \frac{\rho_{A_2} R}{a_2 \left(C_{A_1}\right)_{fl} k'} \quad (3.115)$$

$$\frac{t}{t_{total}} = 1-\left(1-X_{A_2}\right)^{1/3} \tag{3.116}$$

Here also, Figure 3.4b can be used to identify the controlling resistance. If the reaction time for a given conversion varies linearly with the particle size, the process is limited by chemical reaction proper. Otherwise, it is controlled by mass transfer in fluid film.

EXAMPLE 3.8

For the noncatalytic gas–solid reaction

$$A_1(g) + A_2(s) \rightarrow P(g)$$

derive an expression for t/t_{total} when A_2 is in the form of a solid cylindrical particle of radius R and length L with $L \gg R$ and the process is controlled by mass transfer of A_1 in the gas. Gas flow is in the Stokes regime.

Solution

Since $L \gg R$, and $a_2 = 1$

$$\left(-r_{A_2}\right)' = \frac{1}{2\pi r_{CS} L}\left(-\frac{dN_{A_2}}{dt}\right) = \left(-r_{A_1}\right)' = \frac{\left(C_{A_1}\right)_{fl}}{1/k_{fl}} \tag{3.8.1}$$

In the Stokes regime, from Equation 3.98, $k_{fl} = D_{fl}/r_{CS}$, so from Equation 3.8.1,

$$\frac{1}{2\pi r_{CS} L}\left(-\frac{dN_{A_2}}{dt}\right) = \frac{\left(C_{A_1}\right)_{fl}}{r_{CS}/D_{fl}}$$

$$\Rightarrow \quad \frac{\rho_{A_2}}{2}\left(-\frac{d(r_{CS}^2)}{dt}\right) = \frac{\left(C_{A_1}\right)_{fl}}{1/D_{fl}}$$

$$\Rightarrow \quad -\rho_{A_2} r_{CS}\frac{dr_{CS}}{dt} = \frac{\left(C_{A_1}\right)_{fl}}{1/D_{fl}}$$

$$\Rightarrow \quad -\rho_{A_2}\int_R^{r_{CS}} r_{CS}\, d(r_{CS}) = \left(C_{A_1}\right)_{fl} D_{fl}\int_0^t dt$$

$$\Rightarrow \quad \frac{\rho_{A_2}\left(R^2 - r_{CS}^2\right)}{2} = \left(C_{A_1}\right)_{fl} D_{fl} t$$

$$\Rightarrow \quad t = \frac{\rho_{A_2}\left(R^2 - r_{CS}^2\right)}{2\left(C_{A_1}\right)_{fl} D_{fl}} \tag{3.8.2}$$

From Equation 3.7.2,

$$t = \frac{\rho_{A_2} R^2}{2\left(C_{A_1}\right)_{fl} D_{fl}} X_{A_2} \tag{3.8.3}$$

At complete conversion,

$$t_{total} = \frac{\rho_{A_2} R^2}{2\left(C_{A_1}\right)_{fl} D_{fl}} \tag{3.8.4}$$

Hence,

$$\frac{t}{t_{total}} = X_{A_2} \tag{3.8.5}$$

3.3.2 Fluid–Fluid Noncatalytic Reactions

Consider the irreversible reaction

$$A_1(I) + A_2(II) \rightarrow P_1(II) \tag{3.117}$$

where I and II represent two fluid phases, either both liquids or one liquid and the other gas. The overall process can be visualized as follows:

1. Migration of A_1 from the bulk of the fluid phase I to the interface between the two phases.
2. Migration of A_1 from the interface between the two phases toward the bulk of fluid phase II.
3. Chemical reaction.
4. Migration of A_2 from the bulk of the fluid phase II toward the interface between the two phases.

Before we deliberate on the rate of the above process, let us first consider the case of simple mass transfer where there is no chemical reaction but only transfer of A_1 from phase I to phase II. Therefore, we need to consider only Step 1 and Step 2. At steady state

$$\left(-\frac{dN_{A_1}}{dt}\right)_{\text{Step 1}} = \left(-\frac{dN_{A_1}}{dt}\right)_{\text{Step 2}} = \left(-\frac{dN_{A_1}}{dt}\right) \tag{3.118}$$

For Step 1,

$$\left(-\frac{dN_{A_1}}{dt}\right) = k_I \times Ar_i\left[\left(C_{A_1}\right)_I - \left(C_{A_1}\right)_{iI}\right]$$

$$\Rightarrow \quad \frac{1}{k_I \times Ar_i}\left(-\frac{dN_{A_1}}{dt}\right) = \left(C_{A_1}\right)_I - \left(C_{A_1}\right)_{iI} \tag{3.119}$$

For Step 2,

$$\left(-\frac{dN_{A_1}}{dt}\right) = k_{II} \times Ar_i\left[\left(C_{A_1}\right)_{iII} - \left(C_{A_1}\right)_{II}\right]$$

$$\Rightarrow \quad \frac{1}{k_{II} \times Ar_i}\left(-\frac{dN_{A_1}}{dt}\right) = \left(C_{A_1}\right)_{iII} - \left(C_{A_1}\right)_{II} \qquad (3.120)$$

At the interface, the two phases are essentially in equilibrium. If the equilibrium concentrations of A_1 in the two phases are related by Henry's law, so that

$$\left(C_{A_1}\right)_{iI} = H\left(C_{A_1}\right)_{iII} \qquad (3.121)$$

From Equations 3.120 and 3.121,

$$\frac{1}{k_{II} \times Ar_i}\left(-\frac{dN_{A_1}}{dt}\right) = \frac{\left(C_{A_1}\right)_{iI}}{H} - \left(C_{A_1}\right)_{II}$$

$$\Rightarrow \quad \frac{H}{k_{II} \times Ar_i}\left(-\frac{dN_{A_1}}{dt}\right) = \left(C_{A_1}\right)_{iI} - H\left(C_{A_1}\right)_{II} \qquad (3.122)$$

Adding Equations 3.119 and 3.122,

$$\left[\frac{1}{k_I \times Ar_i} + \frac{H}{k_{II} \times Ar_i}\right]\left(-\frac{dN_{A_1}}{dt}\right) = \left(C_{A_1}\right)_I - H\left(C_{A_1}\right)_{II}$$

$$\Rightarrow \quad \left(-\frac{dN_{A_1}}{dt}\right) = \frac{\left(C_{A_1}\right)_I - H\left(C_{A_1}\right)_{II}}{[1/k_I \times Ar_i + H/k_{II} \times Ar_i]} \qquad (3.123)$$

The two terms in the denominator on the RHS of Equation 3.123 represent the resistances to migration of A_1 in Steps 1 and 2, respectively. In the event of chemical reaction 3.117 taking place in phase II some of A_1 is consumed in the reaction. This is compensated by a faster transfer of A_1 from the interface to the site of reaction. In effect, the transfer coefficient k_{II} is enhanced and the resistance to migration of A_1 in Step 2 is reduced. This enhancement in k_{II} is quantified in the parameter E, known as the enhancement factor, so that the transfer coefficient to be used in the rate expression is now Ek_{II} It is defined as

$$E = \frac{\text{Rate of migration of } A_1 \text{ from interface to the bulk of phase II with reaction}}{\text{Rate of migration of } A_1 \text{ from interface to the bulk of phase II without reaction}}$$

Similarly, some of A_2 is consumed in the reaction and this is compensated by the migration of A_2 from the bulk of phase II toward the reaction site. Now for reaction 3.117, the steps involving A_1 are Steps 1 through 3.

For Step 1, we already have Equation 3.119. For Step 2, the modified equation would be

$$\frac{H}{Ek_{II} \times Ar_i}\left(-\frac{dN_{A_1}}{dt}\right) = \left(C_{A_1}\right)_{iI} - H\left(C_{A_1}\right)_{II} \tag{3.124}$$

For Step 3,

$$\left(-\frac{dN_{A_1}}{dt}\right) = k\left(C_{A_1}\right)_{II} C_{A_2} V_{II}$$

$$\Rightarrow \frac{1}{kC_{A_2}V_{II}}\left(-\frac{dN_{A_1}}{dt}\right) = \left(C_{A_1}\right)_{II}$$

$$\Rightarrow \frac{H}{kC_{A_2}V_{II}}\left(-\frac{dN_{A_1}}{dt}\right) = H\left(C_{A_1}\right)_{II} \tag{3.125}$$

Adding Equations 3.119, 3.124, and 3.125,

$$\left[\frac{1}{k_I \times Ar_i} + \frac{H}{Ek_{II} \times Ar_i} + \frac{H}{kC_{A_2}V_{II}}\right]\left(-\frac{dN_{A_1}}{dt}\right) = \left(C_{A_1}\right)_I$$

$$\Rightarrow \left(-\frac{dN_{A_1}}{dt}\right) = \frac{\left(C_{A_1}\right)_I}{[1/k_I \times Ar_i + H/Ek_{II} \times Ar + H/kC_{A_2}V_{II}]} \tag{3.126}$$

Based on Equation 3.126, the overall rate of the process can be expressed on the basis of unit area of the interface. In such cases,

$$(-r_{A_1})' = \frac{1}{Ar_i}\left(-\frac{dN_{A_1}}{dt}\right) = \frac{\left(C_{A_1}\right)_I}{[(1/k_I) + (H/Ek_{II}) + (HAr_i/kC_{A_2}V_{II})]} \tag{3.127}$$

The rate can also be expressed on the basis of unit volume of phase II, in which case

$$-r_{A_1} = \frac{1}{V_{II}}\left(-\frac{dN_{A_1}}{dt}\right) = \frac{\left(C_{A_1}\right)_I}{\left[(V_{II}/k_I \times Ar_i) + (HV_{II}/Ek_{II} \times Ar_i) + (H/kC_{A_2})\right]} \tag{3.128}$$

Alternatively, the rate can also be expressed on the basis of unit volume of the entire reacting system (both phases I and II). We then have

$$\left(-r_{A_1}\right)_V = \frac{1}{V}\left(-\frac{dN_{A_1}}{dt}\right) = \frac{\left(C_{A_1}\right)_I}{[[(V/k_I \times Ar_i) + (HV/Ek_{II} \times Ar_i) + (HV/kC_{A_2}V_{II})]} \tag{3.129}$$

In Equations 3.127, 3.128, and 3.129, the three terms in the denominator on the RHS of the equation represent the resistance to the process due to mass transfer of A_1 in phase I, mass transfer of A_1 in phase II, and the chemical reaction, respectively. Depending on the relative values of the parameters, there can be instances where one of these resistances far exceeds the other two. In that case, the particular step will act as the rate-limiting step for the entire process.

EXAMPLE 3.9

Ammonia from air is absorbed in water through the reaction

$$NH_3(g) + H_2O(l) \rightarrow NH_4OH \tag{3.9.1}$$

in a gas–liquid-contacting equipment at 20°C and 1 atm total pressure. At the point under consideration in the equipment the air contains 10% ammonia by volume. The combined mole fraction of unconverted ammonia and NH_4OH in the liquid film is 0.02. The following data are given:

$H = 0.02$ Pa · m^3/mol, specific surface area = 25 m^2/m^3 reactor volume, $k_g = 4.5 \times 10^{-4}$ m/h, $k_l = 2$ m/h, $k = 1 \times 10^5$ m^3/(mol · h), volume fraction liquid in the reactor = 0.8, and enhancement factor = 100.

Calculate the local rate of absorption of ammonia based on unit volume of the reactor and identify the rate-limiting step, if any.

SOLUTION

Here, A_1 is NH_3 and A_2 is H_2O. Phase I is gas and phase II is liquid

$$\left(C_{A_1}\right)_I = \left(C_{NH_3}\right)_g = \frac{p_{NH_3}}{RT} = \frac{1 \times 0.1}{8.2 \times 10^{-5} \times 273} = 4.46 \text{ mol/m}^3$$

$$C_{A_2} = C_{H_2O} = \frac{1000 \times 1000}{18}(1 - 0.02) = 54444.45 \text{ mol/m}^3$$

Also given is $Ar_i/V = 25$ m^2/m^3, $k_I = k_g = 4.5 \times 10^{-4}$ m/h, and $k_{II} = k_l = 2$ m/h

$$H = 0.02 \text{ Pa} \cdot \text{m}^3/\text{mol} = \frac{0.02}{8.2 \times 10^{-5} \times 293} = 0.832, \quad \frac{V_{II}}{V} = 0.8$$

From Equation 3.129,

$$\left(-r_{A_1}\right)_V = \frac{\left(C_{A_1}\right)_I}{[V / k_I \times Ar_i + HV / Ek_{II} \times Ar + HV / kC_{A_2} V_{II}]}$$

$$= \frac{4.46}{[1/4.5 \times 10^{-4} \times 25 + 0.832/100 \times 2 \times 25 + 0.832/1 \times 10^5 \times 54444.45 \times 0.8]} \tag{3.9.2}$$

$$\Rightarrow \quad \left(-r_{A_1}\right)_V = \frac{4.46}{[88.88 + 0.00016 + 1.91\times10^{-10}]} \tag{3.9.3}$$

$$\Rightarrow \quad \left(-r_{A_1}\right)_V = 0.05\ \text{mol/m}^3\cdot\text{h}$$

From Equation 3.9.3, it is also evident that the resistance due to mass transfer of ammonia in the gas phase far exceeds the other two resistances. Accordingly, we can say that it is the rate-limiting step.

The key issue associated with the use of Equation 3.126 is the estimation of the enhancement factor, *E*. In the above discussion we have presented the general case of fluid–fluid noncatalytic reaction with the two phases identified as I and II. But the systems, commercially most important, are the ones where phase I is a gas and phase II is a liquid. Technical literature is full of mathematical models that attempt to describe this phenomenon. Each model comes with its own set of assumptions and carries subtle variations. A brief discussion is presented here which brings out the basic approach to enhancement factor estimation.

The first attempt at estimation of the enhancement factor was due to Hatta in 1932 (Hatta, 1932), who analyzed the simplest of possible gas–liquid reactions based on the two-film theory. For the combined mass transfer and chemical reaction process

$$A_1(I) \rightarrow A_1(II) \rightarrow P(II) \tag{3.130}$$

we need to solve the following differential equation obtained from mass balance over the liquid film:

$$D_{A_1}\frac{d^2\left(C_{A_1}\right)_{II}}{dx^2} = k\left(C_{A_1}\right)_{II} \quad 0 \le x \le \delta \tag{3.131}$$

The equation is solved for the boundary conditions

$$\text{At } x = 0: \left(C_{A_1}\right)_{II} = \left(C_{A_1}\right)_{iII} \tag{3.132}$$

$$\text{and at } x = \delta: D_{A_1}\frac{d\left(C_{A_1}\right)_{II}}{dx} = k\left(C_{A_1}\right)_{II} V_{II} \tag{3.133}$$

resulting in

$$J_{A_1} = \frac{D_{A_1}}{\delta}\left(C_{A_1}\right)_{iII} Ha\frac{Ha\left(AL-1\right)+\tanh Ha}{Ha\left(AL-1\right)\tanh Ha+1} \tag{3.134}$$

In this equation, *Ha* is the Hatta modulus and AL is the Hinterland ratio. These are defined as

$$Ha = \delta\sqrt{\frac{k}{D_{A_1}}} = \frac{\left(k\,D_{A_1}\right)^{1/2}}{k_{II}}$$

$$AL = \frac{1-\varepsilon}{Ar_i\delta} = \frac{(1-\varepsilon)k_{II}}{Ar_iD_{A_1}} \tag{3.135}$$

For most of the practical gas–liquid-contacting equipment AL, the ratio of the total liquid volume to the liquid film volume at the interface is very large. Therefore, as a realistic approximation, Equation 3.134 is simplified to

$$J_{A_1} = \frac{D_{A_1}}{\delta}\left(C_{A_1}\right)_{iII}\frac{Ha}{\tanh Ha} = k_{II}\left(C_{A_1}\right)_{iII}\frac{Ha}{\tanh Ha} \tag{3.136}$$

This provides for the enhancement factor

$$E = \frac{Ha}{\tanh Ha} \tag{3.137}$$

For reaction 3.117, the governing differential equation would be

$$D_{A_1}\frac{d^2\left(C_{A_1}\right)_{II}}{dx^2} = k\left(C_{A_1}\right)_{II}C_{A_2} \quad 0 \le x \le \delta \tag{3.138}$$

Because it includes a second dependent variable C_{A_2}, we need additional equations for the reactant species A_2 and simultaneously solve them. Analytical solutions for such systems of equations are not possible and approximations are often resorted to. The approximate method of van Krevelen and Hoftijzer (van Krevelen and Hoftijzer, 1948; van Swaaij and Versteeg, 1992) is the one that is widely used. It defines a very small zone within the liquid film near the interface where the concentration of A_2 is essentially constant so that we have a pseudo-first-order reaction. The approximate concentration profiles are presented in Figure 3.5. The enhancement factor can be computed with the following equation:

$$E = \frac{Ha\sqrt{(E_\infty - E)/(E_\infty - 1)}}{\tanh(Ha\sqrt{(E_\infty - E)/(E_\infty - 1)})} \tag{3.139}$$

It has been found that the enhancement factor depends on two dimensionless parameters, namely, the Hatta modulus, *Ha*, and the maximum enhancement factor, E_∞. For the general case where the chemical reaction is nth order with respect to A_2, the Hatta modulus is defined as (Hikita and Asai, 1964)

$$Ha = \frac{\left(k\left(C_{A_2}\right)_{II}^{n}D_{A_1}\right)^{1/2}}{k_{II}} \tag{3.140}$$

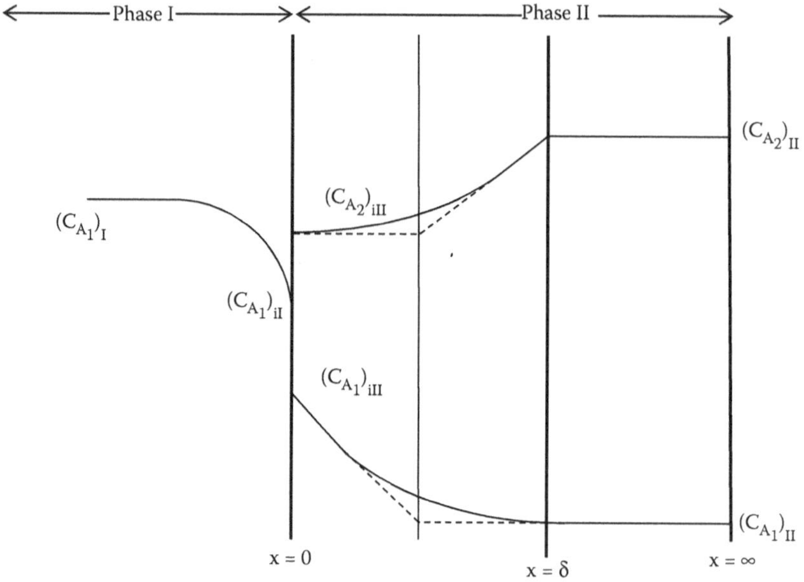

FIGURE 3.5 Concentration profiles in fluid–fluid noncatalytic reaction. Solid lines show actual concentration profiles and the dashed lines show concentration profiles approximated by Van Krevelen and Hoftijzer. (Adapted from *Chem. Eng. Sci.*, 47, van Swaaij, W.P.M. and Versteeg, G.F., Mass transfer accompanied with complex reversible chemical reactions in gas-liquid systems. An overview. 3181, Copyright 1992, with permission from Elsevier.)

And, for reaction 3.117,

$$Ha = \frac{\left(k\left(C_{A_2}\right)_{II} D_{A_1}\right)^{1/2}}{k_{II}} \tag{3.141}$$

The maximum enhancement factor, E_∞, is approximated as

$$E_\infty \cong \left(1 + \frac{D_{A_2}\left(C_{A_2}\right)_{II} H}{D_{A_1}\left(C_{A_1}\right)_{iI}}\right) \tag{3.142}$$

E_∞ can be physically interpreted as the limit to enhancement factor due to an infinitely fast chemical reaction. From Equation 3.139, the values of the enhancement factor for some of the limiting values of Ha and E_∞ are as follows:

When $E_\infty < Ha/5$,

$$E \cong E_\infty - \frac{E_\infty^2(E_\infty - 1)}{Ha^2} \cong E_\infty \tag{3.143}$$

When $E_\infty > 5Ha$,

$$E \cong Ha\left(1 - \frac{(Ha-1)}{2E_\infty}\right) \cong Ha \tag{3.144}$$

When $Ha \leq 1$,

$$E \cong 1 + \frac{Ha^2}{3} \tag{3.145}$$

These values of enhancement factor are then to be used in conjunction with Equations 3.127, 3.128, or 3.129.

Expressions for enhancement factor have also been derived using the penetration theory as well as the surface renewal theory (Danckwerts, 1955). However, there is not much variation in the predicted values of enhancement factor predicted by these theories and that provided by the two-film theory. Equation 3.139 is, therefore, adequate for the determination of enhancement factor required for engineering calculations.

For a general reaction

$$a_1A_1(I) + a_2A_2(II) \rightarrow P_1(II) \tag{3.146}$$

we can still use Equation 3.139 for the estimation of enhancement factor. For this purpose, Hikita and Asai proposed the Hatta modulus to be calculated as

$$Ha = \frac{\sqrt{[2k/(a_1+1)] \times [(C_{A_1})_{iII}^{a_1-1}/H^{a_1-1}] \times \left(C_{A_2}\right)_{II}^{a_2} \times D_{A_1})}}{k_{II}} \tag{3.147}$$

The main difficulty in the estimation of enhancement factor using Ha and E_∞ is to measure the quantity $\left(C_{A_1}\right)_{iI}$, the interfacial concentration on the gas side. This introduces much inaccuracy in the calculated value of enhancement factor.

EXAMPLE 3.10

For the reactive absorption case of Example 3.9, estimate the enhancement factor for the following situation:

$$D_{NH_3} = 2.5 \times 10^{-6}\ m^2/h, \quad D_{H_2O} = 1 \times 10^{-10}\ m^2/h, \quad \left(C_{NH_3}\right)_{iI} = 2\ mol/m^3$$

Solution

From Example 3.9,

$$C_{A_2} = C_{H_2O} = 54444.45\ mol/m^3$$

$$k_{\mathrm{II}} = k_l = 2\ \mathrm{m/h}$$

$$\mathrm{H} = 0.02\ \mathrm{Pa \cdot m^3/mol} = \frac{0.02}{8.2 \times 10^{-5} \times 293} = 0.832$$

$$\mathrm{k} = 1 \times 10^5\ \mathrm{m^3/mol \cdot h}$$

From Equation 3.141,

$$Ha = \frac{\left(\mathrm{k}\left(\mathrm{C}_{\mathrm{A}_2}\right)_{\mathrm{II}} D_{\mathrm{NH_3}}\right)^{1/2}}{k_{\mathrm{II}}} = \frac{(1 \times 10^5 \times 54444.45 \times 2.5 \times 10^{-6})^{1/2}}{2} = 58.33$$

From Equation 3.142,

$$E_\infty \cong \left(1 + \frac{D_{\mathrm{H_2O}} \mathrm{C}_{\mathrm{H_2O}} \mathrm{H}}{D_{\mathrm{NH_3}} \left(\mathrm{C}_{\mathrm{NH_3}}\right)_{\mathrm{iI}}}\right) = 1 + \frac{1 \times 10^{-10} \times 54444.45 \times 0.832}{2.5 \times 10^{-7} \times 2} = 9.06$$

This gives

$$E_\infty < \frac{Ha}{5}$$

and

$$E \cong E_\infty = 9.06$$

3.3.3 Fluid–Solid Catalytic Reactions

Owing to their overwhelming commercial importance, the fluid–solid catalytic reactions have drawn much attention. Solid catalysts enable us to carry out chemical transformations of gaseous or liquid reactants that are otherwise unviable. The entire downstream processing in petroleum refining and subsequent petrochemicals is very strongly dependent on solid-catalyzed reactions. These lead to tremendous value addition to the original feedstock. Vegetable oils are hydrogenated over nickel catalysts.

Let us consider the simplest of the fluid–solid catalytic reactions

$$\mathrm{A_1(fl)} \rightarrow \mathrm{P_1(fl)} \tag{3.148}$$

The overall process may be visualized as consisting of the following steps:

1. Migration of A_1 from the bulk of the fluid to the external surface of the catalyst.
2. Migration of A_1 from the external surface of the catalyst to the active sites through the pores within the catalyst.
3. Chemisorption of A_1 on the active sites of the catalyst.
4. Chemical transformation of A_1 into products.

5. Desorption of P_1 from the active sites.
6. Migration of P_1 from the active sites to the external surface of the catalyst through the pores within the catalyst.
7. Migration of P_1 from the external surface of the catalyst to the bulk of the fluid.

Among these steps, Steps 1 and 7 involve mass transfer within the fluid phase where no reaction is taking place. Their rates can be determined with the help of empirical correlations such as Equations 3.61 and 3.63. For the other steps, the analysis is presented in the following paragraphs.

3.3.3.1 Langmuir–Hinshelwood–Hougen–Watson Kinetic Model

Among the above-listed steps, Steps 3–5 are the ones that are responsible for the chemical transformation of A_1 to P_1. A mathematical analysis of the rate of these transformations is provided by the Langmuir–Hinshelwood–Hougen–Watson kinetic model (Hougen and Watson, 1943). The classical Langmuir adsorption model is based on the following hypotheses:

1. Monolayer of adsorbate species on the catalytic surface.
2. All adsorption sites are energetically equivalent.
3. No interaction between adsorbate species that are adsorbed.

Steps 3–5 of the overall process can be schematically represented by the following:
Adsorption step:

$$A_1 + S_v \underset{k_2}{\overset{k_1}{\rightleftharpoons}} A_1 \bullet S_v \tag{3.149}$$

Chemical transformation step:

$$A_1 \bullet S_v \underset{k_4}{\overset{k_3}{\rightleftharpoons}} P_1 \bullet S_v \tag{3.150}$$

Desorption step:

$$P_1 \bullet S_v \underset{k_6}{\overset{k_5}{\rightleftharpoons}} P_1 + S_v \tag{3.151}$$

The net rate of disappearance of A_1,

$$-r_{A_1} = k_1 C_{A_1} C_{S_v} - k_2 C_{A_1 \bullet S_v} \tag{3.152}$$

$$\Rightarrow \quad -r_{A_1} = k_1 C_{A_1} C_{S_v} - \frac{k_1}{K_{ad}} C_{A_1 \bullet S_v}$$

$$\Rightarrow \quad -r_{A_1} = k_1 \left(C_{A_1} C_{S_v} - \frac{C_{A_1 \bullet S_v}}{K_{ad}} \right) \tag{3.153}$$

From the pseudo-steady-state approximation,

$$r_{A_1\bullet S_v} = k_1C_{A_1}C_{S_v} - k_2C_{A_1\bullet S_v} - k_3C_{A_1\bullet S_v} + k_4C_{P_1\bullet S_v} = 0$$

$$\Rightarrow \quad k_1C_{A_1}C_{S_v} - (k_2 + k_3)C_{A_1\bullet S_v} + k_4C_{P_1\bullet S_v} = 0$$

$$\Rightarrow \quad (k_2 + k_3)C_{A_1\bullet S_v} = k_1C_{A_1}C_{S_v} + k_4C_{P_1\bullet S_v}$$

$$\Rightarrow \quad (k_2 + k_3)C_{A_1\bullet S_v} = k_1C_{A_1}C_{S_v} + k_4C_{P_1\bullet S_v} \tag{3.154}$$

Similarly,

$$r_{P_1\bullet S_v} = k_3C_{A_1\bullet S_v} - k_4C_{P_1\bullet S_v} - k_5C_{P_1\bullet S_v} + k_6C_{P_1}C_{S_v} = 0$$

$$\Rightarrow \quad k_3C_{A_1\bullet S_v} - (k_4 + k_5)C_{P_1\bullet S_v} + k_6C_{P_1}C_{S_v} = 0$$

$$\Rightarrow \quad (k_4 + k_5)C_{P_1\bullet S_v} = k_3C_{A_1\bullet S_v} + k_6C_{P_1}C_{S_v}$$

$$\Rightarrow \quad C_{P_1\bullet S_v} = \frac{k_3C_{A_1\bullet S_v} + k_6C_{P_1}C_{S_v}}{(k_4 + k_5)}$$

$$\Rightarrow \quad C_{P_1\bullet S_v} = k^{I}C_{A_1\bullet S_v} + k^{II}C_{P_1}C_{S_v} \tag{3.155}$$

where $k^{I} = k_3/(k_4 + k_5)$, and $k^{II} = k_6/(k_4 + k_5)$
Substituting this value in Equation 3.154

$$(k_2 + k_3)C_{A_1\bullet S_v} = k_1C_{A_1}C_{S_v} + k_4k^{I}C_{A_1\bullet S_v} + k_4k^{II}C_{P_1}C_{S_v}$$

$$\Rightarrow \quad (k_2 + k_3 - k_4k^{I})C_{A_1\bullet S_v} = k_1C_{A_1}C_{S_v} + k_4k^{II}C_{P_1}C_{S_v}$$

$$\Rightarrow \quad C_{A_1\bullet S_v} = k^{III}C_{A_1}C_{S_v} + k^{IV}C_{P_1}C_{S_v} \tag{3.156}$$

where $k^{III} = k_1/(k_2 + k_3 - k_4k^{I})$, and $k^{IV} = k_4k^{II}/(k_2 + k_3 - k_4k^{I})$
From the active site balance for the catalyst,

$$C_{S_t} = C_{S_v} + C_{A_1\bullet S_v} + C_{P_1\bullet S_v} \tag{3.157}$$

Using Equations 3.155 and 3.156,

$$C_{S_t} = C_{S_v} + k^{III}C_{A_1}C_{S_v} + k^{IV}C_{P_1}C_{S_v} + k^{I}C_{A_1\bullet S_v} + k^{II}C_{P_1}C_{S_v}$$

$$\Rightarrow \quad C_{S_t} = C_{S_v} + k^{III}C_{A_1}C_{S_v} + (k^{II} + k^{IV})C_{P_1}C_{S_v} + k^{I}\left(k^{III}C_{A_1}C_{S_v} + k^{IV}C_{P_1}C_{S_v}\right)$$

$$\Rightarrow \quad C_{S_t} = C_{S_v}\left[1 + k^{III}C_{A_1} + (k^{II} + k^{IV})C_{P_1} + k^{I}\left(k^{III}C_{A_1} + k^{IV}C_{P_1}\right)\right]$$

$$\Rightarrow \quad C_{S_v} = C_{S_t} / \left[\left(1 + k^{III}C_{A_1} + (k^{II} + k^{IV}\right)C_{P_1} + k^{I}\left(k^{III}C_{A_1} + k^{IV}C_{P_1}\right)\right] \tag{3.158}$$

From Equations 3.153, 3.156, and 3.158,

$$-r_{A_1} = \frac{k_1 C_{A_1} C_{S_t}}{\left[1 + k^{III} C_{A_1} + \left(k^{II} + k^{IV}\right) C_{P_1} + k^{I}\left(k^{III} C_{A_1} + k^{IV} C_{P_1}\right]\right.}$$

$$- k_2\left(k^{III} C_{A_1} C_{S_v} + k^{IV} C_{P_1} C_{S_v}\right)$$

$$\Rightarrow \quad -r_{A_1} = \frac{k_1 C_{A_1} C_{S_t}}{\left[1 + (k^{III} + k^{I}k^{III}) C_{A_1} + (k^{II} + k^{IV} + k^{I}k^{IV}) C_{P_1}\right]}$$

$$- \frac{C_{S_t} k_2 (k^{III} C_{A_1} + k^{IV} C_{P_1})}{\left[1 + (k^{III} + k^{I}k^{III}) C_{A_1} + (k^{II} + k^{IV} + k^{I}k^{IV}) C_{P_1}\right]}$$

$$\Rightarrow \quad -r_{A_1} = \frac{C_{S_t}\left[(k_1 - k_2 k^{III}) C_{A_1} - k_2 k^{IV} C_{P_1}\right]}{\left[1 + (k^{III} + k^{I}k^{III}) C_{A_1} + (k^{II} + k^{IV} + k^{I}k^{IV}) C_{P_1}\right]}$$

$$\Rightarrow \quad -r_{A_1} = \frac{C_{S_t}\left[k^{V} C_{A_1} - k^{VI} C_{P_1}\right]}{\left[1 + k^{VII} C_{A_1} + k^{VIII} C_{P_1}\right]} \tag{3.159}$$

where

$$k^{V} = k_1 - k_2 k^{III}$$

$$k^{VI} = k_2 k^{IV}$$

$$k^{VII} = k^{III} + k^{I}k^{III}$$

and

$$k^{VIII} = k^{II} + k^{IV} + k^{I}k^{IV}$$

In Equation 3.159, the rate is dependent on three parameters, namely, the fluid-phase concentrations of A_1 and P_1 (the reactant and the product) and the total concentration of active sites on the catalyst. The difficulty in the use of Equation 3.159 lies in the estimation of the constants k^{V}, k^{VI}, k^{VII}, and k^{VIII}, which are themselves functions of different rate constants. The equation is simplified if one of the steps in reaction 3.149–3.151 is slow compared to the other two steps so that this slow step acts as the rate-limiting step. Let us examine the possible cases.

Step (3.149), that is, the adsorption of A_1 is rate limiting:

Since steps (3.150) and (3.151) are fast, they can be considered to attain equilibrium, so that

$$k_3 C_{A_1 \bullet S_v} = k_4 C_{P_1 \bullet S_v}$$

$$\Rightarrow \quad C_{A_1 \bullet S_v} = \frac{C_{P_1 \bullet S_v}}{K_{rxn}} \tag{3.160}$$

And

$$k_5 C_{P_1 \bullet S_v} = k_6 C_{P_1} C_{S_v}$$

$$\Rightarrow \quad C_{P_1 \bullet S_v} = \frac{C_{P_1} C_{S_v}}{K_{des}} \tag{3.161}$$

From Equations 3.157, 3.160, and 3.161,

$$C_{S_t} = C_{S_v} + \frac{C_{P_1 \bullet S_v}}{K_{rxn}} + \frac{C_{P_1} C_{S_v}}{K_{des}}$$

$$\Rightarrow \quad C_{S_t} = C_{S_v} + \frac{C_{P_1} C_{S_v}}{K_{rxn} K_{des}} + \frac{C_{P_1} C_{S_v}}{K_{des}}$$

$$\Rightarrow \quad C_{S_t} = C_{S_v}\left(1 + \frac{C_{P_1}}{K_{rxn} K_{des}} + \frac{C_{P_1}}{K_{des}}\right)$$

$$\Rightarrow \quad C_{S_v} = \frac{C_{S_t}}{(1 + C_{P_1}/K_{rxn} K_{des} + C_{P_1}/K_{des})} \tag{3.162}$$

Using Equations 3.152, 3.161, and 3.162,

$$-r_{A_1} = k_1 \frac{C_{A_1} C_{S_t}}{\left(1 + C_{P_1}/K_{rxn} K_{des} + C_{P_1}/K_{des}\right)} - k_2 \frac{C_{P_1} C_{S_v}}{K_{rxn} K_{des}}$$

$$\Rightarrow \quad -r_{A_1} = \frac{k_1 C_{A_1} C_{S_t}}{\left(1 + C_{P_1}/K_{rxn} K_{des} + C_{P_1}/K_{des}\right)} - \frac{k_2 C_{P_1} C_{S_t}}{K_{rxn} K_{des}\left(1 + C_{P_1}/K_{rxn} K_{des} + C_{P_1}/K_{des}\right)}$$

$$\Rightarrow \quad -r_{A_1} = \frac{K_{rxn} K_{des} k_1 C_{A_1} C_{S_t}}{\left(K_{rxn} K_{des} + C_{P_1} + K_{rxn} C_{P_1}\right)} - \frac{k_2 C_{P_1} C_{S_t}}{\left(K_{rxn} K_{des} + C_{P_1} + K_{rxn} C_{P_1}\right)}$$

$$\Rightarrow \quad -r_{A_1} = \frac{K_{rxn} K_{des} k_1 C_{A_1} C_{S_t}}{\left(K_{rxn} K_{des} + C_{P_1} + K_{rxn} C_{P_1}\right)} - \frac{k_1 C_{P_1} C_{S_t}}{K_{ads}\left(K_{rxn} K_{des} + C_{P_1} + K_{rxn} C_{P_1}\right)}$$

$$\Rightarrow \quad -r_{A_1} = \frac{K k_1 C_{A_1} C_{S_t} - k_1 C_{P_1} C_{S_t}}{K_{ads}\left(K_{rxn} K_{des} + C_{P_1} + K_{rxn} C_{P_1}\right)}$$

$$\Rightarrow \quad -r_{A_1} = \frac{K k_1 C_{S_t}\left(C_{A_1} - C_{P_1}/K\right)}{K_{ads}\left(K_{rxn} K_{des} + C_{P_1} + K_{rxn} C_{P_1}\right)}$$

$$\Rightarrow \quad -r_{A_1} = \frac{k_1 C_{S_t}\left(C_{A_1} - C_{P_1}/K\right)}{\left(1 + (K_{ads}/K) C_{P_1} + C_{P_1}/K_{des}\right)} \tag{3.163}$$

where $K = K_{ads} K_{rxn} K_{des}$

Step (3.150), that is, the chemical transformation of A_1 is rate limiting:

In this case, the rate of the overall process will be equal to the rate of this individual step for which,

$$r_{overall} = r_{P_1\bullet S_v} = k_3 C_{A_1\bullet S_v} - k_4 C_{P_1\bullet S_v} \tag{3.164}$$

From the equilibrium of Step (3.149),

$$\begin{aligned} & k_1 C_{A_1} C_{S_v} = k_2 C_{A_1\bullet S_v} \\ => \quad & C_{A_1\bullet S_v} = K_{ads} C_{A_1} C_{S_v} \end{aligned} \tag{3.165}$$

From Equations 3.157, 3.161, and 3.165,

$$\begin{aligned} & C_{S_t} = C_{S_v} + K_{ads} C_{A_1} C_{S_v} + \frac{C_{P_1} C_{S_v}}{K_{des}} \\ \Rightarrow \quad & C_{S_t} = C_{S_v}\left(1 + K_{ads} C_{A_1} + \frac{C_{P_1}}{K_{des}}\right) \\ \Rightarrow \quad & C_{S_v} = \frac{C_{S_t}}{(1 + K_{ads} C_{A_1} + C_{P_1}/K_{des})} = \frac{K_{des} C_{S_t}}{(K_{des} + K_{ads} K_{des} C_{A_1} + C_{P_1})} \end{aligned} \tag{3.166}$$

From Equations 3.164, 3.165, 3.161, and 3.166,

$$\begin{aligned} & r_{overall} = k_3 K_{ads} C_{A_1} C_{S_v} - k_4 \frac{C_{P_1} C_{S_v}}{K_{des}} \\ \Rightarrow \quad & r_{overall} = \frac{k_3 K_{ads} K_{des} C_{A_1} C_{S_t} - k_4 C_{P_1} C_{S_t}}{(K_{des} + K_{ads} K_{des} C_{A_1} + C_{P_1})} \\ \Rightarrow \quad & r_{overall} = \frac{k_3 K_{ads} C_{A_1} C_{S_t} - (k_4/K_{des}) C_{P_1} C_{S_t}}{(1 + K_{ads} C_{A_1} + C_{P_1}/K_{des})} \\ \Rightarrow \quad & r_{overall} = \frac{k_3 K_{ads} C_{S_t}(C_{A_1} - (k_4/k_3 K_{ads} K_{des}) C_{P_1})}{(1 + K_{ads} C_{A_1} + C_{P_1}/K_{des})} \\ \Rightarrow \quad & r_{overall} = \frac{k_3 K_{ads} C_{S_t}(C_{A_1} - C_{P_1}/K)}{(1 + K_{ads} C_{A_1} + C_{P_1}/K_{des})} \end{aligned} \tag{3.167}$$

Step (3.151), that is, the desorption of P_1 is rate limiting:

$$r_{overall} = k_5 C_{P_1\bullet S_v} - k_6 C_{P_1} C_{S_v} \tag{3.168}$$

From Equations 3.157, 3.165, and 3.160,

$$C_{S_t} = C_{S_v} + K_{ads}C_{A_1}C_{S_v} + K_{rxn}K_{ads}C_{A_1}C_{S_v}$$

$$\Rightarrow \quad C_{S_t} = C_{S_v}\left(1 + K_{ads}C_{A_1} + K_{rxn}K_{ads}C_{A_1}\right)$$

$$\Rightarrow \quad C_{S_v} = \frac{C_{S_t}}{\left(1 + K_{ads}C_{A_1} + K_{rxn}K_{ads}C_{A_1}\right)} \tag{3.169}$$

From Equations 3.168, 3.160, and 3.165,

$$r_{overall} = k_5K_{rxn}K_{ads}C_{A_1}C_{S_v} - k_6C_{P_1}C_{S_v}$$

$$\Rightarrow \quad r_{overall} = \frac{k_5K_{rxn}K_{ads}C_{A_1}C_{S_t}}{\left(1 + K_{ads}C_{A_1} + K_{rxn}K_{ads}C_{A_1}\right)} - \frac{k_6C_{P_1}C_{S_t}}{\left(1 + K_{ads}C_{A_1} + K_{rxn}K_{ads}C_{A_1}\right)}$$

$$\Rightarrow \quad r_{overall} = \frac{k_5K_{rxn}K_{ads}C_{A_1}C_{S_t} - k_6C_{P_1}C_{S_t}}{\left(1 + K_{ads}C_{A_1} + K_{rxn}K_{ads}C_{A_1}\right)}$$

$$\Rightarrow \quad r_{overall} = \frac{k_6KC_{A_1}C_{S_t} - k_6C_{P_1}C_{S_t}}{\left(1 + K_{ads}C_{A_1} + K_{rxn}K_{ads}C_{A_1}\right)}$$

$$\Rightarrow \quad r_{overall} = \frac{k_6C_{S_t}\left(KC_{A_1} - C_{P_1}\right)}{\left(1 + K_{ads}C_{A_1} + K_{rxn}K_{ads}C_{A_1}\right)} \tag{3.170}$$

The rate expressions derived above, seemingly for the simplest of the solid-catalyzed reaction, are complicated even when we consider one of the steps to be rate controlling. For stoichiometrically more complex reactions, the derived equations are still more complicated. They have limited utility in engineering calculations. Simpler empirical equations are often available. These equations attempt to fit the experimental data into nth-order rate equations, which are convenient to use yet reasonably reliable.

We find from Equations 3.159, 3.163, 3.167, and 3.170 that the rate of the process is dependent, besides the catalyst characteristics as determined by C_{S_t}, on the concentrations of the reactant A_1, and the product P_1, C_{A_1}, and C_{P_1}, respectively. These concentrations are the ones prevailing within the core of the catalyst at the reaction site and may be different from the concentrations of these species in the bulk fluid. From the steps listed at the beginning of Section 3.2.3, Steps 1 and 2 restrict the entry and migration of A_1 within the pore. If one of these steps is controlling, C_{A_1} will be less than the bulk concentration of A_1. We now examine the effect of these mass transfer steps on the rate of solid–fluid catalytic reactions.

3.3.3.2 Effect of Mass Transfer within the Catalyst Pore

In all, the rate expressions presented in the preceding section, the rate is found to be linearly dependent on the parameter C_{S_t}. Heterogeneous catalytic reactions take place on the active sites present on the surface of the catalyst. For a given mass (or volume) of the catalyst, the higher the surface area, the higher will be the value of the parameter C_{S_t} and the higher will be the rate. Therefore, it is imperative for the catalyst manufacturers to produce catalysts with as high a specific surface area (area per unit mass or volume) as possible. This is done by making the catalyst particles porous. Commercially available catalysts have specific surface areas varying from a few m^2/g to several hundred m^2/g. Needless to say, the external surface area of the catalyst particle is negligible in comparison to the internal surface area. Along with the specific surface area, the average pore size and the pore size distribution are other important parameters. The interior of the catalyst particle thus consists of an intricate network of pores. These pores are neither straight nor of a uniform regular cross section.

3.3.3.2.1 *Intrapore Mass Transfer in Straight Pore in Combination with First-Order Reaction*

Before getting into these complexities and the mass transfer restrictions posed by them, we start our discussion of mass transfer within the catalyst pore with the analysis of a hypothetical straight pore of a uniform circular cross-section area closed at one end. Such a pore may be thought to be running along the thickness of a catalyst particle in the shape of a disk with the edges blocked for diffusion. A schematic is shown in Figure 3.6. Further, we assume that the chemical reaction is first order. As the molecules of A_1 move through the pore, they progressively undergo chemical transformation on the inner wall of the pore. Therefore, the resistance due to mass transfer and the resistance due to chemical reaction do not act in series and cannot be added. They must be simultaneously accounted for. For the differential section of the pore shown in the figure, a mass balance for species A_1 under steady-state conditions gives

Input = Output + disappearance by chemical reaction

$$\Rightarrow \quad -D_{pore}\pi r^2\left(dC_{A_1}/dx\right)_x = -D_{pore}\pi r^2\left(dC_{A_1}/dx\right)_{x+dx} + \left(\pi r^2 dx\right)kC_{A_1}$$

$$\Rightarrow \quad D_{pore}\left(dC_{A_1}/dx\right)_{x+dx} - D_{pore}\left(dC_{A_1}/dx\right)_x = \left(dx\right)kC_{A_1}$$

$$\Rightarrow \quad \frac{D_{pore}\left(dC_{A_1}/dx\right)_{x+dx} - D_{pore}\left(dC_{A_1}/dx\right)_x}{dx} = kC_{A_1} \tag{3.171}$$

In the limit $dx \to 0$, Equation 3.171 reduces to

$$D_{pore}\frac{d^2C_{A_1}}{dx^2} = kC_{A_1}$$

$$\Rightarrow \quad \frac{d^2C_{A_1}}{dx^2} - \frac{k}{D_{pore}}C_{A_1} = 0 \tag{3.172}$$

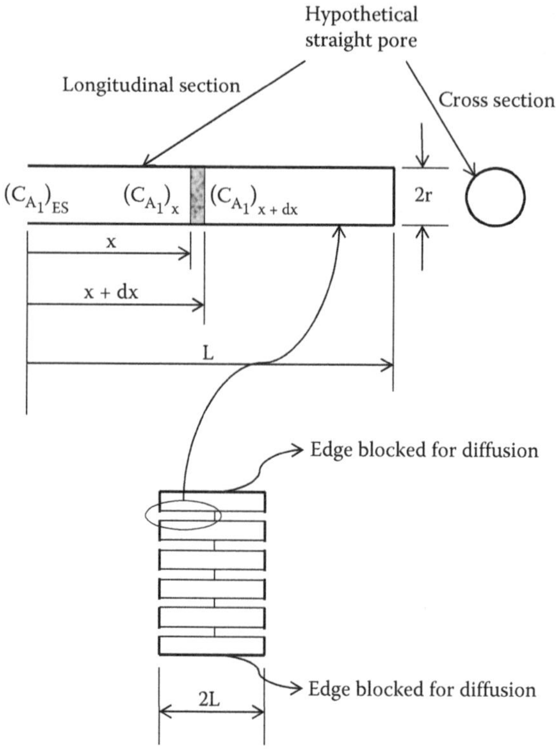

FIGURE 3.6 Schematic of a hypothetical straight pore of uniform circular cross-section area closed at one end. The pore is the part of a catalyst particle in the shape of a disk with the edges blocked for diffusion.

The boundary conditions to this governing differential equation are

$$\text{At } x = 0: C_{A_1} = \left(C_{A_1}\right)_{ES}$$

$$\text{At } x = L: \frac{dC_{A_1}}{dx} = 0$$

With these boundary conditions, the solution to Equation 3.172 is given by

$$C_{A_1} = \frac{\left(C_{A_1}\right)_{ES} e^{-mL}}{e^{mL} + e^{-mL}} e^{mx} + \frac{\left(C_{A_1}\right)_{ES} e^{mL}}{e^{mL} + e^{-mL}} e^{-mx}$$

$$\Rightarrow \quad C_{A_1} = \left(C_{A_1}\right)_{ES} \frac{\left[e^{m(L-x)} + e^{-m(L-x)}\right]}{e^{mL} + e^{-mL}} = \left(C_{A_1}\right)_{ES} \frac{\cosh mL(1 - x/L)}{\cosh mL} \qquad (3.173)$$

where

$$m = \sqrt{\frac{k}{D_{pore}}}$$

The parameter mL, a dimensionless quantity, is known as the Thiele modulus (Thiele, 1939), represented as *Th*. From Equation 3.173, the local concentration of A_1 at any point within the pore is dependent on *Th* and x/L, the fractional distance within the pore. Since the local rate of chemical reaction is kC_{A_1}, this rate is also dependent on *Th* and x/L. The net reaction rate within the pore under the restrictive influence of mass transfer in the pore, as observed from outside, will be the mean of all these local rates. Thus, we have

$$\left(-r_{A_1}\right)_{obs} = k\left(C_{A_1}\right)_{mean} = \frac{k\left(C_{A_1}\right)_{ES}}{L}\int_0^L \frac{\cosh mL(1-x/L)}{\cosh mL}dx$$

$$\Rightarrow \quad \left(-r_{A_1}\right)_{obs} = \frac{k\left(C_{A_1}\right)_{ES}}{mL}\frac{\sinh mL}{\cosh mL} = \frac{k\left(C_{A_1}\right)_{ES}\tanh mL}{mL} = \frac{k\left(C_{A_1}\right)_{ES}\tanh Th}{Th} \tag{3.174}$$

Had there been no resistance to mass transfer due to the pore, the concentration of A_1 would have been $\left(C_{A_1}\right)_{ES}$ everywhere inside the pore. Under this condition, the reaction rate would be

$$\left(-r_{A_1}\right)_{no\ MT\ res} = k\left(C_{A_1}\right)_{ES} \tag{3.175}$$

We define the parameter effectiveness factor as

$$\xi = \frac{\text{Rate of reaction in the presence of resistance to mass transfer due to pore}}{\text{Rate of reaction without resistance to mass transfer due to pore}} \tag{3.176}$$

$$\Rightarrow \quad \xi = \frac{\left(-r_{A_1}\right)_{obs}}{\left(-r_{A_1}\right)_{no\ MT\ res}} \tag{3.177}$$

The effectiveness factor as defined by Equations 3.176 and 3.177 is valid for any reaction stoichiometry and kinetics as well as for any pore geometry. For the particular reaction and the pore under consideration

$$\xi = \frac{\left(-r_{A_1}\right)_{obs}}{\left(-r_{A_1}\right)_{no\ MT\ res}} = \frac{k\left(C_{A_1}\right)_{ES}\tanh Th}{Th} \times \frac{1}{k\left(C_{A_1}\right)_{ES}} = \frac{\tanh Th}{Th} \tag{3.178}$$

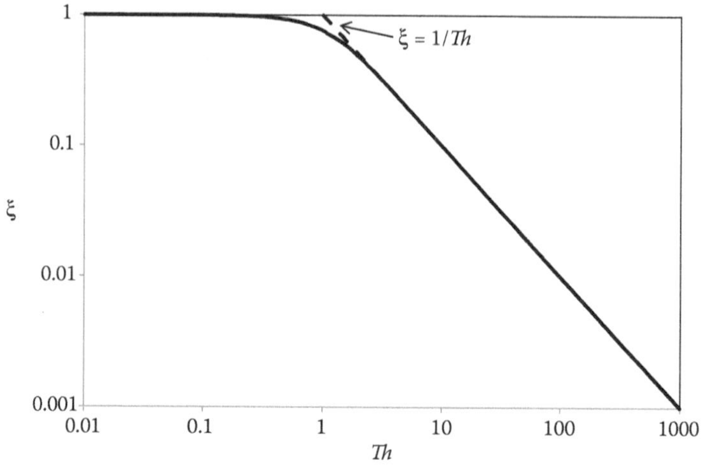

FIGURE 3.7 Effectiveness factor as a function of Thiele modulus for a hypothetical straight pore of uniform circular cross-section area.

The value of effectiveness factor is always less than 1 for the porous catalyst. Equation 3.178 is shown graphically as ξ as a function of Th as a log–log plot in Figure 3.7. From Equation 3.178 when $Th \rightarrow 0$, $\xi \rightarrow 1$. Similarly, from Equation 3.178 when $Th \rightarrow \infty$, $\xi \rightarrow 1/Th$. This is depicted in Figure 3.7 as the curve for ξ as a function of Th asymptotically approaches the lines for $\xi = 1$, and $\xi = 1/Th$. Practically, for values of Th less than about 0.5, the value of $\xi = 1$, and for values of Th more than about 5, the value of $\xi \cong 1/Th$. For the intermediate values of Th, ξ can be determined with Equation 3.178 or Figure 3.7.

From the above discussion, it appears that the porous nature of the catalyst is a disadvantage as the reaction rate within the pore is less than that on the external surface of the catalyst. This apparent anomaly is resolved if we go into the fundamental definition of the reaction rate. Let us consider an assembly of porous catalyst particle of total mass m. For this assembly of porous catalyst particles:

$$\frac{1}{Ar_{\text{pore}}}\left(-\frac{dN_{A_1}}{dt}\right)_{\text{pore}} = \left(-r_{A_1}\right)_{\text{obs}} \tag{3.179}$$

$$\Rightarrow \left(-\frac{dN_{A_1}}{dt}\right)_{\text{pore}} = Ar_{\text{pore}}\left(-r_{A_1}\right)_{\text{obs}} \tag{3.180}$$

Now let us consider another assembly of nonporous catalyst particles of the same total mass m. In this case, the area available for reaction is only the external area of the catalyst particles. For this assembly of nonporous catalyst particles:

$$\frac{1}{Ar_{\text{ex}}}\left(-\frac{dN_{A_1}}{dt}\right)_{\text{nonpore}} = \left(-r_{A_1}\right)_{\text{no MT res}} \tag{3.181}$$

$$\Rightarrow \left(-\frac{dN_{A_1}}{dt}\right)_{\text{nonpore}} = Ar_{ex}\left(-r_{A_1}\right)_{\text{no MT res}} \quad (3.182)$$

Now we compare the RHSs of Equations 3.180 and 3.182. From Equation 3.178, $\left(-r_{A_1}\right)_{obs}$ is less than $\left(-r_{A_1}\right)_{\text{no MT res}}$ by a factor of ξ. However, for the same mass m, the area of the pores in the porous catalyst particles Ar_{pore} is several orders of magnitude larger than the area Ar_{ex} for the nonporous catalyst particles. As a result

$$\left(-\frac{dN_{A_1}}{dt}\right)_{\text{pore}} \gg \left(-\frac{dN_{A_1}}{dt}\right)_{\text{nonpore}} \quad (3.183)$$

This is the reason we favor porous catalysts over nonporous ones.

Equations 3.173 through 3.178 have been derived for a hypothetical straight cylindrical pore in a catalyst particle in the shape of a disk with the edges blocked for diffusion. Similar equations can be derived for other geometries of the porous catalyst particle, like a sphere or a cylinder with both ends blocked for diffusion. From these results combined with experimental findings, Aris (1957) proposed a general expression for the Thiele modulus as

$$Th = \frac{V_p}{Ar_{ex}}\sqrt{\frac{k}{D_{pore}}} \quad (3.184)$$

With *Th* as expressed by Equation 3.184, the plot for ξ versus *Th* for all the three geometries almost coincide with each other with the superimposition virtually perfect for *Th* less than about 0.5 and more than about 5 as shown in Figure 3.8.

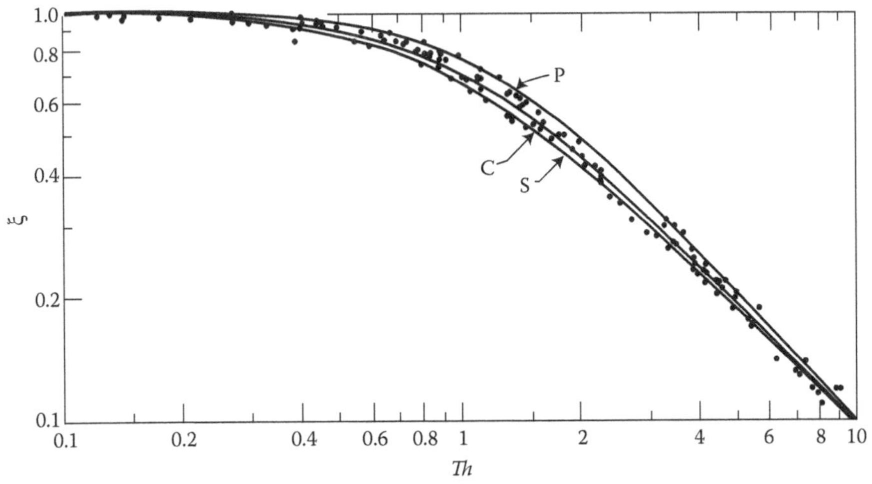

FIGURE 3.8 Effectiveness factor, ξ, versus generalized Thiele modulus, *Th*. Curve P for slab, curve C for cylinder, and curve S for sphere. (Reprinted from *Chem. Eng. Sci.*, 24, Rester, S. and Aris, R., Communications on the theory of diffusion and reaction—II. The effect of shape on the effectiveness factor. 793, Copyright 1969, with permission from Elsevier.)

EXAMPLE 3.11

The gas–solid catalytic reaction

$$A \rightarrow B$$

was carried out at 600°C and 1.2 bar pressure with spherical catalyst particles of 0.25 cm diameter. The molecular weight of A is 120. The mean pore radius of the catalyst is 2.5×10^{-7} cm, having a pore volume of 0.3 cm^3/g, at the given pressure. The diffusion is Knudsen type. The rate constant was found to be 0.3 $cm^3/g \cdot s$. Assuming a tortuosity factor of 2.0, calculate the effectiveness factor of the catalyst.

Solution

For Knudsen-type diffusion, the effective diffusivity within the catalyst pores may be determined as (McCabe et al., 1993)

$$D_{pore} = 9700 r \sqrt{\frac{T}{M_A}} \quad \text{(quantities in CGS units)} \tag{3.11.1}$$

$$\Rightarrow \quad D_{pore} = 9700 \times 2.5 \times 10^{-7} \sqrt{\frac{(600+273)}{120}} = 6.54 \times 10^{-3}\ \mathrm{cm^2/s}$$

The volume-based rate constant is

$$k = \frac{0.3\,\mathrm{cm^3/g \cdot s}}{0.3\,\mathrm{cm^3/g}} = 1\,\mathrm{s^{-1}}$$

From Equation 3.184,

$$Th = \frac{V_p}{A r_{ex}} \sqrt{\frac{k}{D_{pore}}} = \frac{4/3\pi r_p^3}{4\pi r_p^2} \sqrt{\frac{k}{D_{pore}}} = \frac{r_p}{3} \sqrt{\frac{k}{D_{pore}}} = \frac{0.25}{2 \times 3} \sqrt{\frac{1}{6.54 \times 10^{-3}}}$$

$$\Rightarrow \quad Th = 0.515$$

From Equation 3.178,

$$\xi = \frac{\tanh Th}{Th} = \frac{\tanh 0.515}{0.515} = 0.92$$

3.3.3.2.2 Reactions Other Than First Order

With the basic conceptual understanding of the effectiveness factor, we will now generalize this concept to practical situations. A detailed derivation of the equations presented may be found in the works of Aris and Bischoff (Aris, 1965; Bischoff, 1965). In a nutshell, the generalization lies in the formulation of a general Thiele modulus that is applicable to all particle geometries and all reaction orders. Further, it is assumed that the diffusivity within the pore is constant everywhere, which may not always be true. For a reversible reaction:

$$Th_{gen} = \frac{V_p\left(-r_{A_1}\right)_{no\ MT\ res}}{Ar_{ex}\sqrt{2D_{pore}}}\sqrt{\frac{1}{\int_{(C_{A_1})_{eq}}^{(C_{A_1})_{ES}}\left(-r_{A_1}\right)d\left(C_{A_1}\right)}} \tag{3.185}$$

And, for an irreversible reaction:

$$Th_{gen} = \frac{V_p\left(-r_{A_1}\right)_{no\ MT\ res}}{Ar_{ex}\sqrt{2D_{pore}}}\sqrt{\frac{1}{\int_{0}^{(C_{A_1})_{ES}}\left(-r_{A_1}\right)d\left(C_{A_1}\right)}} \tag{3.186}$$

With this generalized Thiele modulus, we still have $\xi \cong 1$, for small values of Th_{gen}, and $\xi \cong 1/Th_{gen}$, for large values of Th_{gen}. In general, the effectiveness factor is

$$\xi = \frac{Ar_{ex}\sqrt{2D_{pore}}}{V_p\left(-r_{A_1}\right)_{no\ MT\ res}}\sqrt{\int_{(C_{A_1})_{cen}}^{(C_{A_1})_{ES}}\left(-r_{A_1}\right)d\left(C_{A_1}\right)} \tag{3.187}$$

For intermediate ranges of Thiele modulus experimental measurement of the effectiveness factor is still the most viable option because $\left(C_{A_1}\right)_{cen}$ is not measurable.

3.3.3.2.3 *Falsified Kinetics*

If we use Equations 3.177 and 3.187 for an nth-order irreversible reaction taking place under strong resistance to intrapore mass transfer, we have (Froment and Bischoff, 1979; Levenspiel, 1999)

$$\xi = \frac{\left(-r_{A_1}\right)_{obs}}{\left(-r_{A_1}\right)_{no\ MT\ res}} = \frac{Ar_{ex}\sqrt{2D_{pore}}}{V_p\left(-r_{A_1}\right)_{no\ MT\ res}}\sqrt{\int_{0}^{(C_{A_1})_{ES}}\left(-r_{A_1}\right)d\left(C_{A_1}\right)}$$

$$\Rightarrow \left(-r_{A_1}\right)_{obs} = \frac{Ar_{ex}\sqrt{2D_{pore}}}{V_p}\sqrt{\int_{0}^{(C_{A_1})_{ES}} k\left(C_{A_1}\right)^n d\left(C_{A_1}\right)}$$

$$\Rightarrow \left(-r_{A_1}\right)_{obs} = \frac{Ar_{ex}}{V_p}\sqrt{\frac{2kD_{pore}}{n+1}}\left(C_{A_1}\right)_{ES}^{(n+1)/2} \quad n \neq 1 \tag{3.188}$$

Thus, one would find as if the reaction is one of (n + 1)/2th order. From Equation 3.188 the observed specific reaction rate constant for the nth-order irreversible reaction under strong resistance to intrapore mass transfer, in combination with Arrhenius temperature dependence of rate constant, would be

$$k_{obs} = \frac{Ar_{ex}}{V_p}\sqrt{\frac{2k_0e^{-E_a/RT}D_{pore}}{n+1}}$$

$$\Rightarrow \quad k_{obs} = \frac{Ar_{ex}}{V_p}\sqrt{\frac{2k_0D_{pore}}{n+1}}\,e^{-\frac{E_a}{2RT}} \tag{3.189}$$

Therefore, the observed activation energy for the reaction would be

$$(E_a)_{obs} = \frac{E_a}{2} \tag{3.190}$$

If both the rate constant and the diffusivity exhibit Arrhenius temperature dependence, the observed rate constant would be

$$k_{obs} = \frac{Ar_{ex}}{V_p}\sqrt{\frac{2k_0e^{-E_a/RT}(k_0)_De^{-E_{aD}/RT}}{n+1}}$$

$$\Rightarrow \quad k_{obs} = \frac{Ar_{ex}}{V_p}\sqrt{\frac{2k_0(k_0)_D}{n+1}}\,e^{-(E_a+E_{aD})/2RT} \tag{3.191}$$

Therefore, the observed activation energy when both the reaction and diffusion are temperature dependent would be

$$(E_a)_{obs} = \frac{E_a + E_{aD}}{2} \tag{3.192}$$

These apparent deviations of observed parameters from the true ones in the presence of strong resistance to intrapore mass transfer are referred to as falsified kinetics.

3.3.3.2.4 Predicting Strong Resistance to Intrapore Mass Transfer

In Equations 3.185 through 3.189, the quantity V_p/Ar_{ex} has the dimensions of length and may be called the characteristic length of the catalyst particle. When the process is not limited by intrapore mass transfer the effectiveness factor is 1, which is independent of the various parameters appearing in Equation 3.188. In such case, the observed rate is the same as that when there is no mass transfer resistance. If different experimental runs are carried out with the same bulk concentration of reactant but particles of different characteristic lengths, the observed rate will be the same in each case. On the other hand, if the process is limited by intrapore mass transfer, then Equation 3.188 predicts that the observed rate is inversely proportional to the characteristic length or directly proportional to Ar_{ex}/V_p. If observed rates for

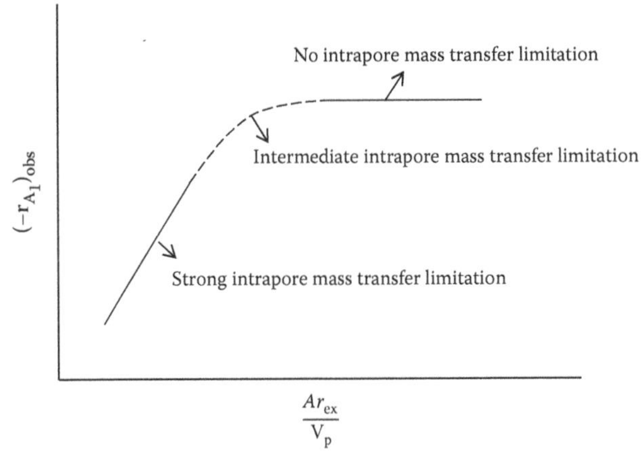

FIGURE 3.9 Experimental observation for intrapore mass transfer limitation.

different experimental runs, carried out with the same bulk concentration of reactant but particles of different characteristic lengths, conform to the inversely proportional relation, then the process is strongly limited by intrapore mass transfer. These two limiting situations are presented in Figure 3.9 with solid lines. In the intermediate regime, the observed rate does increase with decrease in characteristic length but the dependence is not linear as depicted by the dashed line. Therefore, experiments with catalyst particles of different sizes can tell us to what extent the process is limited by intrapore mass transfer.

3.3.3.2.5 *Weisz–Prater Criterion*

For a first-order reaction, Weisz and Prater (1954) proposed the following criterion for predicting the presence of strong intrapore mass transfer. For a first-order reaction

$$Th = L\sqrt{\frac{k}{D_{pore}}}$$

$$\Rightarrow \quad k = \frac{Th^2 D_{pore}}{L^2} \tag{3.193}$$

From Equations 3.178 and 3.193,

$$(-r_{A_1})_{obs} = \frac{k(C_{A_1})_{ES}\tanh Th}{Th} = k(C_{A_1})_{ES}\,\xi$$

$$\Rightarrow \quad (-r_{A_1})_{obs} = \frac{Th^2 D_{pore}}{L^2}(C_{A_1})_{ES}\,\xi$$

$$\Rightarrow \quad \frac{(-r_{A_1})_{obs}L^2}{D_{pore}(C_{A_1})_{ES}} = Th^2\xi \tag{3.194}$$

We know that if there is no resistance to intrapore mass transfer $Th \ll 1$, and $\xi = 1$. So, in the absence of resistance to intrapore mass transfer, we have

$$Th^2\xi = \frac{\left(-r_{A_1}\right)_{obs} L^2}{D_{pore}\left(C_{A_1}\right)_{ES}} \ll 1 \tag{3.195}$$

This is known as the Weisz–Prater criterion for the solid–fluid catalytic reaction to be free from intrapore mass transfer resistance. The advantage of using the criterion is that all the quantities $\left(-r_{A_1}\right)_{obs}$, L, D_{pore}, and $\left(C_{A_1}\right)_{ES}$ could be experimentally determined. The criterion has been extended to a general case by other than first-order reaction by Bischoff (1967). For a general case, we have

$$\left(-r_{A_1}\right)_{obs} = kf\left\{\left(C_{A_1}\right)_{ES}\right\}\xi \tag{3.196}$$

And, from Equation 3.186,

$$Th_{gen} = \frac{V_p kf\left\{\left(C_{A_1}\right)_{ES}\right\}}{Ar_{ex}\sqrt{2D_{pore}}}\sqrt{\frac{1}{\int_0^{\left(C_{A_1}\right)_{ES}} kf\left\{\left(C_{A_1}\right)\right\}d\left(C_{A_1}\right)}}$$

$$\Rightarrow\quad k = \frac{2Th_{gen}^2 D_{pore} Ar_{ex}^2 \int_0^{\left(C_{A_1}\right)_{ES}} f\left\{\left(C_{A_1}\right)\right\}d\left(C_{A_1}\right)}{V_p^2\left[f\left\{\left(C_{A_1}\right)_{ES}\right\}\right]^2} \tag{3.197}$$

Using this value in Equation 3.196,

$$\left(-r_{A_1}\right)_{obs} = \frac{2Th_{gen}^2 D_{pore} Ar_{ex}^2 \int_0^{\left(C_{A_1}\right)_{ES}} f\left\{\left(C_{A_1}\right)\right\}d\left(C_{A_1}\right)}{V_p^2\, f\left\{\left(C_{A_1}\right)_{ES}\right\}}\xi$$

$$\Rightarrow\quad \frac{\left(-r_{A_1}\right)_{obs} V_p^2 f\left\{\left(C_{A_1}\right)_{ES}\right\}}{2D_{pore} Ar_{ex}^2 \int_0^{\left(C_{A_1}\right)_{ES}} f\left\{\left(C_{A_1}\right)\right\}d\left(C_{A_1}\right)} = Th_{gen}^2\xi \tag{3.198}$$

So, the criterion for no resistance to intrapore mass transfer becomes

$$\frac{\left(-r_{A_1}\right)_{obs} V_p^2 f\left\{\left(C_{A_1}\right)_{ES}\right\}}{2D_{pore} Ar_{ex}^2 \int_0^{\left(C_{A_1}\right)_{ES}} f\left\{\left(C_{A_1}\right)\right\}d\left(C_{A_1}\right)} \ll 1 \tag{3.199}$$

For nth-order irreversible reaction, the criterion can be given as

$$\frac{\left(-r_{A_1}\right)_{obs} V_p^2 \left(C_{A_1}\right)_{Es}^n}{2D_{pore} A r_{ex}^2 \int_0^{\left(C_{A_1}\right)_{ES}} \left(C_{A_1}\right)^n d\left(C_{A_1}\right)} \ll 1$$

$$\Rightarrow \quad \frac{(n+1)\left(-r_{A_1}\right)_{obs} V_p^2}{2D_{pore} A r_{ex}^2 \left(C_{A_1}\right)_{ES}} \ll 1 \tag{3.200}$$

3.3.3.2.6 *Combination of Resistances*

For situations when a particular step is not controlling the overall process, we must consider the resistance offered by each of the steps for estimating the overall rate of the process. This is not very difficult for a first-order reaction.

For mass transfer within the fluid:

$$\left(-\frac{dN_{A_1}}{dt}\right)_{fl} = k_{fl} A r_{ex}\left[\left(C_{A_1}\right)_{fl} - \left(C_{A_1}\right)_{ES}\right]$$

$$\Rightarrow \quad \frac{1}{k_{fl} A r_{ex}}\left(-\frac{dN_{A_1}}{dt}\right)_{fl} = \left(C_{A_1}\right)_{fl} - \left(C_{A_1}\right)_{ES} \tag{3.201}$$

For mass transfer and reaction within the pore, from Equation 3.174

$$\left(-\frac{dN_{A_1}}{dt}\right)_{pore} = \frac{V_p k \left(C_{A_1}\right)_{ES} \tanh Th}{Th}$$

$$\Rightarrow \quad \frac{Th}{V_p k \tanh Th}\left(-\frac{dN_{A_1}}{dt}\right)_{pore} = \left(C_{A_1}\right)_{ES} \tag{3.202}$$

At steady state, $\left(-dN_{A_1}/dt\right)_{fl} = \left(-dN_{A_1}/dt\right)_{pore}$ so we can drop the subscripts and add Equations 3.201 and 3.202. Then,

$$\left[\frac{1}{k_{fl} A r_{ex}} + \frac{Th}{V_p k \tanh Th}\right]\left(-\frac{dN_{A_1}}{dt}\right) = \left(C_{A_1}\right)_{fl}$$

$$\Rightarrow \quad \left(-\frac{dN_{A_1}}{dt}\right) = \frac{\left(C_{A_1}\right)_{fl}}{\dfrac{1}{k_{fl} A r_{ex}} + \dfrac{Th}{V_p k \tanh Th}} = \frac{\left(C_{A_1}\right)_{fl}}{\dfrac{1}{k_{fl} A r_{ex}} + \dfrac{1}{V_p k\, \xi}} \tag{3.203}$$

The overall rate can now be expressed either in terms of the external surface area of the catalyst particle or the volume of the catalyst particle. These are

$$\left(-r_{A_1}\right)_{Ar_{ex}} = \frac{1}{Ar_{ex}}\left(-\frac{dN_{A_1}}{dt}\right) = \frac{\left(C_{A_1}\right)_{fl}}{1/k_{fl} + Ar_{ex}/V_p k\,\xi} \tag{3.204}$$

$$\left(-r_{A_1}\right)_{V_p} = \frac{1}{V_p}\left(-\frac{dN_{A_1}}{dt}\right) = \frac{\left(C_{A_1}\right)_{fl}}{V_p/k_{fl}Ar_{ex} + 1/k\,\xi} \tag{3.205}$$

These equations have striking similarity with Equations 3.127 and 3.129 for fluid–fluid noncatalytic reactions. For the fluid–fluid noncatalytic reactions, the significance of enhancement factor lies in its modifying effect on the resistance offered by the second fluid phase. Similarly, the effectiveness factor in solid–fluid catalytic reaction modifies the resistance offered by the pore. The two parameters differ in the sense that while enhancement factor reduces the resistance (enhances the rate), the effectiveness factor increases the resistance. This is evident from the limiting values of these two parameters. The enhancement factor is always greater than or equal to one. On the other hand, the effectiveness factor is always less than or equal to one.

3.3.3.3 Catalyst Deactivation

It is found that the catalyst loses its activity to catalyze a particular reaction over a period of time. This phenomenon is known as catalyst deactivation. The activity of a catalyst at any time may be defined as

$$\chi = \frac{-r_{A_1}}{\left(-r_{A_1}\right)_0} \tag{3.206}$$

$$\Rightarrow \quad -r_{A_1} = \chi\left(-r_{A_1}\right)_0 \tag{3.207}$$

The timescale over which a catalyst deactivates may be of the order of a few seconds to a few years. The decline in the activity of a catalyst may be attributed to the following mechanisms (Bartholomew, 2001):

1. Mechanical
   a. Attrition or crushing. Attrition refers to loss of catalyst material due to abrasion. Catalyst particles may also be crushed under mechanical stresses so that their internal surface area is reduced.
   b. Fouling. This refers to physical deposition of substances from the fluid phase on the catalyst surface and inside the pores.
2. Thermal
   a. Sintering. Thermal stresses may lead to irreversible structural change in the crystalline structure of the active material of the catalyst that alters the chemical nature of the surface. As a result, active sites are no

longer present on the catalyst surface. Thermal stresses may also lead to the collapse of the pore structure of the catalyst support. These effects are collectively termed *sintering*.

3. Chemical
   a. Poisoning. Presence of such chemical species—known as *catalyst poison*—in the feed which compete with the reactant chemical species for adsorption on the active sites on the catalyst surface. At any instant of time, a part of the active sites are masked by the poison and become unavailable for the reactant chemical species to adsorb and react. In some cases, the adsorbed poison not only masks the active site on which it is adsorbed but also makes the adjacent sites ineffective by altering their electronic configuration.
   b. Coking. Some of the product chemical species remain irreversibly adsorbed on the active sites so that these sites are no longer available for the adsorption of reactant chemical species. Sometimes the product chemical species are so large that they block the entry of reactant chemical species into the pores of the catalyst. *Coke* refers to the unwanted product that masks the active sites and/or blocks the catalyst pores. The name derives its origin from petroleum refining and petrochemical industries where carbonaceous products—traditionally known as coke—are formed as a result of undesired side reactions during catalytic reactions of hydrocarbons.

3.3.3.3.1 Deactivation by Poisoning

From Equation 3.159, the rate of the solid-catalyzed reaction is proportional to the total number of active sites present on the catalyst surface, C_{S_t}. This is the case when all the active sites are unmasked and available for the reaction. If, due to the irreversible adsorption of the poison, $C_{S_{posn}}$ sites are no longer available for the desired chemical reaction, the active sites available for the reaction are only $C_{S_t} - C_{S_{posn}}$. As a result, the rate of reaction in the presence of poison, from Equation 3.159, would be

$$-r_{A_1} = \frac{\left(C_{S_t} - C_{S_{posn}}\right)\left[k^{V}C_{A_1} - k^{VI}C_{P_1}\right]}{\left[1 + k^{VII}C_{A_1} + k^{VIII}C_{P_1}\right]} \tag{3.208}$$

The rate for the fresh catalyst with no poisoning would be

$$\left(-r_{A_1}\right)_0 = \frac{C_{S_t}\left[k^{V}C_{A_1} - k^{VI}C_{P_1}\right]}{\left[1 + k^{VII}C_{A_1} + k^{VIII}C_{P_1}\right]} \tag{3.209}$$

Combining Equations 3.206, 3.208, and 3.209

$$\chi_{posn} = \frac{\left(C_{S_t} - C_{S_{posn}}\right)}{C_{S_t}} = 1 - \frac{C_{S_{posn}}}{C_{S_t}} \tag{3.210}$$

According to Wheeler (1951) the quantity $C_{S_{posn}}$ is directly proportional to the concentration of poison in the fluid stream, so that

$$C_{S_{posn}} = \kappa C_{posn} \tag{3.211}$$

Equation 3.210 then reduces to

$$\chi_{posn} = 1 - \frac{\kappa C_{posn}}{C_{S_t}} = 1 - \kappa' C_{posn} \tag{3.212}$$

The quantity κ' is known as the susceptibility of the catalyst, which is a measure of the sensitivity of the catalyst toward a particular poison. Equation 3.212 predicts a linear variation of activity with the poison concentration in the fluid. This type of poisoning is known as *nonselective poisoning*. If the poison preferentially adsorbs on the most active sites, we get *selective poisoning* in which the susceptibility decreases with increasing poison concentration. The reverse behavior where the susceptibility increases with poison concentration is known as *antiselective poisoning*.

In Equations 3.211 and 3.212, C_{posn} is the actual concentration of the poison inside the catalyst pore. When the intrapore mass transfer resistance is negligible, this concentration is equal to the bulk concentration of the poison in the fluid stream. However, in the presence of strong resistance to intrapore mass transfer, the concentration inside the pore is less than the bulk concentration. In the limit when $Th_0 \to \infty$ the catalyst activity, $\chi_{posn} \to \sqrt{1 - \kappa' C_{posn}}$.

3.3.3.3.2 *Deactivation by Coking*

In relation to catalyst deactivation, coke refers to carbonaceous substances that are formed during solid-catalyzed reactions. These substances may chemisorb onto the active sites of the catalyst and make them unavailable for the desired chemical reaction. Alternatively, these substances, owing to their molecular size and geometry, block the catalyst pores and prevent the reactant molecules access to the active sites within the pores. These carbonaceous substances may be relatively pure carbon such as graphite or high molecular weight condensed hydrocarbons. The formation of coke itself may not be detrimental to the catalytic reaction if the substance formed is not strongly chemisorbed and is carried away by the fluid stream. In such cases, the reaction is termed *coke insensitive*. On the other hand, if the coke is strongly chemisorbed on the active sites and does not react further, the catalytic activity is diminished and the reaction is said to be *coke sensitive*.

Analogous to Equation 3.210, the catalyst activity in the presence of coking may be expressed as

$$\chi_{coke} = \frac{\left(C_{S_t} - C_{S_{coke}}\right)}{C_{S_t}} = 1 - \frac{C_{S_{coke}}}{C_{S_t}} \tag{3.213}$$

The utility of Equation 3.213 depends on the ability to express the quantity $C_{S_{coke}}$, which is not measurable, in terms of the weight percent of coke on catalyst which is a measurable quantity. Further, it is necessary to express the weight percent of coke as a function of reaction time in order to have a usable kinetic equation. Complete unanimity is lacking in the understanding of deactivation kinetics by coking. Voorhies (1945) empirically related the weight percent of coke on the catalyst with the reaction time as

$$C_{coke} = Kt^{n} \tag{3.214}$$

The value of the exponent n has been found to be close to 0.5. Froment and Bischoff (1961, 1962) have empirically related the catalytic activity to the coke content of the catalyst as

$$\chi_{coke} = e^{-\alpha C_{coke}} \tag{3.215}$$

Or,

$$\chi_{coke} = \frac{1}{1+\alpha C_{coke}} \tag{3.216}$$

3.3.3.3.3 *Deactivation by Sintering*

Sintering collectively refers to a dozen odd thermally induced changes in the catalyst particle that alters its crystalline and porous structure thereby rendering it ineffective for catalysis. In most cases, it is irreversible and a slowly occurring process (Bartholomew, 2001; Forzatti and Lietti, 1999). In supported metal catalysts, sintering is mainly due to agglomeration and coalescence of small metal crystallites into larger ones. This reduces the surface area per unit volume and hence the number of active sites. Temperature exponentially increases the sintering rate in the initial stages. The rate of sintering is also dependent on the composition of the fluid phase. In general, the higher the melting point of the active metal in the catalyst, the more stable it is against sintering. Catalyst promoters also affect the sintering behavior by affecting the metal mobility on the catalyst support. Fuentes and Gamas (1991) have proposed the following general power law expression (GPLE) for the kinetics of catalyst deactivation by sintering:

$$-\frac{d}{dt}\left[\frac{a'}{(a')_0}\right] = k_{sint}\left[\frac{a'}{(a')_0} - \frac{(a')_{eq}}{(a')_0}\right]^{m} \tag{3.217}$$

Experimental observations have shown that the value of the exponent m known as the *sintering order*, is either 1 or 2, that is, deactivation by sintering is either a first-order or a second-order process. $(a')_{eq}$ is the active area of the catalyst after an infinite time of sintering. The activity of a catalyst subject to sintering is expressed as (Bartholomew, 1993)

$$\chi_{sint} = \left[\frac{a'}{(a')_0}\right]^s \quad (3.218)$$

where s is a positive exponent that can even have noninteger values.

EXERCISES

3.1 For the nonelementary reaction

$$C_2H_6 \rightarrow C_2H_4 + H_2$$

The mechanism and the observed rate equation are given in Table 3.1. Show that the proposed mechanism correctly explains the observed rate. List the necessary assumptions that you take.

3.2 For the decomposition of ozone in stratosphere, the following mechanism is proposed:

$$ClO + ClO \rightarrow (ClO)_2$$

$$(ClO)_2 \xrightarrow{h\upsilon} ClO_2 + Cl$$

$$ClO_2 \rightarrow Cl + O_2$$

$$Cl + O_3 \rightarrow ClO + O_2$$

Derive a rate equation compatible with the proposed mechanism in accordance with the pseudo-steady-state approximation.

3.3 From Michaelis–Menten mechanism as given below

$$E + S \xrightarrow{k_1} E.S$$

$$E.S \xrightarrow{k_2} E + S$$

$$E.S \xrightarrow{k_3} P + E$$

derive a rate equation for the enzyme–substrate reaction.

3.4 For the gas-phase nonelementary reaction

$$H_2 + Br_2 \rightarrow 2HBr$$

the reaction rate is expressed as

$$-r_{H_2} = \frac{k_1 C_{H_2} C_{Br_2}^{3/2}}{\left(k_2 C_{Br_2} + C_{HBr}\right)}$$

The reaction is carried out with initial concentrations of the species as $(C_{H_2})_0$, $(C_{Br_2})_0$, and $(C_{HBr})_0$, respectively. Express the reaction rate in terms of the initial concentrations and the conversion of H_2.

3.5 Repeat Exercise 3.4 if $(C_{H_2})_0 = (C_{Br_2})_0$, and $(C_{HBr})_0 = 0$.

3.6 In Exercise 3.4, if $-r_{H_2}$ has the units of mol/m^3 · h what are the units of k_1 and k_2?

3.7 The gaseous elementary reaction

$$A + 2B \rightarrow 2C + 2D$$

is taking place in a constant pressure reaction system. If we start with a mixture of A and B in the mole ratio of 1:2, develop the rate equation for the reaction in terms of the conversion of A.

3.8 Repeat Exercise 3.7 if the starting molar ratio of A and B is 1:3.

3.9 Repeat Exercise 3.7 if we start with a mixture of A, B, and an inert gas in the molar ratio of 1:3:1.

3.10 For the multiple reaction

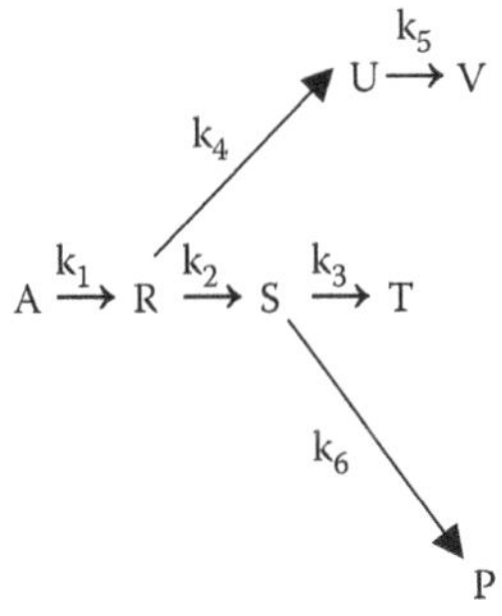

write down the constituent series and parallel reactions and the reaction rates for different species.

3.11 Repeat the problem of Example 3.7 for the case when the process is controlled by migration of reactant A_1 through the layer of P_2 to the surface of unreacted A_2.

3.12 Repeat the problem of Example 3.7 for the case when the process is controlled by chemical reaction proper.

3.13 Repeat the problem of Example 3.7 if the solid particle is in the shape of a flat plate of thickness L with the edges closed for reaction and the process controlled by mass transfer of A_1 in the fluid phase.

3.14 Repeat Exercise 3.13 for the case when the process is controlled by migration of reactant A_1 through the layer of P_2 to the surface of unreacted A_2.

3.15 Repeat Exercise 3.13 for the case when the process is controlled by chemical reaction proper.

3.16 In an absorption column for the absorption of CO_2 in methylmonoethanolamine, the following data are obtained at a height of 0.25 m from the bottom of the column (data adapted from Liu et al., 2014):

Column operating temperature 20°C

Column pressure 1 atm

Bulk gas composition: 6.95% CO_2 rest N_2
Interface CO_2 concentration 2.72×10^{-3} mol/L
Concentration of amine 1.64 mol/L
Gas-phase mass transfer coefficient 6.2×10^{-7} kmol/(m^2 · s · kPa)
Liquid-phase mass transfer coefficient 1.2×10^{-5} m/s
Diffusivity of CO_2 in liquid 6.5×10^{-16} m^2/s
Diffusivity of amine in liquid 8.33×10^{-10} m^2/s
Henry's law constant 2580 m^3 · kPa/kmol
The reaction rate between CO_2 and absorbent may be expressed as

$$-r_{CO_2} = 4418 \times C_{CO_2} C_{amine} \text{ kmol/(m}^3 \cdot \text{s)}$$

Calculate the enhancement factor.

3.17 Determine the limiting resistance and the rate of mass transfer for the absorption of CO_2 described in Exercise 3.16. The interfacial area in the absorption column is reported to be 600 m^2/m^3 (Liu et al., 2014). The volume fraction of liquid in the column may be taken as 0.7.

3.18 In an absorption column for the absorption of H_2S in aqueous NaOH, the following data are obtained at a particular location in the absorption column (data adapted from Biard and Couvert, 2013):
Column temperature 293 K
Bulk gas concentration of H_2S: 1 mg/m^3
Interface H_2S concentration 8.1×10^{-7} mol/L
Concentration of NaOH 1 mol/L
Gas-phase mass transfer coefficient 1×10^{-2} m/s
Liquid-phase mass transfer coefficient 1×10^{-4} m/s
Diffusivity of H_2S in liquid 1.75×10^{-9} m^2/s
Diffusivity of NaOH in liquid 2.09×10^{-9} m^2/s
Henry's law constant 864 m^3 · kPa/kmol
The reaction is reversible and may be expressed as

$$H_2S + OH^- \rightleftharpoons HS^- + H_2O$$

with

$$-r_{H_2S} = 6.24 \times 10^3 C_{H_2S} \text{ mol/(L} \cdot \text{s)}$$

Calculate the enhancement factor.

3.19 Using the Langmuir–Hinshelwood–Hougen–Watson kinetic model, derive a rate equation for the solid- catalyzed gas reaction represented by the overall reaction

$$A + B \rightarrow C$$

when the adsorption of A is rate limiting. Both A and B react only when adsorbed on the catalyst surface.

3.20 Repeat Exercise 3.19 if the chemical reaction step is rate controlling.

3.21 Repeat Exercise 3.19 if the desorption of C is rate limiting.

3.22 Repeat Example 3.11 if the catalyst particles are cylindrical pellets of 0.35 cm diameter with the side edges blocked.

3.23 The cracking of n-hexane to light olefins is found to be first order in hexane concentration. In an experimental study n-hexane was cracked over nanocrystalline ZSM-5 zeolite catalyst with the characteristic dimension of the crystals being 150 nm, and a density of 1780 kg/m^3. The catalyst mass-based reaction rate constant was determined to be 1.04×10^{-2} $m^3/(kg \cdot s)$ at 923 K. Under the reaction conditions the effective diffusivity of n-hexane within the catalyst particle is 1.3×10^{-11} m^2/s. Estimate the effectiveness factor of the catalyst. (Data adapted from Konno et al., 2012).

3.24 In a separate experiment for the cracking of n-hexane spherical composite catalyst pellets of 3 μm diameter were prepared by mixing zeolite crystals with silica alumina. The volume-based rate constant for the reaction is reported as 4.54 s^{-1}. The effective diffusivity in the catalyst pellet is 1×10^{-10} m^2/s. Calculate the effectiveness factor. (Data adapted from Smirniotis and Ruckenstein, 1993).

3.25 In Exercise 3.24, what will be the effectiveness factor if the catalyst pellet is in the form of a cylindrical disk of diameter 3 μm and thickness 2 μm with the round edges sealed for diffusion?

REFERENCES

Aris, R., On shape factors of irregular particles—I. *Chem. Eng. Sci.*, 6, 262, 1957.

Aris, R., A normalization for the Thiele modulus. *Ind. Eng. Chem. Fund.*, 4, 227, 1965.

Arrhenius, S.A., Uber die reaktionsgeschwindigkeit bei der inversion von rohrzucker durch sauren. *Z. Physik. Chem.*, 4, 226, 1889.

Atkins, P. and de Paula, J., *Physical Chemistry*, 8th ed., Freeman, New York, 830, 2006.

Bartholomew, C.H., Sintering kinetics of supported metals: New perspectives from a unifying GPLE treatment. *App. Catal. A: Gen.*, 107, 1, 1993.

Bartholomew, C.H., Mechanisms of catalyst deactivation. *App. Catal. A: Gen.*, 212, 17, 2001.

Biard, P.-F. and Couvert, A., Overview of mass transfer enhancement factor determination for acidic and basic compounds absorption in water. *Chem. Eng. J.*, 222, 444, 2013.

Bischoff, K.B., Effectiveness factors for general reaction rate forms. *AIChE J.*, 11, 351, 1965.

Bischoff, K.B., An extension of the general criterion for importance of pore diffusion with chemical reactions. *Chem. Eng. Sci.*, 22, 525, 1967.

Compendium of Chemical Terminology Gold Book, Version 2.3.3, IUPAC, 1459, 2014. Available online; http://goldbook.iupac.org/PDF/goldbook.pdf.

Danckverts, P.V., Gas absorption accompanied by chemical reaction. *AIChE J.*, 1, 456, 1955.

Forzatti, P., and Lietti, L., Catalyst deactivation. *Catal. Today*, 52, 165, 1999.

Froessling, N., The evaporation of falling drops. *Gerland Beitr. Geophys.*, 52, 170 1938.

Froment, G.F. and Bischoff, K.B., Non-steady state behaviour of fixed bed catalytic reactors due to catalyst fouling. *Chem. Eng. Sci.*, 16, 189, 1961.

Froment, G.F. and Bischoff, K.B., Kinetic data and product distribution from fixed bed catalytic reactors subject to catalyst fouling. *Chem. Eng. Sci.*, 17, 105, 1962.

Froment, G.F. and Bischoff, K.B., *Chemical Reactor Analysis and Design*, John Wiley & Sons, New York, 186, 1979.

Fuentes, G.A. and Gamas, E.D., in Bartholomew, C.H., and Butt, J.B. (eds.), *Catalyst Deactivation*, Elsevier, Amsterdam, 637, 1991.

Hatta, S., On the absorption velocity of gases by liquids. *Tech. Rep. Tohoku Imperial Univ.*, 10, 119, 1932.

Hikita, H., and Asai, S., Gas absorption with (m,n)th order irreversible chemical reaction. *Int. Chem. Eng.*, 4, 332, 1964.

Hougen, O.A. and Watson, K.M., Solid catalysts and reaction rates. *Ind. Eng. Chem.*, 35, 529, 1943.

Konno, H., Okamura, T., Kawahara, T., Nakasaka, Y., Tago, T., and Masuda, T., Kinetics of n-hexane cracking over ZSM-5 zeolites—Effect of crystal size on effectiveness factor and catalyst lifetime. *Chem. Eng. J.*, 207–208, 490, 2012.

Levenspiel, O., *Chemical Reaction Engineering*, 3rd ed., John Wiley & Sons, New York, 390, 1999.

Liu, Y., Li, H., Wei, G., Zhang, H., Li, X., and Jia, Y., Mass transfer performance of CO_2 absorption by alkanolamine aqueous solution for biogas purification. *Sep. Purif. Technol.*, 133, 476, 2014.

Logan, S.R., The origin and status of the Arrhenius equation. *J. Chem. Edu.*, 59, 279, 1982.

Lund, E.W., Guldberg and Waage and the law of mass action. *J. Chem. Edu.*, 42, 548, 1965.

McCabe, W.L., Smith, J.C., and Harriott, P., *Unit Operations of Chemical Engineering*, 5th ed., McGraw-Hill, New York, 839, 1993.

Ranz, W.E., Friction and transfer coefficients for single particles and packed beds. *Chem. Eng. Prog.*, 48, 247, 1952.

Rester, S. and Aris, R., Communications on the theory of diffusion and reaction—II. The effect of shape on the effectiveness factor. *Chem. Eng. Sci.*, 24, 793, 1969.

Smirniotis, P.G. and Ruckenstein, E., The activity of composite catalysts: Theory and experiments for spherical and cylindrical single pellets. *Chem. Eng. Sci.*, 48, 585, 1993.

Thiele, E.W., Relation between catalytic activity and size of particle. *Ind. Eng. Chem.*, 31, 916, 1939.

van Krevelen, D.W. and Hoftijzer, P.J., Kinetics of gas-liquid reactions part I. General theory. *Rec. Trav. Chim. Pays-Bas*, 67, 563 1948.

van Swaaij, W.P.M. and Versteeg, G.F., Mass transfer accompanied with complex reversible chemical reactions in gas-liquid systems. An overview. *Chem. Eng. Sci.*, 47, 3181, 1992.

Voorhies, A., Carbon formation in catalytic cracking. *Ind. Eng. Chem.*, 37, 318, 1945.

Weisz, P.B. and Prater, C.D., Interpretation of measurements in experimental catalysis. *Adv. Catal.*, 6, 143, 1954.

Wheeler, A., Reaction rates and selectivity in catalyst pores. *Adv. Catal.*, 3, 249, 1951.

Yagi, S. and Kunii, D., Fluidized-solids reactors with continuous solids feed—I. Residence time of particles in fluidized beds; II. Conversion for overflow and carryover particles; III. Conversion in experimental fluidized-solids reactors. *Chem. Eng. Sci.*, 16, 364, 372, 380, 1961.

4 Thermodynamics of Chemical Reaction

As outlined in Chapter 1, chemical transformations may result in change in energy of the reacting system. It is imperative to study these energy changes in relation to the progress of the reaction. It is all the more important as energy changes are often associated with temperature changes, which have direct bearing on kinetics. On a more fundamental basis, energy transformations decide the feasibility of a reaction and the extent to which a reaction can proceed.

4.1 HEAT OF REACTION

During a chemical reaction, existing bonds are broken and new ones are formed, which leads to the formation of products from the reactants. It means that the spatial arrangement of nuclei and the electronic configurations around these nuclei are altered during a chemical reaction. The change in enthalpy during a chemical reaction is called *enthalpy of reaction* or more commonly the *heat of reaction* (Atkins and de Paula, 2006; DeVoe, 2015). For the general chemical reaction as expressed by the balanced chemical Equation 2.8, from a simple enthalpy balance we have

$$\Delta H_r = \text{Sum of enthalpies of products} - \text{Sum of enthalpies of reactants}$$

$$\Rightarrow \quad \Delta H_r = \sum_{j=1}^{l} p_j H_{P_j} - \sum_{i=1}^{n} a_i H_{A_i} \tag{4.1}$$

In terms of *internal energy*

$$\Delta H_r = \Delta U_r + \left(\sum_{j=1}^{l} p_j - \sum_{i=1}^{n} a_i \right) RT \tag{4.2}$$

When the reaction is carried out in such a way that all the products and the reactants are in their standard state, the resulting heat of reaction is known as the *standard heat of reaction*. The modified equation can then be written as

$$\Delta H_r^0 = \sum_{j=1}^{l} p_j H_{P_j}^0 - \sum_{i=1}^{n} a_i H_{A_i}^0 \tag{4.3}$$

Equations 4.1 and 4.2 provide the change in enthalpy when a_1 moles of the reactant A_1, reacts with a_2 moles of the reactant A_2, a_3 moles of the reactant A_3, and so on to form p_1 moles of the product P_1, p_2 moles of the product P_2, and so on. For calculations in chemical reaction engineering, it is often useful to express the heat of reaction in terms of one mole of the limiting reactant. It may be expressed as

$$(\Delta H_r)_{LR} = \sum_{j=1}^{l} \frac{p_j}{a_{LR}} H_{P_j} - \sum_{i=1}^{n} \frac{a_i}{a_{LR}} H_{A_i} \tag{4.4}$$

and

$$\left(\Delta H_r^0\right)_{LR} = \sum_{j=1}^{l} \frac{p_j}{a_{LR}} H_{P_j}^0 - \sum_{i=1}^{n} \frac{a_i}{a_{LR}} H_{A_i}^0 \tag{4.5}$$

The enthalpies of the reactant and product chemical species, sometimes referred to as the enthalpy of formation, are calculated from the mean bond enthalpies. Some mean values of bond enthalpies and enthalpies of formation of some common chemical species are given in Tables 4.1 and 4.2.

EXAMPLE 4.1

Calculate the heat of reaction for combustion of ethane at 298 K and 1 atm pressure expressed as per mole of ethane.

Solution

The balanced chemical equation for the combustion of ethane would be

$$2C_2H_6(g) + 7O_2(g) \rightarrow 4CO_2(g) + 6H_2O(l) \tag{4.1.1}$$

From Table 4.2, at 298 K,

$$H_{CO_2(g)} = -393.51 \text{ kJ/mol}, \quad H_{H_2O(l)} = -285.83 \text{ kJ/mol}$$

$$H_{O_2(g)} = 0 \text{ (element in standard state)}, \quad H_{C_2H_6(g)} = -84.68 \text{ kJ/mol}$$

From reaction 4.1.1,

$$a_{C_2H_6(g)} = 2, \quad p_{CO_2(g)} = 4, \quad p_{H_2O(l)} = 6$$

From Equation 4.4, at 298 K,

$$(\Delta H_r)_{C_2H_6} = \frac{p_{CO_2(g)}}{a_{C_2H_6(g)}} H_{CO_2(g)} + \frac{p_{H_2O(l)}}{a_{C_2H_6(g)}} H_{H_2O(l)} - \frac{a_{C_2H_6(g)}}{a_{C_2H_6(g)}} H_{C_2H_6(g)}$$

$$\Rightarrow \quad (\Delta H_r)_{C_2H_6} = \frac{4}{2}(-393.51) + \frac{6}{2}(-285.83) - \frac{2}{2}(-84.68) = -1559.83 \text{ kJ/mol}$$

TABLE 4.1
Values of Some Mean Bond Enthalpies at 298 K

Bond	Bond Enthalpy (kJ/mol)
Br–Br	192
Br–Cl	218
C–Br	276
C–C	347
C=C	611
C≡C	837
C–Cl	330
C–F	439
C–H	414
C–I	238
C–N	293
C=N	615
C≡N	891
C–O	351
C=O	741
C–S	259
Cl–Br	218
Cl–Cl	243
H–Br	368
H–Cl	431
H–F	569
H–H	435.8
H–I	297
N–H	389
N–N	159
N=N	418
N≡N	946
N–O	201
O–Cl	205
O–H	464
O–O	138
O=O	498

Source: Darwent, B.., *Natl. Stand. Ref. Data Ser.–Natl. Bur. Stand.*, 31, 9, 1970.

When a balanced chemical equation is written along with the associated heat of reaction, it is known as a thermochemical equation. Thus, the thermochemical form of Equation 2.8, Chapter 2 can be written as

$$\sum_{i=1}^{n} a_i A_i \rightarrow \sum_{j=1}^{l} p_j P_j \quad \Delta H_r^0 = \text{value} \tag{4.6}$$

TABLE 4.2
Values of Enthalpies of Formation of Some Common Chemical Species at 298 K

Chemical Species	Enthalpy of Formation (kJ/mol)
$Al_2O_3(s)$	−1675.7
$Br_2(g)$	30.91
$CaCO_3(s)$	−1206.92
$CaO(s)$	−635.09
$CH_4(g)$	−74.81
$C_2H_4(g)$	52.26
$C_2H_6(g)$	−84.68
$C_4H_{10}(g)$	−126.15
$C_6H_6(l)$	49
$CH_3OH(l)$	−238.86
$C_2H_5OH(l)$	−277.69
$CO(g)$	−110.53
$CO_2(g)$	−393.51
$H_2O(g)$	−241.82
$H_2O(l)$	−285.83
$HCl(g)$	−92.31
$NaCl(s)$	−411.15
$NH_3(g)$	−46.11
$NO(g)$	90.25
$NO_2(g)$	33.18
$SO_2(g)$	−296.83
$SO_3(g)$	−395.72

Source: Lide, D.R., *CRC Handbook of Chemistry and Physics*, CRC Press, Boca Raton, FL, 2005; Stull, D.R., Prophet, H., *Natl. Stand. Ref. Data Ser.–Natl. Bur. Stand.*, 37, 13, 1971.

In Equation 4.1, if $\sum_{j=1}^{l} p_j H_{P_j} > \sum_{i=1}^{n} a_i H_{A_i}$, ΔH_r is positive. The sum of the enthalpies of all the products exceeds the sum of the enthalpies of all the reactants so heat must be supplied during the reaction. If heat is not supplied from outside, the required heat to drive the reaction comes at the expense of the sensible heat of the reaction and the products. Consequently, the temperature of the reacting system drops during the reaction. Such reactions are called endothermic reactions. Conversely, if $\sum_{j=1}^{l} p_j H_{P_j} < \sum_{i=1}^{n} a_i H_{A_i}$, ΔH_r is negative. The reaction proceeds with the evolution of heat and if this heat is not taken out of the system, the temperature of the reacting system increases during the reaction. These reactions are called exothermic reactions.

4.1.1 Heat of Reaction at Different Temperatures

Commonly, the heat of reaction is measured at 298 K. Frequently, it is required to know the heat of reaction at other temperatures as well (Atkins and de Paula, 2006). This can be done by realizing the fact that enthalpy is a state property. Thus, the change in enthalpy depends only on the initial and final states of the system and not the path by which the system goes from the initial to the final state. Consider the following schematic:

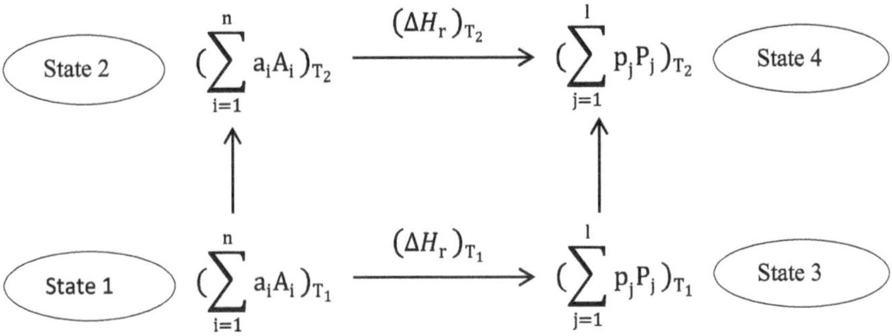

If the system goes from State 1 to State 4, then the change in enthalpy would be the same whether in doing so it passes through State 2 or State 3. If the system goes from State 1 to State 4 through State 2, we have

$$H_{\text{State1}} + \sum_{i=1}^{n} a_i (C_p)_{A_i} (T_2 - T_1) + (\Delta H_r)_{T_2} = H_{\text{State4}}$$

$$\Rightarrow \quad H_{\text{State4}} - H_{\text{State1}} = \sum_{i=1}^{n} a_i (C_p)_{A_i} (T_2 - T_1) + (\Delta H_r)_{T_2} \tag{4.7}$$

Similarly, if the system goes from State 1 to State 4 through State 3, we have

$$H_{\text{State4}} - H_{\text{State1}} = (\Delta H_r)_{T_1} + \sum_{j=1}^{1} p_j (C_p)_{P_j} (T_2 - T_1) \tag{4.8}$$

Since the left-hand sides of Equations 4.7 and 4.8 are the same, their RHSs must also be equal, and we can write

$$\sum_{i=1}^{n} a_i (C_p)_{A_i} (T_2 - T_1) + (\Delta H_r)_{T_2} = (\Delta H_r)_{T_1} + \sum_{j=1}^{1} p_j (C_p)_{P_j} (T_2 - T_1)$$

$$\Rightarrow \quad (\Delta H_r)_{T_2} = (\Delta H_r)_{T_1} + \sum_{j=1}^{1} p_j (C_p)_{P_j} (T_2 - T_1) - \sum_{i=1}^{n} a_i (C_p)_{A_i} (T_2 - T_1)$$

$$\Rightarrow \quad (\Delta H_r)_{T_2} = (\Delta H_r)_{T_1} + (T_2 - T_1) \left[\sum_{j=1}^{1} p_j (C_p)_{P_j} - \sum_{i=1}^{n} a_i (C_p)_{A_i} \right] \tag{4.9}$$

The following situations are possible:

1. $T_2 > T_1$ and $[\sum_{j=1}^{l} p_j (C_p)_{P_j} > \sum_{i=1}^{n} a_i (C_p)_{A_i}]$ then $(\Delta H_r)_{T_2} > (\Delta H_r)_{T_1}$
2. $T_2 > T_1$ and $[\sum_{j=1}^{l} p_j (C_p)_{P_j} < \sum_{i=1}^{n} a_i (C_p)_{A_i}]$ then $(\Delta H_r)_{T_2} < (\Delta H_r)_{T_1}$
3. $T_2 < T_1$ and $[\sum_{j=1}^{l} p_j (C_p)_{P_j} > \sum_{i=1}^{n} a_i (C_p)_{A_i}]$ then $(\Delta H_r)_{T_2} < (\Delta H_r)_{T_1}$
4. $T_2 < T_1$ and $[\sum_{j=1}^{l} p_j (C_p)_{P_j} < \sum_{i=1}^{n} a_i (C_p)_{A_i}]$ then $(\Delta H_r)_{T_2} > (\Delta H_r)_{T_1}$
5. $\sum_{j=1}^{l} p_j (C_p)_{P_j} = \sum_{i=1}^{n} a_i (C_p)_{A_i}$, $(\Delta H_r)_{T_2} = (\Delta H_r)_{T_1}$ for all temperatures

EXAMPLE 4.2

Calculate the heat of reaction of ethane at 600°C given that molar heat capacities at constant pressure of liquid water, carbon dioxide gas, oxygen gas, and ethane are 4.184, 37.5, 29.1, and 53.2 J/mol, respectively.

Solution

$$T_1 = 298\ \text{K}, \quad T_2 = 600 + 273 = 873\ \text{K}$$

From Example 4.1 $(\Delta H_r)_{298} = -1559.83\ \text{kJ/mol} = -1559.83 \times 10^3\ \text{J/mol}$

$$(C_p)_{H_2O(l)} = 4.184\ \text{J/mol}, (C_p)_{CO_2(g)} = 37.5\ \text{J/mol}, (C_p)_{O_2(g)} = 29.1\ \text{J/mol}$$

$$(C_p)_{C_2H_6(g)} = 53.2\ \text{J/mol}$$

Again, from Example 4.1 $a_{C_2H_6(g)} = 2$, $a_{O_2} = 7$, $p_{CO_2(g)} = 4$, $p_{H_2O(l)} = 6$
From Equation 4.9

$$(\Delta H_r)_{873} = (\Delta H_r)_{298} + (873 - 298)[4 \times 37.5 + 6 \times 4.184 - 2 \times 53.2 - 7 \times 29.1]$$

$$\Rightarrow \quad (\Delta H_r)_{873} = -1559.83 \times 10^3 + 575[150 + 25.104 - 106.4 - 203.7]$$

$$\Rightarrow \quad (\Delta H_r)_{873} = -1559.83 \times 10^3 - 77622.7 = -1637.45 \times 10^3\ \text{J/mol} = -1637.45\ \text{kJ/mol}$$

4.2 REACTION EQUILIBRIUM

In the previous chapter, we used terms like "reversible reaction," "irreversible reaction," and "equilibrium constant" while dealing with the rates of some of the heterogeneous reactions. We now evolve a conceptual basis for these ideas and the mathematics involving them so that they can be used in solving the problems of reaction engineering.

4.2.1 Thermodynamic Interpretation of Reaction Equilibrium

In this book we have frequently used the phrase "reacting system." Let us now examine the criterion for the reacting system to be at equilibrium. All systems tend to acquire the state of minimum energy and maximum entropy to be at equilibrium and reacting systems are no exception. This seemingly conflicting notion is resolved by the concept of free energy, which may be defined at constant pressure and temperature and constant volume and temperature. For most of the reacting systems (not all, however), the pressure and temperature are constant. Under these conditions, the Gibbs free energy is defined as (Denbigh, 1981)

$$G = H - \mathrm{T}S \tag{4.10}$$

$$\Rightarrow \quad \mathrm{d}G = \mathrm{d}H - \mathrm{Td}S - S\mathrm{dT}$$

In addition, at constant temperature, dT is zero, so

$$\mathrm{d}G = \mathrm{d}H - \mathrm{Td}S \tag{4.11}$$

The tendency of the reacting system is to minimize its free energy and the state at which the free energy is at its minimum is the state of equilibrium (Seider and Widagdo, 1996; Neron et al., 2012). The spontaneous response of any reacting system shifted from the state of equilibrium would be to try to attain equilibrium by lowering its free energy. Therefore, for a chemical reaction to be spontaneous, it should be associated with a decrease in free energy. Thus, the criterion for a reaction to be spontaneous is (Ismail, 2013)

$$\mathrm{d}G = \mathrm{d}H - \mathrm{Td}S < 0 \tag{4.12}$$

Once the equilibrium is attained, there is no more incentive for the system to change its state and

$$\mathrm{d}G = \mathrm{d}H - \mathrm{Td}S = 0 \tag{4.13}$$

At equilibrium, the sum of free energies of all the reacting chemical species is equal to the sum of free energies of all the product chemical species.

EXAMPLE 4.3

For the gaseous reaction

$$2C_2H_6(g) + 7O_2(g) \rightarrow 4CO_2(g) + 6H_2O\ (l)$$

Calculate the Gibbs free energy change at 298 K and 1 atm pressure for the reaction if the standard entropies of formation of $C_2H_6(g)$, $O_2(g)$, $CO_2(g)$, and $H_2O(l)$ are 229.6, 205.14, 213.74, and 69.91 J/K · mol, respectively.

Solution

$$(\Delta S_r)_{C_4H_{10}} = \frac{p_{CO_2(g)}}{a_{C_2H_6(g)}} S_{CO_2(g)} + \frac{p_{H_2O(l)}}{a_{C_2H_6(g)}} S_{H_2O(l)} - \frac{a_{C_2H_6(g)}}{a_{C_2H_6(g)}} S_{C_2H_6(g)} - \frac{a_{O_2(g)}}{a_{C_2H_6(g)}} S_{O_2(g)}$$

$$\Rightarrow \quad (\Delta S_r)_{C_4H_{10}} = \frac{4}{2}\times 213.74 + \frac{6}{2}\times 69.91 - \frac{2}{2}\times 229.6 - \frac{7}{2}\times 205.14 = -310.38\ \mathrm{J/K \cdot mol}$$

From Example 4.1,

$$(\Delta H_r)_{C_2H_6} = -1559.83 \text{ kJ/mol}$$

From Equation 4.11,

$$(\Delta G_r)_{C_2H_6} = (\Delta H_r)_{C_4H_{10}} - T(\Delta S_r)_{C_4H_{10}} = -1559.83 \times 10^3 - 298(-310.38)$$

$$\Rightarrow \quad (\Delta G_r)_{C_2H_6} = -1467.34 \times 10^3 \text{ J/mol} = -1467.34 \text{ kJ/mol}$$

Gibbs free energy change is negative so the reaction is spontaneous.

4.2.2 Kinetic Interpretation of Reaction Equilibrium

Thermodynamics profess the cessation of all intent for change in the reacting system at equilibrium. At the macroscopic level, we cannot observe any change in the system at equilibrium. For a reacting system, it means no change in the composition of the system (IUPAC Gold Book, 2014). Does it mean that all the activity at the microscopic level ceases at equilibrium? Kinetics has an answer to this paradox for reacting systems. Even when the reacting system is at equilibrium, its microscopic world is buzzing with activity. But the net effect of all these microscopic activities averages out to be zero at the macroscopic level. Averaged over a measurable time span, the number of events leading to the formation of products from the reactants is equal to the number of events transforming the products back to the reactants. This gives us the concept of forward reaction and backward reaction. The equilibrium is merely the state when the rate of the forward reaction is the same as that of the backward reaction. In theory, every reaction is reversible, that is, given suitable conditions the products can revert back to the reactants. However, many reactions are practically irreversible in the sense that the backward reaction is infinitesimally slow in comparison to the forward reaction. All elementary reactions are by definition reversible.

Consider the general chemical reaction represented by the balanced chemical Equation 2.8, Chapter 2 to be rigorously a reversible reaction. We rewrite this equation as

$$\sum_{i=1}^{n} a_i A_i \rightleftharpoons \sum_{j=1}^{l} p_j P_j \qquad (4.14)$$

Rate of the forward reaction

$$r_{fwd} = k_1 \prod_{i=1}^{n} C_{A_i}^{a_i} \qquad (4.15)$$

Rate of the backward reaction

$$r_{bwd} = k_2 \prod_{j=1}^{l} C_{P_j}^{p_j} \tag{4.16}$$

At equilibrium,

$$r_{fwd} = r_{bwd}$$

$$\Rightarrow \quad k_1 \prod_{i=1}^{n} \left(C_{A_i}\right)_{eq}^{a_i} = k_2 \prod_{j=1}^{l} \left(C_{P_j}\right)_{eq}^{p_j}$$

$$\Rightarrow \quad K_C = \frac{k_1}{k_2} = \frac{\prod_{j=1}^{l} \left(C_{P_j}\right)_{eq}^{p_j}}{\prod_{i=1}^{n} \left(C_{A_i}\right)_{eq}^{a_i}} \tag{4.17}$$

From Equations 3.14 and 3.15 (Chapter 3),

$$C_{A_i} = \frac{\left(C_{A_i}\right)_0}{1+\varepsilon_{LR}X_{A_{LR}}} - \frac{a_i}{a_{LR}}\left\{\frac{\left(C_{A_{LR}}\right)_0}{1+\varepsilon_{LR}X_{A_{LR}}} - \frac{\left(C_{A_{LR}}\right)_0\left(1-X_{A_{LR}}\right)}{1+\varepsilon_{LR}X_{A_{LR}}}\right\}$$

$$\Rightarrow \quad C_{A_i} = \frac{\left(C_{A_i}\right)_0}{1+\varepsilon_{LR}X_{A_{LR}}} - \frac{a_i}{a_{LR}}\left\{\frac{\left(C_{A_{LR}}\right)_0 X_{A_{LR}}}{1+\varepsilon_{LR}X_{A_{LR}}}\right\}$$

$$\Rightarrow \quad C_{A_i} = \frac{a_{LR}\left(C_{A_i}\right)_0 - a_i\left(C_{A_{LR}}\right)_0 X_{A_{LR}}}{a_{LR}\left(1+\varepsilon_{LR}X_{A_{LR}}\right)} \tag{4.18}$$

Similarly,

$$C_{P_j} = \frac{a_{LR}\left(C_{P_j}\right)_0 + p_j\left(C_{A_{LR}}\right)_0 X_{A_{LR}}}{a_{LR}\left(1+\varepsilon_{LR}X_{A_{LR}}\right)} \tag{4.19}$$

From Equations 4.17 through 4.19,

$$K_C = \left[a_{LR}\left(1+\varepsilon_{LR}X_{A_{LR}}\right)\right]^{\left(\Sigma_{i=1}^{n} a_i - \Sigma_{j=1}^{l} p_j\right)} \times \frac{\prod_{j=1}^{l}\left[a_{LR}\left(C_{P_j}\right)_0 + p_j\left(C_{A_{LR}}\right)_0 X_{A_{LR}}\right]_{eq}^{p_j}}{\prod_{i=1}^{n}\left[a_{LR}\left(C_{A_i}\right)_0 - a_i\left(C_{A_{LR}}\right)_0 X_{A_{LR}}\right]_{eq}^{a_i}} \tag{4.20}$$

For constant volume systems, $\varepsilon_{LR} = 0$, and Equation 4.20 reduces to

$$K_C = [a_{LR}]^{\left(\Sigma_{i=1}^{n} a_i - \Sigma_{j=1}^{l} p_j\right)} \times \frac{\prod_{j=1}^{l}\left[a_{LR}\left(C_{P_j}\right)_0 + p_j\left(C_{A_{LR}}\right)_0 X_{A_{LR}}\right]_{eq}^{p_j}}{\prod_{i=1}^{n}\left[a_{LR}\left(C_{A_i}\right)_0 - a_i\left(C_{A_{LR}}\right)_0 X_{A_{LR}}\right]_{eq}^{a_i}} \tag{4.21}$$

For constant volume systems with no products initially present, the equation further reduces to

$$K_C = [a_{LR}]^{\left(\Sigma_{i=1}^{n} a_i - \Sigma_{j=1}^{l} p_j\right)} \times \frac{\prod_{j=1}^{l}\left[p_j\left(C_{A_{LR}}\right)_0 X_{A_{LR}}\right]_{eq}^{p_j}}{\prod_{i=1}^{n}\left[a_{LR}\left(C_{A_i}\right)_0 - a_i\left(C_{A_{LR}}\right)_0 X_{A_{LR}}\right]_{eq}^{a_i}} \tag{4.22}$$

For gaseous-reacting systems, concentration and pressure are related. If the gases behave like ideal gases, from Equation 4.17

$$K_C = \frac{\prod_{j=1}^{l}\left(C_{P_j}\right)_{eq}^{p_j}}{\prod_{i=1}^{n}\left(C_{A_i}\right)_{eq}^{a_i}} = (RT)^{\left(\Sigma_{i=1}^{n} a_i - \Sigma_{j=1}^{l} p_j\right)} \frac{\prod_{j=1}^{l}\left(p_{P_j}\right)_{eq}^{p_j}}{\prod_{i=1}^{n}\left(p_{A_i}\right)_{eq}^{a_i}} \tag{4.23}$$

$$\Rightarrow \quad K_C = (RT)^{\left(\Sigma_{i=1}^{n} a_i - \Sigma_{j=1}^{l} p_j\right)} K_p$$

where

$$K_p = \frac{\prod_{j=1}^{l}\left(p_{P_j}\right)_{eq}^{p_j}}{\prod_{i=1}^{n}\left(p_{A_i}\right)_{eq}^{a_i}} \tag{4.24}$$

EXAMPLE 4.4

For the reaction

$$N_2(g) + 3H_2(g) \rightleftharpoons 2NH_3(g)$$

$K_p = 41$ at 400 K. If the reaction is carried out at constant temperature and pressure of 400 K and 10 atm with an initial mixture of N_2 and H_2 in 1:3 molar ratio, find out the composition of the reaction mixture at equilibrium.

Solution

From the reaction

$$a_{N_2} = 1, \quad a_{H_2} = 3, \quad a_{NH_3} = 2$$

From Equation 4.23

$$K_C = (RT)^{\left(\sum_{i=1}^{n} a_i - \sum_{j=1}^{l} p_j\right)} K_p = (0.08 \times 400)^{(1+3-2)} \times 41 = 41984$$

In the initial mixture:
Mole fraction of N_2, $x_{N_2} = 0.25$, mole fraction of H_2, $x_{H_2} = 0.75$
Partial pressure of N_2, $p_{N_2} = 0.25 \times 10 = 2.5$ atm
Partial pressure of H_2, $p_{H_2} = 0.75 \times 10 = 7.5$ atm
From Equation 2.26, Chapter 2 Initial concentration of N_2, $(C_{N_2})_0 = \frac{p_{N_2}}{RT} = \frac{2.5}{0.08 \times 298} = 0.1048$ mol/L

Initial concentration of H_2, $(C_{H_2})_0 = \frac{p_{H_2}}{RT} = \frac{7.5}{0.08 \times 298} = 3(C_{N_2})_0 = 0.3146$ mol/L

Let the conversion of N_2 at equilibrium be $(X_{N_2})_{eq}$ and the conversion of H_2 at equilibrium be $(X_{H_2})_{eq}$. From the reaction stoichiometry and the initial molar ratio of reactants we have $(X_{N_2})_{eq} = (X_{H_2})_{eq}$.

From Equation 2.44 (Chapter 2),

$$\varepsilon_{N_2} = \varepsilon_{H_2} = \frac{(N_{N_2})_0 [2-1-3]}{(N_{N_2})_0 + (N_{H_2})_0} = \frac{-2}{1+3} = -0.5$$

Then, from Equation 2.45 (Chapter 2),

$$(C_{N_2})_{eq} = \frac{(C_{N_2})_0 \left[1-(X_{N_2})_{eq}\right]}{1+\varepsilon_{N_2}(X_{N_2})_{eq}} \tag{4.4.1}$$

$$(C_{H_2})_{eq} = \frac{(C_{H_2})_0 \left[1-(X_{N_2})_{eq}\right]}{1+\varepsilon_{N_2}(X_{N_2})_{eq}} = 3(C_{N_2})_{eq} \tag{4.4.2}$$

$$(C_{NH_3})_{eq} = \frac{2(C_{N_2})_0 (X_{N_2})_{eq}}{1+\varepsilon_{N_2}(X_{N_2})_{eq}} \tag{4.4.3}$$

From Equation 4.17,

$$K_C = \frac{\prod_{j=1}^{l} (C_{P_j})_{eq}^{p_j}}{\prod_{i=1}^{n} (C_{A_i})_{eq}^{a_i}} = \frac{(C_{NH_3})_{eq}^2}{(C_{N_2})_{eq}(C_{H_2})_{eq}^3} = \frac{(C_{NH_3})_{eq}^2}{(C_{N_2})_{eq} \times 27 \times (C_{N_2})_{eq}^3}$$

$$\Rightarrow \quad K_C = \frac{(C_{NH_3})_{eq}^2}{27 \times (C_{N_2})_{eq}^4} = \frac{4 \times (C_{N_2})_0^2 (X_{N_2})_{eq}^2 \times \left[1+\varepsilon_{N_2}(X_{N_2})_{eq}\right]^2}{27 \times (C_{N_2})_0^4 \times \left[1-(X_{N_2})_{eq}\right]^4}$$

$$= \frac{4 \times (X_{N_2})_{eq}^2 \times \left[1+\varepsilon_{N_2}(X_{N_2})_{eq}\right]^2}{27 \times (C_{N_2})_0^2 \times \left[1-(X_{N_2})_{eq}\right]^4}$$

$$\Rightarrow \quad \frac{41984 \times 27 \times 0.1048 \times 0.1048}{4} = \frac{(X_{N_2})_{eq}^2 \times \left[1 + \varepsilon_{N_2}(X_{N_2})_{eq}\right]^2}{\left[1 - (X_{N_2})_{eq}\right]^4}$$

$$\Rightarrow \quad \frac{(X_{N_2})_{eq}^2 \times \left[1 - 0.5(X_{N_2})_{eq}\right]^2}{\left[1 - (X_{N_2})_{eq}\right]^4} = 3112.51$$

$$\Rightarrow \quad (X_{N_2})_{eq} = 0.906 \quad \text{(solving by iteration)}$$

From Equation 4.4.1,

$$(C_{N_2})_{eq} = \frac{(C_{N_2})_0\left[1 - (X_{N_2})_{eq}\right]}{1 + \varepsilon_{N_2}(X_{N_2})_{eq}} = \frac{0.1048\,[1 - 0.906]}{1 - 0.5 \times 0.906} = 0.018 \text{ mol/L}$$

From Equation 4.4.2,

$$(C_{H_2})_{eq} = 3(C_{N_2})_{eq} = 3 \times 0.018 = 0.054 \text{ mol/L}$$

From Equation 4.4.3,

$$(C_{NH_3})_{eq} = \frac{2(C_{N_2})_0(X_{N_2})_{eq}}{1 + \varepsilon_{N_2}(X_{N_2})_{eq}} = \frac{2 \times 0.1048 \times 0.906}{1 - 0.5 \times 0.906} = 0.347 \text{ mol/L}$$

4.2.2.1 Factors Affecting Equilibrium Position in Chemical Reaction

The equilibrium position for a reversible reaction is dependent on the thermodynamic parameters temperature, pressure, and the composition of the reacting system. These may be elaborated as follows:

Temperature dependence of equilibrium position

Temperature dependence of equilibrium constant is expressed by the van't Hoff equation for gaseous-reacting systems. The equation in its differential form gives

$$\frac{d\ \ln K_p}{dT} = \frac{\Delta H_r^0}{RT^2} \tag{4.25}$$

The equation can be integrated between temperatures T_1 and T_2 to give

$$\int_{(K_p)_1}^{(K_p)_2} d\ \ln K_p = \int_{T_1}^{T_2} \frac{\Delta H_r^0}{RT^2}\, dT$$

$$\Rightarrow \quad \ln\frac{(K_p)_2}{(K_p)_1} = \frac{\Delta H_r^0}{R}\left[\frac{1}{T_1} - \frac{1}{T_2}\right] \tag{4.26}$$

The following situations are possible:

a. ΔH_r^0 is positive (endothermic reaction); $T_2 > T_1$, then $(K_p)_2 > (K_p)_1$. For endothermic reaction increase in temperature results in increase of the equilibrium constant. Therefore, from Equations 4.23 and 4.24, the position of equilibrium shifts toward the products with increase in temperature and vice versa.
b. ΔH_r^0 is negative (exothermic reaction); $T_2 > T_1$, then $(K_p)_2 < (K_p)_1$. For exothermic reaction increase in temperature results in decrease of the equilibrium constant. Therefore, from Equations 4.23 and 4.24, the position of equilibrium shifts toward the reactants with increase in temperature and vice versa.

From Equation 4.23,

$$K_C = (RT)^{\left(\sum_{i=1}^{n} a_i - \sum_{j=1}^{l} p_j\right)} K_p$$

$$\Rightarrow \quad K_p = (RT)^{\left(\sum_{j=1}^{l} p_j - \sum_{i=1}^{n} a_i\right)} K_C \tag{4.27}$$

From Equation 4.26,

$$\ln\frac{(K_p)_2}{(K_p)_1} = \frac{\Delta H_r^0}{R}\left[\frac{1}{T_1} - \frac{1}{T_2}\right]$$

$$\Rightarrow \quad \ln\frac{(K_C)_2}{(K_C)_1}\left(\frac{T_2}{T_1}\right)^{\left(\sum_{j=1}^{l} p_j - \sum_{i=1}^{n} a_i\right)} = \frac{\Delta H_r^0}{R}\left[\frac{1}{T_1} - \frac{1}{T_2}\right]$$

$$\Rightarrow \quad \ln\frac{(K_C)_2}{(K_C)_1} + \left(\sum_{j=1}^{l} p_j - \sum_{i=1}^{n} a_i\right)\ln\frac{T_2}{T_1} = \frac{\Delta H_r^0}{R}\left[\frac{1}{T_1} - \frac{1}{T_2}\right] \tag{4.28}$$

Pressure dependence of equilibrium position
If the total pressure of the reacting system is constant, from Equation 4.24,

$$K_p = \frac{\prod_{j=1}^{l}\left(p_{P_j}\right)_{eq}^{p_j}}{\prod_{i=1}^{n}\left(p_{A_i}\right)_{eq}^{a_i}}$$

$$\Rightarrow \quad K_p = \frac{\prod_{j=1}^{l}\left(x_{P_j}p\right)_{eq}^{p_j}}{\prod_{i=1}^{n}\left(x_{A_i}p\right)_{eq}^{a_i}}$$

$$\Rightarrow \quad K_p = p^{\left(\sum_{j=1}^{l} p_j - \sum_{i=1}^{n} a_i\right)} \frac{\prod_{j=1}^{l} \left(x_{P_j}\right)_{eq}^{p_j}}{\prod_{i=1}^{n} \left(x_{A_i}\right)_{eq}^{a_i}}$$

$$\Rightarrow \quad \frac{\prod_{j=1}^{l} \left(x_{P_j}\right)_{eq}^{p_j}}{\prod_{i=1}^{n} \left(x_{A_i}\right)_{eq}^{a_i}} = K_p p^{\left(\sum_{i=1}^{n} a_i - \sum_{j=1}^{l} p_j\right)} \tag{4.29}$$

At a given temperature, K_p is constant, so depending on the total pressure of the reacting system, the following situations are possible:

1. If $\sum_{i=1}^{n} a_i > \sum_{j=1}^{l} p_j$, increasing the total pressure of the reacting system will increase the term $p^{\left(\sum_{i=1}^{n} a_i - \sum_{j=1}^{l} p_j\right)}$; hence, the RHS of Equation 4.29 will increase. To maintain the equality, the numerator in the left-hand side of the equation must increase and the denominator must decrease. Thus, the position of the equilibrium shifts toward the products. A contrary effect will be observed if the total pressure is decreased.
2. If $\sum_{i=1}^{n} a_i < \sum_{j=1}^{l} p_j$, increasing the total pressure of the reacting system will decrease the term $p^{\left(\sum_{i=1}^{n} a_i - \sum_{j=1}^{l} p_j\right)}$; hence, the RHS of Equation 4.29 will decrease. To maintain the equality, the numerator in the left-hand side of the equation must decrease and the denominator must increase. Thus, the position of the equilibrium shifts toward the reactants. A contrary effect will be observed if the total pressure is decreased.

Composition dependence of equilibrium position

From Equation 4.20, K_C depends, besides the initial concentrations of reactants and products, on ϵ_{LR}, and $X_{A_{LR}}$ at equilibrium. From Equation 2.44, Chapter 2, reproduced here:

$$\varepsilon_{LR} = \frac{\left(N_{A_{LR}}\right)_0 \left[\sum_{j=1}^{l} (p_j/a_{LR}) - \sum_{i=1}^{n} (a_i/a_{LR})\right]}{\sum_{i=1}^{n} \left(N_{A_i}\right)_0 + N_{inerts}}$$

The following cases are possible for given initial concentrations of reactants and products:

1. If $\sum_{j=1}^{l} p_j = \sum_{i=1}^{n} a_i$, ε_{LR} is zero irrespective of N_{inerts} and from Equation 4.21 K_C, and hence the position of equilibrium, remains unchanged by the addition of inerts.

2. If $\sum_{j=1}^{l} p_j > \sum_{i=1}^{n} a_i$, ε_{LR} is positive and decreases in value with increase in N_{inerts}. From Equation 4.20, in order to keep K_C unchanged, $X_{A_{LR}}$ at equilibrium will increase. Hence, the position of equilibrium will shift toward the products.
3. If $\sum_{j=1}^{l} p_j < \sum_{i=1}^{n} a_i$, ε_{LR} is negative and increases in value with increase in N_{inerts}. From Equation 4.20, in order to keep K_C unchanged, $X_{A_{LR}}$ at equilibrium will decrease. Hence, the position of equilibrium will shift toward the reactants.

For the initial concentrations of reactants and products:

1. If more reactants are added initially, the position of equilibrium shifts toward products.
2. If more products are added initially, the position of equilibrium shifts toward reactants.

EXAMPLE 4.5

For the reaction of Example 4.4, $\Delta H_r^0 = -91.8$ kJ/mol. Find the equilibrium constant K_p at 298 K.

SOLUTION

From Example 4.4 $K_p = 41$ at 400 K. Therefore,

$$T_1 = 400\,\text{K}, \quad (K_p)_1 = 41$$

$$T_2 = 298\,\text{K}$$

From Equation 4.26,

$$\ln\frac{(K_p)_2}{(K_p)_1} = \frac{\Delta H_r^0}{R}\left[\frac{1}{T_1} - \frac{1}{T_2}\right]$$

$$\Rightarrow \quad \ln\frac{(K_p)_2}{41} = \frac{-91.8\times10^3}{8.314}\left[\frac{1}{400} - \frac{1}{298}\right] = 9.44$$

$$\Rightarrow \quad \frac{(K_p)_2}{41} = 12581.72$$

$$\Rightarrow \quad (K_p)_2 = 515850.5$$

4.2.3 Relating Thermodynamics and Kinetics at Equilibrium

Thermodynamics looks at chemical equilibrium from the point of view of free energy change while kinetics provides for an equilibrium constant for the chemical equilibrium. Since both deal with the same physical phenomenon, it is natural to expect a relationship between them (Awasthi et al., 2013). For a multicomponent system such as our reacting system under consideration,

$$\Delta G_r = \sum_{j=1}^{1} p_j \mu_{P_j} - \sum_{i=1}^{n} a_i \mu_{A_i}$$

$$\Rightarrow \quad \Delta G_r = \sum_{j=1}^{1} p_j \left(\mu_{P_j}^0 + RT\ \ln a_{P_j}\right) - \sum_{i=1}^{n} a_i \left(\mu_{A_i}^0 + RT\ \ln a_{A_i}\right)$$

$$\Rightarrow \quad \Delta G_r = \sum_{j=1}^{1} p_j \mu_{P_j}^0 - \sum_{i=1}^{n} a_i \mu_{A_i}^0 + \sum_{j=1}^{1} p_j RT\ \ln a_{P_j} - \sum_{i=1}^{n} a_i RT\ \ln a_{A_i}$$

$$\Rightarrow \quad \Delta G_r = \Delta G_r^0 + RT\left(\sum_{j=1}^{1} \ln a_{P_j}^{p_j} - \sum_{i=1}^{n} \ln a_{A_i}^{a_i}\right)$$

$$\Rightarrow \quad \Delta G_r = \Delta G_r^0 + RT\left(\ln \prod_{j=1}^{1} a_{P_j}^{p_j} - \ln \prod_{i=1}^{n} a_{A_i}^{a_i}\right)$$

$$\Rightarrow \quad \Delta G_r = \Delta G_r^0 + RT\ \ln \frac{\prod_{j=1}^{1} a_{P_j}^{p_j}}{\prod_{i=1}^{n} a_{A_i}^{a_i}} \tag{4.30}$$

At equilibrium, $\Delta G_r = 0$, and

$$0 = \Delta G_r^0 + RT\ \ln \frac{\prod_{j=1}^{1} \left(a_{P_j}\right)_{eq}^{p_j}}{\prod_{i=1}^{n} \left(a_{A_i}\right)_{eq}^{a_i}}$$

$$\Rightarrow \quad \Delta G_r^0 = -RT\ \ln K_a \tag{4.31}$$

This is the required relationship between the standard free energy change of the reaction and the equilibrium constant. In Equation 4.31, the equilibrium constant is expressed in terms of the activities of the individual components. For ideal gases, K_a can be substituted with K_p so that

$$\Delta G_r^0 = -RT\ \ln K_p \tag{4.32}$$

EXERCISES

4.1 The thermochemical equation for the combustion of carbon at 298 K is

$$C(s) + O_2(g) \rightarrow CO_2(g) \quad \Delta H_r = -393.5\ \text{kJ/mol}$$

How much heat will be released in the combustion of 1 tonne carbon if the combustion is 95% complete?

4.2 Our cells oxidize glucose to CO_2 and water to generate energy. How much energy will be generated by an organism if it completely oxidizes 50 g glucose? The enthalpies of formation, at 298 K, of glucose(s), CO_2(g), and H_2O(l) are −1268, −393.5, and −285.8 kJ/mol, respectively.

4.3 From the data given in Example 4.5, calculate the heat of formation of NH_3(g) at 298 K.

4.4 From the data and results of Exercises 4.2 and 4.3, calculate the ΔH_r for the reaction for synthesis of urea from ammonia and carbon dioxide at 298 K if the heat of formation of urea at 298 K is—333.1 kJ/mol.

4.5 Repeat Exercise 4.4 to calculate ΔH_r if the synthesis is carried out at 700 K. The molar heat capacities at constant pressure are NH_3(g)—35.1 J/(mol.K), CO_2(g)—37.5 J/(mol.K), urea(s)—94 J/(mol.K), and water vapor—36 J/(mol.K).

4.6 "Reactions that are exothermic at 298 K can be endothermic at other temperatures." Justify this statement.

4.7 Endothermic reactions require heat to be supplied during reaction yet they are spontaneous. Explain.

4.8 For a gaseous reaction $\Delta H_r = \Delta U_r$. What conclusion can you draw about the reaction?

4.9 The internal energy change for the complete combustion of benzene at 27°C is 3270 kJ/mol. Calculate the enthalpy of reaction for the combustion of benzene at 27°C.

4.10 Given

$$C(s) \rightarrow C(g) \quad \Delta H_r = 714.36 \, kJ/mol,$$

calculate the heat of formation of n-butane and benzene.

4.11 The standard heat of formation of H_2O(g) is −241.82 kJ/mol at 298 K. Calculate the equilibrium constant K_p for the reaction

$$2H_2O(g) \rightleftharpoons 2H_2(g) + O_2(g)$$

at 298 K if the equilibrium constant at 1395 K is 2.1×10^{-13}. Also calculate K_C.

4.12 At 298 K, the standard free energies of formation of N_2O_4(g) and NO_2(g) are 100 and 50 kJ/mol, respectively. Calculate the equilibrium constant K_p for the reaction

$$N_2O_4(g) \rightleftharpoons 2NO_2(g)$$

at 298 K.

4.13 For the reaction of Exercise 4.12, find the composition of the equilibrium mixture if the starting mixture contains N_2O_4 and NO_2 at 0.01 mol/L each.

4.14 For the reversible gaseous reaction

$$A \rightleftharpoons 2B$$

the rate constants for the forward and backward reactions at 300 K are 100 s^{-1} and 10 L/(mol · s). The standard heat of reaction at 300 K is 9 kJ/mol. The activation energy for the forward reaction is 20 kJ/mol. Calculate the activation energy for the backward reaction.

REFERENCES

Atkins, P. and de Paula, J., *Physical Chemistry*, 8th ed., Freeman, New York, 50–56, 2006.

Awasthi, N., Ritschel, T., Lipowsky, R., and Knecht, V., Standard Gibbs energies of formation and equilibrium constants from ab-initio calculations: Covalent dimerization of NO_2 and synthesis of NH_3. *J. Chem. Thermodyn.*, 62, 211, 2013.

Compendium of Chemical Terminology Gold Book, Version 2.3.3, IUPAC, 1459, 2014. Available online; http://goldbook.iupac.org/PDF/goldbook.pdf.

Darwent, B., Bond dissociation energies in simple molecules. *Natl. Stand. Ref. Data Ser.–Natl. Bur. Stand.*, 31, 9, 1970.

Denbigh, K.G., *The Principles of Chemical Equilibrium*, 4th ed., Cambridge University Press, London, 61–74, 1981.

DeVoe, H., *Thermodynamics and Chemistry*, 2nd ed., 319, 2015. Available online; http://www.chem.umd.edu/thermobook.

Ismail, T., *The Thermodynamics of Phase and Reaction Equilibria*, 1st ed., Elsevier, Oxford, 551, 2013.

Lide, D.R. (Ed.), *CRC Handbook of Chemistry and Physics*, Internet Version 2005, CRC Press, Boca Raton, FL, 5–4, 2005.

Neron, A., Lantagne, G., and Marcos, B., Computation of complex and constrained equilibria by minimization of the Gibbs free energy. *Chem. Eng. Sci.*, 82, 260, 2012.

Seider, W.D. and Widagdo, S., Multiphase equilibria of reactive systems. *Fluid Phase Equilib.*, 123, 283, 1996.

Stull, D.R. and Prophet, H., JANAF thermochemical tables, 2nd edition. *Natl. Stand. Ref. Data Ser.–Natl. Bur. Stand.*, 37, 13, 1971.

5 Ideal Isothermal Batch Reactor

5.1 GENERAL MOLE BALANCE EQUATION FOR HOMOGENEOUS REACTIONS

A batch reactor is essentially a vessel that holds the reacting system within its confines while the reactants chemically transform into products. Once filled, there is no entry of material into the vessel and no exit of material out of the vessel. Thus, the reacting system does not exchange material with its surroundings. Their importance and advantage lies in their flexibility and versatility. These are best suited for small production rates and/or multiple products from the same facility. Their major drawback is high operational cost compared to continuous reactors. Batch reactors are valuable in the experimental study of reacting systems before scale up. An ideal batch reactor is one that has no spatial variation of composition. In other words, at any given instant of time, the concentration of each of the chemical species is the same everywhere inside the reactor. Further, an isothermal batch reactor has constant temperature both with respect to time as well as with respect to position within the reactor. These ideal conditions may be approximated but not rigorously met in a real batch reactor. Nevertheless, the study of the hypothetical idealized system provides vital understanding that can be used for the real system as well. For gaseous-reacting systems, the reactor can be operated either at constant volume or at constant pressure. We start our analysis of the ideal isothermal batch reactor by writing a mole balance as follows:

For the limiting reactant species

$$\text{Input} = \text{Output} + \text{Accumulation} + \text{Disappearance due to chemical reaction}$$

By definition, for the batch reactor, input and output are zero. Therefore, we have

$$\text{Accumulation} = -\text{Disappearance due to chemical reaction}$$

$$\Rightarrow \quad \frac{dN_{A_{LR}}}{dt} = r_{A_{LR}} V_{BR} \tag{5.1}$$

From Equation 2.43 (Chapter 2), $V_f/V_0 = 1 + \varepsilon_{LR} X_{A_{LR}}$, so from Equation 5.1

$$\frac{dN_{A_{LR}}}{dt} = r_{A_{LR}} (V_{BR})_0 \left(1 + \varepsilon_{LR} X_{A_{LR}}\right) \tag{5.2}$$

From Equation 2.23 (Chapter 2),

$$N_{A_{LR}} = \left(N_{A_{LR}}\right)_0 - \left(N_{A_{LR}}\right)_0 X_{A_l}$$

$$\Rightarrow \quad \frac{dN_{A_{LR}}}{dt} = -\left(N_{A_{LR}}\right)_0 \frac{dX_{A_{LR}}}{dt} \tag{5.3}$$

From Equations 5.2 and 5.3,

$$-\left(N_{A_{LR}}\right)_0 \frac{dX_{A_{LR}}}{dt} = r_{A_{LR}} (V_{BR})_0 \left(1+\varepsilon_{LR} X_{A_{LR}}\right)$$

$$\Rightarrow \quad \left(C_{A_{LR}}\right)_0 \frac{dX_{A_{LR}}}{dt} = -r_{A_{LR}} \left(1+\varepsilon_{LR} X_{A_{LR}}\right) \tag{5.4}$$

$$\Rightarrow (C_{A_{LR}})_0 \int_0^{X_{A_{LR}}} \frac{dX_{A_{LR}}}{-r_{A_{LR}} \left(1+\varepsilon_{LR} X_{A_{LR}}\right)} = \int_0^t dt = t \tag{5.5}$$

Equations 5.4 and 5.5 are the basic mole balance equations, in differential and integral forms, respectively, for an ideal batch reactor. These can be developed into convenient usable forms under specific conditions.

5.2 CONSTANT VOLUME BATCH REACTOR

For constant volume of the reacting system, $\varepsilon_{LR} = 0$, and Equation 5.4 reduces to

$$\left(C_{A_{LR}}\right)_0 \frac{dX_{A_{LR}}}{dt} = -r_{A_{LR}} \tag{5.6}$$

Similarly, Equation 5.5 reduces to

$$\left(C_{A_{LR}}\right)_0 \int_0^{X_{A_{LR}}} \frac{dX_{A_{LR}}}{-r_{A_{LR}}} = \int_0^t dt = t \tag{5.7}$$

In terms of the concentration of the limiting reactant, from Equation 2.25 (Chapter 2),

$$C_{A_{LR}} = \left(C_{A_{LR}}\right)_0 \left(1 - X_{A_{LR}}\right) = \left(C_{A_{LR}}\right)_0 - \left(C_{A_{LR}}\right)_0 X_{A_{LR}}$$

$$\Rightarrow \quad \frac{dC_{A_{LR}}}{dt} = -\left(C_{A_{LR}}\right)_0 \frac{dX_{A_{LR}}}{dt} \tag{5.8}$$

Combining Equations 5.6 and 5.8,

$$\frac{dC_{ALR}}{dt} = r_{ALR} \tag{5.9}$$

$$\Rightarrow \int_{(C_{ALR})_0}^{C_{ALR}} \frac{dC_{ALR}}{r_{ALR}} = \int_0^t dt = t \tag{5.10}$$

Equations 5.6, 5.7, 5.9, and 5.10 are the performance equations for the constant volume batch reactor and can be used as suited. Other than evaluating the performance of a batch reactor, these equations are useful in estimating the kinetic parameters for a chemical reaction under study as elaborated next.

5.3 SINGLE-REACTANT nTH-ORDER REACTION

Consider a nonelementary reaction

$$A \rightarrow \text{Products}$$

for which the reaction kinetics are to be studied.

To start with, let us assume that the rate of reaction can be expressed by the empirical equation

$$-r_A = kC_A^n \tag{5.11}$$

Combining this with Equation 5.9, we have

$$\begin{aligned} \frac{dC_A}{dt} &= -kC_A^n \\ \Rightarrow \quad -\frac{dC_A}{dt} &= kC_A^n \end{aligned} \tag{5.12}$$

$$\Rightarrow \quad \ln\left(-\frac{dC_A}{dt}\right) = \ln k + n \ln C_A \tag{5.13}$$

Equation 5.13 implies that a plot between $\ln(-dC_A/dt)$ and $\ln C_A$ is a straight line with a slope equal to n and the y-axis intercept equal to ln k. This interpretation is used for the purpose of kinetic study of the reaction. The steps involved are given next.

Step 1

Carry out the reaction in a batch reactor. Take samples of the content of the reactor at different times and determine the instantaneous concentrations of the reactant A. Tabulate the concentration, C_A versus time data.

Step 2

Plot the concentration versus time data on a graph sheet. From the plot, graphically determine the quantity $-dC_A/dt$ at different times (refer to Example 3.2, Chapter 3). Tabulate them in the next column in the table prepared in Step 1.

Step 3

Calculate $\ln C_A$ from the data tabulated from Step 1 and tabulate them in the next column in the table.

Step 4

On a graph sheet plot $\ln(-dC_A/dt)$ against $\ln C_A$. Fit a straight line through the data points so obtained.

Step 5

Determine the slope of the straight line fitted. This slope is equal to "n" the reaction order.

Step 6

Extend the straight line to the $\ln(-dC_A/dt)$ axis. Measure the intercept of the line on this axis. This intercept is equal to ln k. Take the antilog and determine "k" the specific reaction rate constant.

We can also make our calculations based on conversion of the reactant instead of concentration. From Equations 5.8 and 5.12 and Equation 2.25 (Chapter 2),

$$-\frac{dC_A}{dt} = kC_A^n = k(C_A)_0^n(1-X_A)^n$$

$$\Rightarrow \quad (C_A)_0\frac{dX_A}{dt} = k(C_A)_0^n(1-X_A)^n$$

$$\Rightarrow \quad \frac{dX_A}{dt} = k(C_A)_0^{n-1}(1-X_A)^n \tag{5.14}$$

$$\Rightarrow \quad \ln\left(\frac{dX_A}{dt}\right) = \ln k + (n-1)\ln(C_A)_0 + n\ln(1-X_A) \tag{5.15}$$

Equation 5.15 implies that a plot between $\ln(dX_A/dt)$ and $\ln(1 - X_A)$ is a straight line with a slope equal to n and the y-axis intercept equal to $\ln k + (n - 1)\ln(C_A)_0$. This interpretation is used for the purpose of kinetic study of the reaction. The steps involved are given next.

Step 1

Carry out the reaction in a batch reactor. Take samples of the content of the reactor at different times and determine the concentration of the reactant A. Calculate the conversion X_A by Equation 2.25 (Chapter 2). Tabulate the concentration versus time data. In the next column, tabulate the calculated values of X_A.

Step 2

Plot the conversion versus time data on a graph sheet. From the plot graphically determine the quantity dX_A/dt at different times (refer to Example 3.2, Chapter 3). Tabulate them in the next column in the table prepared in Step 1.

Step 3

Calculate $\ln(1 - X_A)$ and $\ln(dX_A/dt)$ from the data tabulated from Step 1 and tabulate them in the next columns in the table.

Step 4

On a graph sheet plot $\ln(dX_A/dt)$ against $\ln(1 - X_A)$. Fit a straight line through the data points so obtained.

Step 5

Determine the slope of the straight line fitted. This slope is equal to "n" the reaction order.

Step 6

Extend the straight line to the $\ln(dX_A/dt)$ axis. Measure the intercept of the line on this axis. This intercept is equal to $\ln k + (n - 1)\ln(C_A)_0$. Determine "k" the specific reaction rate constant.

This procedure is known as the *differential method of analysis* (Gurol and Singer, 1982; Bardow and Marquardt, 2004). The steps listed above are further elaborated with the help of the following example.

EXAMPLE 5.1

From the data given in Example 3.2 (Chapter 3), determine the reaction order and the specific reaction rate constant by the differential method of analysis.

Solution

Step 1

We use the table from Example 3.2 (Chapter 3) where the concentration versus time data is tabulated.

Step 2

We start with Figure EXE3.2 (Chapter 3) where concentration is plotted against time. Draw tangents to the curve at different times and from the slopes of these tangents calculate $-dC_A/dt$ at different times. The procedure is presented in Figure EXE5.1a.

$$\text{At } t = 2\text{ min;}\quad -\frac{dC_A}{dt} = \frac{48}{13.2} = 3.64\text{ mol/(L}\cdot\text{min)}$$

$$\text{At } t = 4\text{ min;}\quad -\frac{dC_A}{dt} = \frac{45}{15.4} = 2.92\text{ mol/(L}\cdot\text{min)}$$

$$\text{At } t = 6\text{ min;}\quad -\frac{dC_A}{dt} = \frac{42-5}{17} = 2.18\text{ mol/(L}\cdot\text{min)}$$

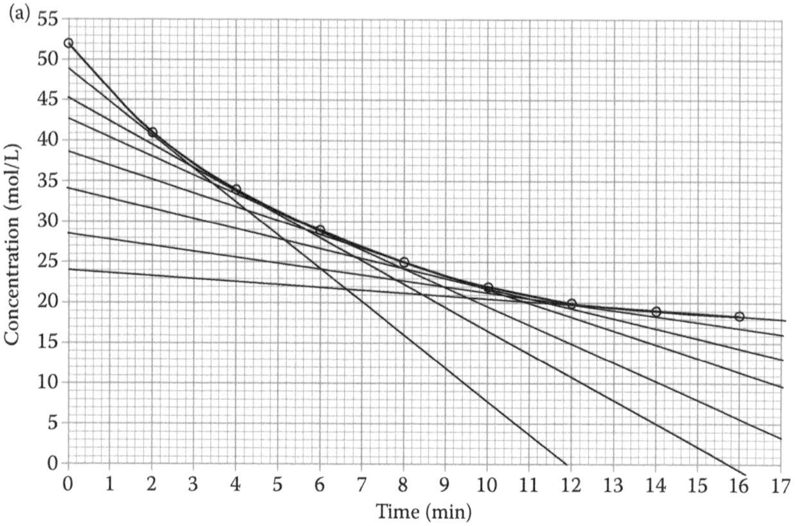

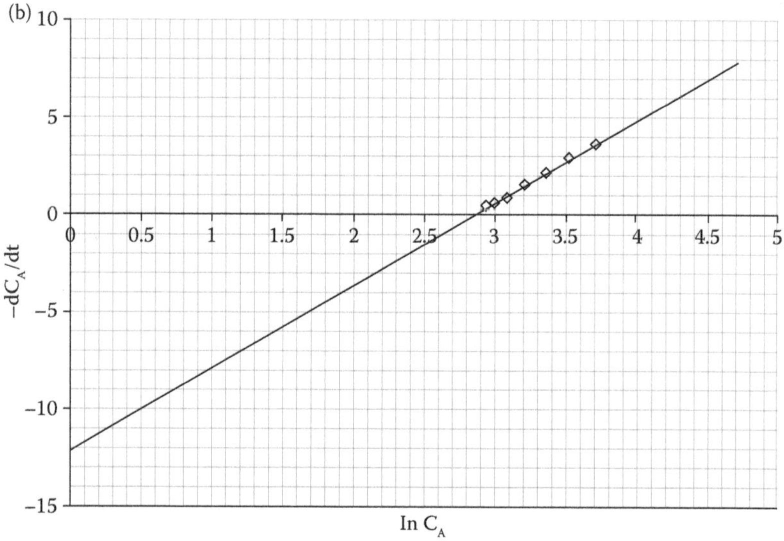

FIGURE EXE5.1

$$\text{At } t = 8 \text{ min}; \quad -\frac{dC_A}{dt} = \frac{37-15}{14.1} = 1.56 \text{ mol/(L} \cdot \text{min)}$$

$$\text{At } t = 10 \text{ min}; \quad -\frac{dC_A}{dt} = \frac{30.2-15}{17} = 0.89 \text{ mol/(L} \cdot \text{min)}$$

$$\text{At } t = 12 \text{ min}; \quad -\frac{dC_A}{dt} = \frac{25.1-15.2}{17} = 0.58 \text{ mol/(L} \cdot \text{min)}$$

TABLE EXE5.1
Calculation Table

t	C_A	$-dC_A/dt$	ln C_A
0	52		3.951244
2	41	3.64	3.713572
4	34	2.92	3.526361
6	29	2.18	3.367296
8	25	1.56	3.218876
10	22	0.89	3.091042
12	20	0.58	2.995732
14	19	0.52	2.944439
16	18.5		2.917771

$$\text{At } t = 14\ \text{min;}\quad -\frac{dC_A}{dt} = \frac{24-15.2}{17} = 0.52\ \text{mol/(L}\cdot\text{min)}$$

The values of $-dC_A/dt$ at different times are tabulated in Table EXE5.1.
Step 3
The calculated values of ln C_A are tabulated in the next column in Table EXE5.1.
Step 4
The values of $-dC_A/dt$ are plotted against ln C_A. This is shown in Figure EXE5.1b.
Step 5
From Figure EXE5.1b, the slope of the straight line = 12/2.9 = 4.14 ≅ 4
Step 6
From Figure EXE5.1b, the y-axis intercept is –12. Therefore,

$$\ln k = -12$$
$$\Rightarrow\quad k = 6.14\times 10^{-6}\ \text{L}^3/(\text{mol}^3\cdot\text{s})$$

We can see from the previous example that a graphical procedure to determine $-dC_A/dt$ involves visual judgment and may lead to error. Alternatively, some numerical differentiation methods may be used with some present-day computer software. In Figure EXE5.1a, the tangents have not been drawn at time 0 and 16 min as this would require the plot to be extrapolated. These are some of the limitations of the differential method of analysis.

From Equation 5.12, we can separate the variables and write

$$-\frac{dC_A}{C_A^n} = k\,dt$$

$$\Rightarrow\quad -\int_{(C_A)_0}^{C_A}\frac{dC_A}{C_A^n} = \int_0^t k\,dt = kt$$

$$\Rightarrow\quad \frac{1}{(n-1)}\left[\frac{1}{C_A^{n-1}} - \frac{1}{(C_A)_0^{n-1}}\right] = kt \quad \text{for } n \neq 1$$

$$\Rightarrow\quad \frac{1}{C_A^{n-1}} = (n-1)kt + \frac{1}{(C_A)_0^{n-1}} \tag{5.16}$$

Equation 5.16 implies that a plot between $(1/C_A^{n-1})$; and t is a straight line with slope equal to (n – 1)k and the y-axis intercept as $1/(C_A)_0^{n-1}$. With this interpretation, we can determine the kinetic parameters of the reaction with an iterative approach. The steps involved are as follows:

Step 1

Carry out the reaction in a batch reactor. Take samples of the content of the reactor at different times and determine the instantaneous concentrations of the reactant A. Tabulate the concentration, C_A versus time data.

Step 2

Assume a value for the reaction order "n" not equal to 1. Compute the values of $1/C_A^{n-1}$ and tabulate these values in the next column in the table prepared in Step 1.

Step 3

Plot the $1/C_A^{n-1}$ versus time data on a graph sheet. If the plotted points fit well into a straight line, the reaction order assumed is correct.

Step 4

Determine the slope of the straight line obtained in Step 3. Calculate the specific reaction rate constant as Slope/(n – 1)

Step 5

If the points plotted in Step 3 do not fit well into a straight line, the reactor order assumed is not correct. Assume a new value for the reaction order "n" not equal to 1 and repeat Steps 2–4. This iteration is carried out until a satisfactory linear fit is obtained in Step 3.

Alternatively, from Equation 2.25 (Chapter 2),

$$C_A = (C_A)_0(1 - X_A)$$

$$\Rightarrow \frac{dC_A}{dt} = -(C_A)_0\frac{dX_A}{dt} \tag{5.17}$$

From Equations 5.14 and 5.17,

$$(C_A)_0\frac{dX_A}{dt} = k(C_A)_0^n(1 - X_A)^n$$

$$\Rightarrow \int_0^{X_A}\frac{dX_A}{(1 - X_A)^n} = (C_A)_0^{n-1}\int_0^t kdt = (C_A)_0^{n-1}kt$$

$$\Rightarrow \frac{1}{(n-1)}[(1 - X_A)^{1-n} - 1] = (C_A)_0^{n-1}kt \quad \text{for } n \neq 1$$

$$\Rightarrow (1 - X_A)^{1-n} = 1 + (n-1)(C_A)_0^{n-1}kt$$

$$\Rightarrow \frac{1}{(C_A)_0^{n-1}(1 - X_A)^{n-1}} = (n-1)kt + \frac{1}{(C_A)_0^{n-1}} \tag{5.18}$$

Equation 5.18 can also be directly obtained from Equation 5.16 by combining with Equation 2.25 (Chapter 2).

So, a plot between $1/(C_A)_0^{n-1}(1-X_A)^{n-1}$ and t is a straight line with slope equal to (n – 1)k and the y-axis intercept as $1/(C_A)_0^{n-1}$. With this interpretation, we can determine the kinetic parameters of the reaction with an iterative approach. The steps involved are as follows:

Step 1

Carry out the reaction in a batch reactor. Take samples of the content of the reactor at different times and determine the instantaneous concentrations of the reactant A. Calculate X_A. Tabulate the X_A versus time data.

Step 2

Assume a value for the reaction order "n" not equal to 1. Compute the values of $1/(C_A)_0^{n-1}(1-X_A)^{n-1}$ and tabulate these values in the next column in the table prepared in Step 1.

Step 3

Plot the $1/(C_A)_0^{n-1}(1-X_A)^{n-1}$ versus time data on a graph sheet. If the plotted points fit well into a straight line, the reaction order assumed is correct.

Step 4

Determine the slope of the straight line obtained in Step 3. Calculate the specific reaction rate constant as Slope/(n – 1).

Step 5

If the points plotted in Step 3 do not fit well into a straight line, the reaction order assumed is not correct. Assume a new value for the reaction order "n" not equal to 1 and repeat Steps 2–4. This iteration is carried out until a satisfactory linear fit is obtained in Step 3.

This procedure is known as the *integral method of analysis* (Kittrel and Mezaki, 1966). Being iterative in nature, it is not very useful in finding the reaction order as several iterations may be needed. However, if the reaction order is known from some other experimental observations, the integral method of analysis may come handy for finding the specific reaction rate constant.

EXAMPLE 5.2

The following data are obtained for a reaction carried out in a constant volume batch reactor

Concentration (mol/L)	15	9.8	7.4	6.1	5.2	4.7	4.3	4	3.7
Time (min)	0	2	4	6	8	10	12	14	16

Check if the reaction is third order. Also, determine the specific reaction rate constant.

Solution

Step 1

We prepare Table EXE5.2 by tabulating the values of time and concentration in the first two columns.

TABLE EXE5.2
Calculation Table

Time (min)	Concentration (mol/L)	$1/C_A^2$
0	15	0.004444
2	9.8	0.010412
4	7.4	0.018262
6	6.1	0.026874
8	5.2	0.036982
10	4.7	0.045269
12	4.3	0.054083
14	4	0.0625
16	3.7	0.073046

Step 2

Next, we compute the values of $1/C_A^2$. These values are tabulated in the next column in Table EXE5.2.

Step 3

Next, we plot $1/C_A^2$ versus t. The plot is shown in Figure EXE5.2. We see that the plotted points fit very well into a straight line. Hence, we conclude that the reaction is third order. This is further confirmed by the value of the y-axis intercept determined as 0.004 from the graph. This value is in close agreement with the value of $1/(C_A)_0^2$.

Step 4

The slope of the line in Figure EXE5.2 is $(0.04 - .004)/8.6 = 0.0042$

Specific reaction rate constant $k = 0.0042/(3 - 1) = 0.0021$ $L^2/(mol^2 \cdot s)$

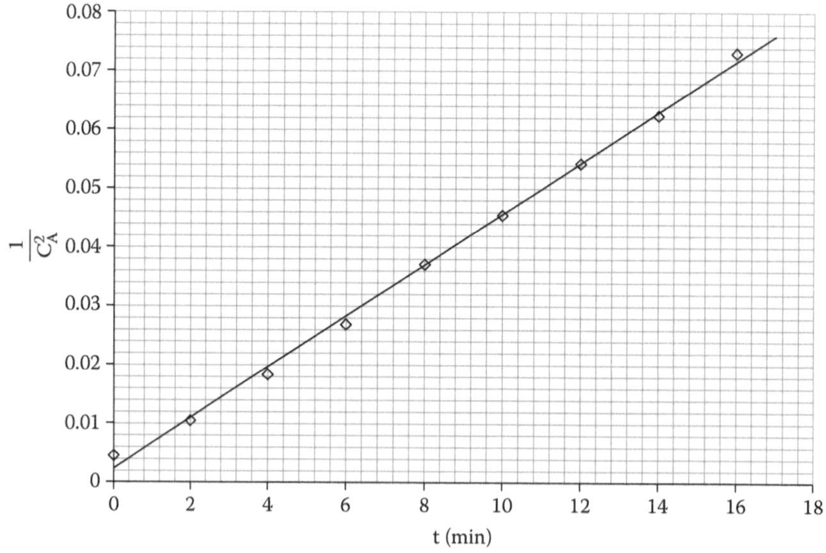

FIGURE EXE5.2

Equation 5.16 is valid for all values of "n" except 1. For a first-order reaction, from Equation 5.12

$$-\frac{dC_A}{dt} = kC_A$$

$$\Rightarrow \quad -\frac{dC_A}{C_A} = kdt$$

$$\Rightarrow \quad -\int_{(C_A)_0}^{C_A} \frac{dC_A}{C_A} = \int_0^t kdt = kt$$

$$\Rightarrow \quad \ln\frac{C_A}{(C_A)_0} = -kt \tag{5.19}$$

$$\Rightarrow \quad \ln C_A = \ln(C_A)_0 - kt \tag{5.20}$$

$$\Rightarrow \quad C_A = (C_A)_0 e^{-kt} \tag{5.21}$$

From Equation 5.20 for a first-order reaction, a plot between $\ln C_A$ and t is a straight line with the slope equal to $-k$ and y-axis intercept equal to $\ln(C_A)_0$. This can be used as a criterion to check if a reaction is first order.

Combining Equation 5.21 with Equation 2.25 (Chapter 2)

$$\begin{aligned} &(C_A)_0(1 - X_A) = (C_A)_0 e^{-kt} \\ \Rightarrow \quad &(1 - X_A) = e^{-kt} \end{aligned} \tag{5.22}$$

$$\Rightarrow \quad \ln(1 - X_A) = -kt \tag{5.23}$$

It follows that for a first-order reaction, a plot between $\ln(1 - X_A)$ and t is a straight line passing through the origin with slope equal to $-k$.

5.4 GENERAL REACTION INVOLVING SEVERAL REACTANTS

Consider the general reaction

$$\sum_{i=1}^{n} a_i A_i \rightarrow \sum_{j=1}^{l} p_j P_j \tag{5.24}$$

Let the rate equation for the reaction be expressed by

$$-r_{A_{LR}} = k\prod_{i=1}^{n} C_{A_i}^{a_i} \tag{5.25}$$

From Equation 3.17 (Chapter 3), for constant volume system with $\varepsilon_{LR} = 0$, we have

$$C_{A_i} = \left(C_{A_i}\right)_0 - \frac{a_i}{a_{LR}}\left\{\left(C_{A_{LR}}\right)_0 - C_{A_{LR}}\right\} \tag{5.26}$$

$$\Rightarrow \quad C_{A_i} = \left(C_{A_i}\right)_0 - \frac{a_i}{a_{LR}}\left(C_{A_{LR}}\right)_0 X_{A_{LR}} \tag{5.27}$$

Substituting the value from Equation 5.26 into Equation 5.25

$$-r_{A_{LR}} = k\prod_{i=1}^{n}\left[\left(C_{A_i}\right)_0 - \frac{a_i}{a_{LR}}\left\{\left(C_{A_{LR}}\right)_0 - C_{A_{LR}}\right\}\right]^{a_i} \tag{5.28}$$

$$\Rightarrow \quad -\frac{dC_{A_{LR}}}{dt} = k\prod_{i=1}^{n}\left[\left(C_{A_i}\right)_0 - \frac{a_i}{a_{LR}}\left\{\left(C_{A_{LR}}\right)_0 - C_{A_{LR}}\right\}\right]^{a_i}$$

$$\Rightarrow \quad -\int_{\left(C_{A_{LR}}\right)_0}^{C_{A_{LR}}} \frac{dC_{A_{LR}}}{\prod_{i=1}^{n}\left[\left(C_{A_i}\right)_0 - (a_i/a_{LR})\left\{\left(C_{A_{LR}}\right)_0 - C_{A_{LR}}\right\}\right]^{a_i}} = \int_0^t k\,dt = kt \tag{5.29}$$

For the special case when $\left(C_{A_i}\right)_0 = (a_i/a_{LR})\left(C_{A_{LR}}\right)_0$ Equation 5.29 reduces to

$$-\int_{\left(C_{A_{LR}}\right)_0}^{C_{A_{LR}}} \frac{dC_{A_{LR}}}{\prod_{i=1}^{n}(a_i/a_{LR})^{a_i}\left(C_{A_{LR}}\right)^{a_i}} = kt$$

$$\Rightarrow \quad -\frac{(a_{LR})^{\Sigma_{i=1}^{n} a_i}}{\prod_{i=1}^{n}(a_i)^{a_i}}\int_{\left(C_{A_{LR}}\right)_0}^{C_{A_{LR}}} \frac{dC_{A_{LR}}}{\left(C_{A_{LR}}\right)^{\Sigma_{i=1}^{n} a_i}} = kt$$

$$\Rightarrow \quad \frac{(a_{LR})^{\Sigma_{i=1}^{n} a_i}}{\left(1-\sum_{i=1}^{n} a_i\right)\prod_{i=1}^{n}(a_i)^{a_i}}\left[\left(C_{A_{LR}}\right)_0^{1-\Sigma_{i=1}^{n} a_i} - \left(C_{A_{LR}}\right)^{\left(1-\Sigma_{i=1}^{n} a_i\right)}\right] = kt \tag{5.30}$$

Equation 5.29 can be used to determine the final concentration of A_{LR} for a given reaction time in the batch reactor or vice versa after integrating the left-hand side. If the number of reactants is more and/or the stoichiometric coefficients, a_1, a_2, etc., are large integers, it is easier to numerically or graphically integrate rather than to look for an analytical solution.

Similarly, substituting the value from Equation 5.27 into Equation 5.25

$$-r_{A_{LR}} = k\prod_{i=1}^{n}\left[(C_{A_i})_0 - \frac{a_i}{a_{LR}}(C_{A_{LR}})_0 X_{A_{LR}}\right]^{a_i}$$

$$\Rightarrow \quad -r_{A_{LR}} = k\prod_{i=1}^{n}(C_{A_{LR}})_0^{a_i}\left[\frac{(C_{A_i})_0}{(C_{A_{LR}})_0} - \frac{a_i}{a_{LR}}X_{A_{LR}}\right]^{a_i}$$

$$\Rightarrow \quad -r_{A_{LR}} = k(C_{A_{LR}})_0^{\Sigma_{i=1}^{n} a_i}\prod_{i=1}^{n}\left[\frac{(C_{A_i})_0}{(C_{A_{LR}})_0} - \frac{a_i}{a_{LR}}X_{A_{LR}}\right]^{a_i}$$

Combining this with Equation 5.6

$$(C_{A_{LR}})_0\frac{dX_{A_{LR}}}{dt} = k(C_{A_{LR}})_0^{\Sigma_{i=1}^{n} a_i}\prod_{i=1}^{n}\left[\frac{(C_{A_i})_0}{(C_{A_{LR}})_0} - \frac{a_i}{a_{LR}}X_{A_{LR}}\right]^{a_i}$$

$$\Rightarrow \quad (C_{A_{LR}})_0^{1-\Sigma_{i=1}^{n} a_i}\int_0^{X_{A_{LR}}}\frac{dX_{A_{LR}}}{\prod_{i=1}^{n}\left[(C_{A_i})_0/(C_{A_{LR}})_0 - (a_i/a_{LR})X_{A_{LR}}\right]^{a_i}} = \int_0^t k\,dt = kt \quad (5.31)$$

Again, for the special case when $(C_{A_i})_0 = (a_i/a_{LR})(C_{A_{LR}})_0$ Equation 5.31 reduces to

$$(C_{A_{LR}})_0^{1-\Sigma_{i=1}^{n} a_i}\int_0^{X_{A_{LR}}}\frac{dX_{A_{LR}}}{\prod_{i=1}^{n}(a_i/a_{LR})^{a_i}\left[1-X_{A_{LR}}\right]^{a_i}} = kt$$

$$\Rightarrow \quad \frac{(a_{LR})^{\Sigma_{i=1}^{n} a_i}}{\prod_{i=1}^{n}(a_i)^{a_i}}(C_{A_{LR}})_0^{1-\Sigma_{i=1}^{n} a_i}\int_0^{X_{A_{LR}}}\frac{dX_{A_{LR}}}{(1-X_{A_{LR}})^{\Sigma_{i=1}^{n} a_i}} = kt$$

$$\Rightarrow \quad \frac{(a_{LR})^{\Sigma_{i=1}^{n} a_i}}{\left(1-\sum_{i=1}^{n} a_i\right)\prod_{i=1}^{n}(a_i)^{a_i}}(C_{A_{LR}})_0^{1-\Sigma_{i=1}^{n} a_i}\left[(1-X_{A_{LR}})^{\left(1-\Sigma_{i=1}^{n} a_i\right)} - 1\right] = -kt \quad (5.32)$$

Equation 5.31 can be used to determine the final conversion of A_{LR} for a given reaction time in the batch reactor or vice versa after integrating the left-hand side.

Again, if the number of reactants is more and/or the stoichiometric coefficients, a_1, a_2, etc., are large integers, it is easier to numerically or graphically integrate rather than to look for an analytical solution.

The general algorithm for engineering calculations can be summarized as follows:

Step 1

Write down the rate equation in terms of the concentrations of the reactants. The rate equation can be of the form of Equation 5.25 or any other form as determined from the experiment.

Step 2

From reaction stoichiometry express the concentration of each reactant in terms of the concentration of the limiting reactant as shown in Equation 5.26 or Equation 5.27.

Step 3

Substitute the concentration terms of each reactant in the rate equation written in Step 1 with the equivalent expression in terms of the limiting reactant as formulated in Step 2. Now, we have the rate equation in terms of the concentration of the limiting reactant alone.

Step 4

Use the rate equation obtained in Step 3 in the performance equation of the batch reactor to determine the desired quantities.

EXAMPLE 5.3

The liquid-phase irreversible reaction

$$A + 2B \rightarrow \text{Products}$$

has a reaction rate $-r_A = 0.02C_AC_B^2$ mol/(L.s). The reaction is carried out in a batch reactor with $(C_A)_0 = 1$ mol/L and $(C_B)_0 = 2.5$ mol/L. Calculate the time required for 50% conversion of A.

SOLUTION

From the reaction stoichiometry and the initial concentrations, it is evident that A is the limiting reactant. From Equation 5.31

$$(C_{ALR})_0^{1-\Sigma_{i=1}^n a_i} \int_0^{X_{ALR}} \frac{dX_{ALR}}{\prod_{i=1}^n [(C_{Ai})_0/(C_{ALR})_0 - (a_i/a_{LR})X_{ALR}]^{a_i}} = kt$$

$$\Rightarrow \quad (C_A)_0^{1-1-2} \int_0^{0.5} \frac{dX_A}{(1-X_A)[(C_B)_0/(C_A)_0 - 2X_A]^2} = 0.02t$$

$$\Rightarrow \quad \int_0^{0.5} \frac{dX_A}{(1-X_A)[2.5-2X_A]^2} = I = 0.02t$$

TABLE EXE5.3
Table for Numerical Integration

X_A	$1/(1-X_A)[2.5-2X_A]^2$
0	0.181818
0.05	0.182749
0.1	0.21004
0.15	0.243072
0.2	0.283447
0.25	0.333333
0.3	0.395726
0.35	0.474834
0.4	0.576701
0.45	0.710227
0.5	0.888889

Next, we numerically integrate the left-hand side as follows (Table EXE5.3):
By Simpson's one-third rule

$$I = \frac{0.05}{3}[0.181818 + 0.888889 + 4(0.182749 + 0.243072$$
$$+0.333333 + 0.474834 + 0.710227)$$
$$+2(0.21004 + 0.283447 + 0.395726 + 0.576701)] = 0.02t$$

$$\Rightarrow \quad t = \frac{0.1963}{0.02} = 9.82 \text{ s}$$

5.5 REVERSIBLE REACTIONS

Consider the general reversible reaction

$$\sum_{i=1}^{n} a_i A_i \rightleftharpoons \sum_{j=1}^{l} p_j P_j \tag{5.33}$$

Here,

$$-r_{A_{LR}} = k_1 \prod_{i=1}^{n} C_{A_i}^{a_i} - k_2 \prod_{j=1}^{l} C_{P_j}^{p_j}$$

$$\Rightarrow \quad -r_{A_{LR}} = k_1 \prod_{i=1}^{n} \left[\left(C_{A_i}\right)_0 - \frac{a_i}{a_{LR}} \left(C_{A_{LR}}\right)_0 X_{A_{LR}} \right]^{a_i}$$
$$- k_2 \prod_{j=1}^{l} \left[\left(C_{P_j}\right)_0 + \frac{p_j}{a_{LR}} \left(C_{A_{LR}}\right)_0 X_{A_{LR}} \right]^{p_j}$$

$$\Rightarrow \quad -r_{A_{LR}} = k_1 \prod_{i=1}^{n} \left(C_{A_{LR}}\right)_0^{a_i} \left[\frac{\left(C_{A_i}\right)_0}{\left(C_{A_{LR}}\right)_0} - \frac{a_i}{a_{LR}} X_{A_{LR}}\right]^{a_i}$$

$$- k_2 \prod_{j=1}^{l} \left(C_{A_{LR}}\right)_0^{p_j} \left[\frac{\left(C_{P_j}\right)_0}{\left(C_{A_{LR}}\right)_0} + \frac{p_j}{a_{LR}} X_{A_{LR}}\right]^{p_j}$$

$$\Rightarrow \quad -r_{A_{LR}} = k_1 \left(C_{A_{LR}}\right)_0^{\Sigma_{i=1}^{n} a_i} \prod_{i=1}^{n} \left[\frac{\left(C_{A_i}\right)_0}{\left(C_{A_{LR}}\right)_0} - \frac{a_i}{a_{LR}} X_{A_{LR}}\right]^{a_i}$$

$$- k_2 \left(C_{A_{LR}}\right)_0^{\Sigma_{j=1}^{l} p_j} \prod_{j=1}^{l} \left[\frac{\left(C_{P_j}\right)_0}{\left(C_{A_{LR}}\right)_0} + \frac{p_j}{a_{LR}} X_{A_{LR}}\right]^{p_j} \tag{5.34}$$

$$\Rightarrow \quad \left(C_{A_{LR}}\right)_0 \frac{dX_{A_{LR}}}{dt} = k_1 \left(C_{A_{LR}}\right)_0^{\Sigma_{i=1}^{n} a_i} \prod_{i=1}^{n} \left[\frac{\left(C_{A_i}\right)_0}{\left(C_{A_{LR}}\right)_0} - \frac{a_i}{a_{LR}} X_{A_{LR}}\right]^{a_i}$$

$$- k_2 \left(C_{A_{LR}}\right)_0^{\Sigma_{j=1}^{l} p_j} \prod_{j=1}^{l} \left[\frac{\left(C_{P_j}\right)_0}{\left(C_{A_{LR}}\right)_0} + \frac{p_j}{a_{LR}} X_{A_{LR}}\right]^{p_j}$$

$$\Rightarrow \quad \int_0^{X_{A_{LR}}} \frac{dX_{A_{LR}}}{k_1 \left(C_{A_{LR}}\right)_0^{\left(\Sigma_{i=1}^{n} a_i\right)^{-1}} \prod_{i=1}^{n} \left[\frac{\left(C_{A_i}\right)_0}{\left(C_{A_{LR}}\right)_0} - \frac{a_i}{a_{LR}} X_{A_{LR}}\right]^{a_i} - k_2 \left(C_{A_{LR}}\right)_0^{\left(\Sigma_{j=1}^{l} p_j\right)^{-1}} \prod_{j=1}^{l} \left[\frac{\left(C_{P_j}\right)_0}{\left(C_{A_{LR}}\right)_0} + \frac{p_j}{a_{LR}} X_{A_{LR}}\right]^{p_j}} = \int_0^t dt = t \tag{5.35}$$

If we integrate the left-hand side of Equation 5.35 between the limits 0 and $\left(X_{A_{LR}}\right)_{eq}$ we can get the time required to achieve equilibrium. $\left(X_{A_{LR}}\right)_{eq}$ is related to the equilibrium constant by Equations 4.21 and 4.22 (Chapter 4) for constant volume systems. Here also, numerical or graphical integration is recommended for systems consisting of many components with large stoichiometric coefficients.

EXAMPLE 5.4

The liquid-phase reversible reaction

$$A \rightleftharpoons B$$

with $k_1 = 0.05\ \text{min}^{-1}$ and $k_2 = 0.02\ \text{min}^{-1}$, is carried out in a batch reactor with initial concentrations of A and B as 1 and 0.1 mol/L, respectively. Determine the time required to achieve equilibrium.

SOLUTION

Equilibrium constant $K_c = (k_1/k_2) = (0.05/0.02) = 2.5$

From Equation 4.17 (Chapter 4),

$$K_C = \frac{(C_B)_{eq}}{(C_A)_{eq}} = \frac{(C_B)_0 + (C_A)_0 (X_A)_{eq}}{(C_A)_0[1-(X_A)_{eq}]}$$

$$\Rightarrow \quad \frac{0.1+(X_A)_{eq}}{[1-(X_A)_{eq}]} = 2.5$$

$$\Rightarrow \quad 0.1+(X_A)_{eq} = 2.5 - 2.5(X_A)_{eq}$$

$$\Rightarrow \quad (X_A)_{eq} = \frac{2.4}{3.5} = 0.68$$

From Equation 5.35,

$$t = \int_0^{0.68} \frac{dX_A}{0.05(1-X_A) - k_2[0.1+X_A]}$$

$$\Rightarrow \quad t = \int_0^{0.68} \frac{dX_A}{0.05(1-X_A) - 0.02[0.1+X_A]}$$

$$\Rightarrow \quad t = \frac{1}{(0.02-0.05)}[\ln\{0.05\times 0.32 - 0.02\times 0.78\} - \ln\{0.05-0.002\}]$$

$$\Rightarrow \quad t = 159.33\,\text{min}$$

5.6 HALF-LIFE FOR IRREVERSIBLE REACTIONS

In a batch reactor, the time required for the reactant concentration to drop to half its initial value is known as the half-life of the reaction. In terms of conversion, it is the time required to achieve 50% conversion. For reactions having only one reactant species, from Equation 5.16

$$\frac{1}{[(C_A)_0/2]^{n-1}} = (n-1)kt_{1/2} + \frac{1}{(C_A)_0^{n-1}} \quad \text{for } n \neq 1$$

$$\Rightarrow \quad t_{1/2} = \frac{1}{(n-1)k(C_A)_0^{n-1}}[2^{n-1}-1]$$

$$\Rightarrow \quad t_{1/2} = \frac{2^{n-1}-1}{(n-1)k}(C_A)_0^{1-n} \quad \text{for } n \neq 1 \tag{5.36}$$

$$\Rightarrow \quad \ln t_{1/2} = \ln\left[\frac{2^{n-1}-1}{(n-1)k}\right] + (1-n)\ln(C_A)_0 \quad \text{for } n \neq 1 \tag{5.37}$$

It follows that a plot between ln $t_{1/2}$ and ln$(C_A)_0$ is a straight line with slope equal to (1 – n) and the y-axis intercept equal to ln[$(2^{n-1}$ – 1)/((n – 1)k)]. This can be used as a diagnostic tool for estimating the kinetic parameters of the reaction. The steps are as follows:

Step 1
Carry out the reaction in a batch reactor with different initial concentrations of the reactant. For each initial concentration, note the time when the concentration drops to half the initial value. Tabulate the data for the initial concentration and $t_{1/2}$.

Step 2
On a log–log graph sheet, plot $(C_A)_0$ on the x-axis and $t_{1/2}$ on the y-axis. Determine the slope of the line. This slope is equal to (1 – n). Calculate n.

Step 3
Determine the y-axis intercept of the line drawn in Step 2. This intercept is equal to ln[2^{n-1} – 1/((n – 1)k)]. Using the value of reaction order determined in Step 2, determine the specific reaction rate constant k.

For a first-order reaction, from Equation 5.19

$$\ln\frac{(C_A)_0/2}{(C_A)_0} = -kt_{1/2}$$

$$\Rightarrow \quad t_{1/2} = \frac{\ln 2}{k} \tag{5.38}$$

Therefore, a distinguishing feature of a first-order reaction is that its half-life is independent of the initial reactant concentration. This can be used as a criterion to check if a reaction is first order. By determining the half-life of a first-order reaction, its specific reaction rate constant can be determined by using Equation 5.38.

For the general reaction involving several reactants, as represented by reaction (5.24), when $\left(C_{A_i}\right)_0 = (a_i/a_{LR})\left(C_{A_{LR}}\right)_0$ we have, from Equation 5.32

$$\frac{(a_{LR})^{\sum_{i=1}^{n} a_i}}{\left(1-\sum_{i=1}^{n} a_i\right)\prod_{i=1}^{n}(a_i)^{a_i}}\left(C_{A_{LR}}\right)_0^{1-\sum_{i=1}^{n} a_i}\left[(1-0.5)^{\left(1-\sum_{i=1}^{n} a_i\right)}-1\right] = -kt_{1/2}$$

$$\Rightarrow \quad t_{1/2} = \frac{(a_{LR})^{\sum_{i=1}^{n} a_i}}{k\left(\sum_{i=1}^{n} a_i - 1\right)\prod_{i=1}^{n}(a_i)^{a_i}}\left(C_{A_{LR}}\right)_0^{1-\sum_{i=1}^{n} a_i}\left[(0.5)^{\left(1-\sum_{i=1}^{n} a_i\right)}-1\right] \tag{5.39}$$

EXAMPLE 5.5

For the reaction

$$A \rightarrow \text{Products}$$

the following data are obtained by carrying out the reaction with different initial concentrations of A.

$(C_A)_0$ (mol/L)	1.22	1.65	2.46	4.48	7.39	11
$t_{1/2}$ (s)	20	36	81	330	897	2440

Determine the order of reaction and the specific reaction rate constant.

SOLUTION

From the given data we prepare Table EXE5.5.

Next we plot $\ln t_{1/2}$ versus $\ln(C_A)_0$ as shown in Figure EXE5.5.

TABLE EXE5.5
Calculation Table

$(C_A)_0$	$t_{1/2}$	$\ln(C_A)_0$	$\ln t_{1/2}$
1.22	20	0.198851	2.995732
1.65	36	0.500775	3.583519
2.46	81	0.900161	4.394449
4.48	330	1.499623	5.799093
7.39	897	2.000128	6.799056
11	2440	2.397895	7.799753

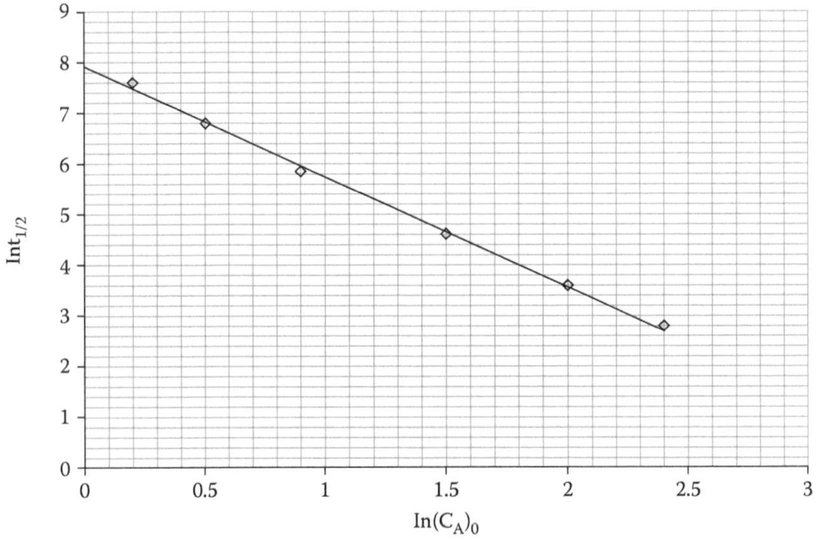

FIGURE EXE5.5

Slope of the line in Figure EXE5.5 is ((2.7–7.9)/2.4) = –2.17
So, 1 – n= –2.17

$$\Rightarrow \quad n = 1 + 2.17 = 3.17 \cong 3$$

y-axis intercept in Figure EXE5.5 is 7.9
So, $2^{n-1} - 1/((n-1)k) = 7.9$

$$\Rightarrow \quad k = \frac{2^2 - 1}{(3-1)7.9} = 0.19\ L^2/(mol^2 \cdot s)$$

5.7 MULTIPLE REACTIONS

5.7.1 Series Reactions

Consider the series reaction represented by the balanced chemical Equation 3.33 (Chapter 3). For convenience, we reproduce the equation here

$$\sum_{i=1}^{n} a_i A_i \rightarrow \sum_{j=1}^{l} p_j P_j \rightarrow \sum_{r=1}^{r} q_r Q_r$$

with the specific reaction rate constants for the two consecutive reaction steps being k_1 and k_2, respectively. Further, we consider that each consecutive reaction step follows the law of mass action and $(C_{A_i})_0 = a_i / a_{LR}(C_{A_{LR}})_0$. Then, from Equation 5.30

$$\frac{(a_{LR})^{\sum_{i=1}^{n} a_i}}{\left(1-\sum_{i=1}^{n} a_i\right)\prod_{i=1}^{n}(a_i)^{a_i}}\left[\left(C_{A_{LR}}\right)_0^{\left(1-\sum_{i=1}^{n} a_i\right)} - \left(C_{A_{LR}}\right)^{\left(1-\sum_{i=1}^{n} a_i\right)}\right] = k_1 t$$

$$\Rightarrow \quad \left(C_{A_{LR}}\right)_0^{\left(1-\sum_{i=1}^{n} a_i\right)} - \left(C_{A_{LR}}\right)^{\left(1-\sum_{i=1}^{n} a_i\right)} = \frac{k_1 t\left(1-\sum_{i=1}^{n} a_i\right)\prod_{i=1}^{n}(a_i)^{a_i}}{(a_{LR})^{\sum_{i=1}^{n} a_i}}$$

$$\Rightarrow \quad \left(C_{A_{LR}}\right) = \left[\left(C_{A_{LR}}\right)_0^{\left(1-\sum_{i=1}^{n} a_i\right)} - \frac{k_1 t\left(1-\sum_{i=1}^{n} a_i\right)\prod_{i=1}^{n}(a_i)^{a_i}}{(a_{LR})^{\sum_{i=1}^{n} a_i}}\right]^{1/\left(1-\sum_{i=1}^{n} a_i\right)} \tag{5.40}$$

For the product P_1,

$$r_{P_1} = \frac{d(C_{P_1})}{dt} = \frac{p_1}{a_{LR}} k_1 \prod_{i=1}^{n} \left(\frac{a_i}{a_{LR}}\right)^{a_i} (C_{A_{LR}})^{a_i} - k_2 \prod_{j=1}^{1} C_{P_j}^{p_j}$$

$$\Rightarrow \frac{d(C_{P_1})}{dt} = \frac{p_1}{a_{LR}} k_1 \prod_{i=1}^{n} \left(\frac{a_i}{a_{LR}}\right)^{a_i} (C_{A_{LR}})^{a_i} - k_2 \prod_{j=1}^{1} \left(\frac{p_j}{p_1}\right)^{p_j} C_{P_1}^{p_j}$$

$$\Rightarrow \frac{d(C_{P_1})}{dt} = \left(\frac{p_1}{a_{LR}}\right) k_1 \left[\frac{(C_{A_{LR}})^{\Sigma_{i=1}^{n} a_i}}{(a_{LR})^{\Sigma_{i=1}^{n} a_i}}\right] \prod_{i=1}^{n} (a_i)^{a_i} - k_2 \left[\frac{(C_{P_1})^{\Sigma_{j=1}^{1} p_j}}{(p_1)^{\Sigma_{j=1}^{1} p_j}}\right] \prod_{j=1}^{1} (p_j)^{p_j} \tag{5.41}$$

Substituting the value of $C_{A_{LR}}$ from Equation 5.40 in Equation 5.41 and solving the resulting differential equation will provide C_{P_1} as a function of time. The analytical solution to this complex differential equation will not be possible in a general sense. However, for some simple reactions, the resulting differential equation can be analytically solved (Hill, 1942; Chien, 1948). Here, we see one such simple example of a series reaction.

Consider the series reaction (Bateman, 1910)

$$A \rightarrow I \rightarrow T \tag{5.42}$$

with the specific reaction rate constants for the two consecutive reaction steps being k_1 and k_2, respectively. The time rate of change of concentration of the species A, I, and T can be written as

$$-\frac{dC_A}{dt} = k_1 C_A \tag{5.43}$$

$$\frac{dC_I}{dt} = k_1 C_A - k_2 C_I \tag{5.44}$$

$$\frac{dC_T}{dt} = k_2 C_I \tag{5.45}$$

From Equation 5.21, the solution to Equation 5.43 is

$$C_A = (C_A)_0 e^{-k_1 t} \tag{5.46}$$

$$\Rightarrow \frac{C_A}{(C_A)_0} = e^{-k_1 t} \tag{5.47}$$

Substituting this value in Equation 5.44,

$$\frac{dC_I}{dt} = k_1(C_A)_0 e^{-k_1 t} - k_2 C_I$$

$$\Rightarrow \quad \frac{dC_I}{dt} + k_2 C_I = k_1(C_A)_0 e^{-k_1 t} \tag{5.48}$$

The solution to this first-order differential equation is

$$C_I e^{\int k_2 dt} = \int k_1(C_A)_0 e^{-k_1 t} e^{\int k_2 dt} dt + \text{constant}$$

$$\Rightarrow \quad C_I e^{k_2 t} = \int k_1(C_A)_0 e^{-k_1 t} e^{k_2 t} dt + \text{constant}$$

$$\Rightarrow \quad C_I e^{k_2 t} = \frac{k_1(C_A)_0 e^{(k_2 - k_1)t}}{(k_2 - k_1)} + \text{constant} \tag{5.49}$$

To determine the constant of integration we use the boundary condition

$$\text{At } t = 0, \quad C_I = (C_I)_0$$

From Equation 5.49, at $t = 0$

$$(C_I)_0 = \frac{k_1(C_A)_0}{(k_2 - k_1)} + \text{constant}$$

$$\text{constant} = (C_I)_0 - \frac{k_1(C_A)_0}{(k_2 - k_1)}$$

Substituting this value in Equation 5.49,

$$C_I e^{k_2 t} = \frac{k_1(C_A)_0 e^{(k_2 - k_1)t}}{(k_2 - k_1)} + (C_I)_0 - \frac{k_1(C_A)_0}{(k_2 - k_1)}$$

$$\Rightarrow \quad C_I = \frac{k_1(C_A)_0 e^{-k_1 t}}{(k_2 - k_1)} + (C_I)_0 e^{-k_2 t} - \frac{k_1(C_A)_0 e^{-k_2 t}}{(k_2 - k_1)} \tag{5.50}$$

$$\Rightarrow \quad C_I = \frac{k_1 C_A)_0}{(k_2 - k_1)} [e^{-k_1 t} - e^{-k_2 t}] + (C_I)_0 e^{-k_2 t}$$

$$\Rightarrow \frac{C_I}{(C_A)_0} = \frac{k_1}{(k_2 - k_1)} [e^{-k_1 t} - e^{-k_2 t}] + \frac{(C_I)_0}{(C_A)_0} e^{-k_2 t} \tag{5.51}$$

To evaluate the time dependence of C_T we may solve Equation5.45. Alternatively, we can make use of Equation 3.37 (Chapter 3) to have

$$(C_A)_0 + (C_I)_0 + (C_T)_0 = C_A + C_I + C_T$$

$$\Rightarrow \quad C_T = (C_A)_0 + (C_I)_0 + (C_T)_0 - C_A - C_I$$

$$\Rightarrow \quad C_T = (C_A)_0 + (C_I)_0 + (C_T)_0 - (C_A)_0 e^{-k_1 t} - \frac{k_1 (C_A)_0}{(k_2 - k_1)}[e^{-k_1 t} - e^{-k_2 t}] - (C_I)_0 e^{-k_2 t}$$

$$\Rightarrow \quad C_T = (C_A)_0 \left\{1 - e^{-k_1 t} - \frac{k_1}{(k_2 - k_1)}[e^{-k_1 t} - e^{-k_2 t}]\right\} + (C_I)_0 (1 - e^{-k_1 t}) + (C_T)_0 \tag{5.52}$$

$$\Rightarrow \frac{C_T}{(C_A)_0} = \left\{1 - e^{-k_1 t} - \frac{k_1}{(k_2 - k_1)}[e^{-k_1 t} - e^{-k_2 t}]\right\} + \frac{(C_I)_0}{(C_A)_0}(1 - e^{-k_1 t}) + \frac{(C_T)_0}{(C_A)_0} \tag{5.53}$$

We define the dimensionless concentrations φ as

$$\varphi = \frac{C}{(C_A)_0}$$

We can rewrite Equations 5.47, 5.51, and 5.53 in terms of the dimensionless concentration and have

$$\varphi_A = e^{-k_1 t} \tag{5.54}$$

$$\varphi_I = \frac{k_1}{(k_2 - k_1)}[e^{-k_1 t} - e^{-k_2 t}] + \frac{(C_I)_0}{(C_A)_0} e^{-k_2 t} \tag{5.55}$$

$$\varphi_T = \left\{1 - e^{-k_1 t} - \frac{k_1}{(k_2 - k_1)}[e^{-k_1 t} - e^{-k_2 t}]\right\} + \frac{(C_I)_0}{(C_A)_0}(1 - e^{-k_1 t}) + \frac{(C_T)_0}{(C_A)_0} \tag{5.56}$$

From Equations 5.54 through 5.56, as $t \to \infty$, $\varphi_A \to 0$, $\varphi_I \to 0$, and $\varphi_T \to [1 + ((C_I)_0 \backslash (C_A)_0) + ((C_T)_0 \backslash (C_A)_0))]$. Additionally, φ_I the dimensionless concentration of species I passes through a maximum. These dimensionless concentrations are graphically shown in Figure 5.1.

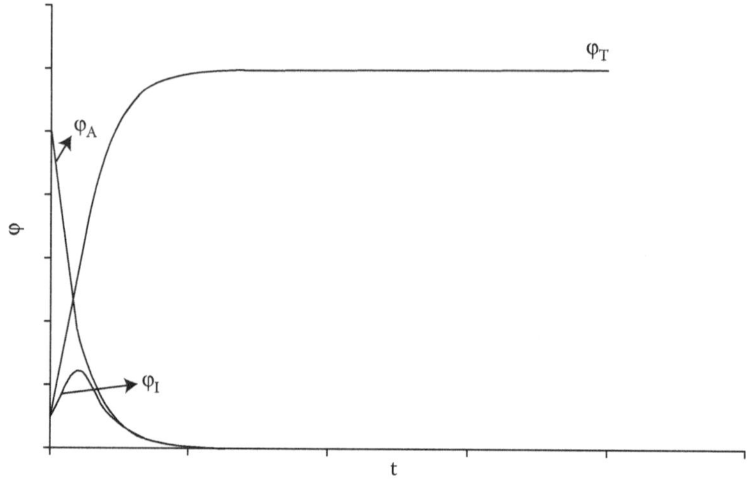

FIGURE 5.1 Dimensionless concentrations for series reaction (5.42).

One of the practical examples of series reaction (5.42) is the two-step dehydration of prostaglandin E_1 methyl ester (Toledo-Velasquez, 1992). For this reaction, the concentration of the intermediate (B) and the final product (C) is shown in Figure 5.2.

Let us now determine the time when we get the maximum of φ_I. From, Equation 5.55, for maximum φ_I

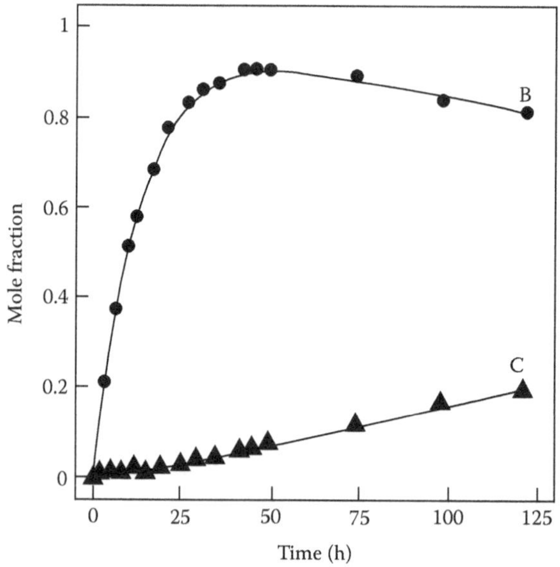

FIGURE 5.2 Concentration profiles of intermediate (B) and final product (C) in the dehydration of prostaglandin E1 methyl ester. (Reprinted from Yeow, Y.L. et al. *Chem. Eng. Sci.*, 58(16), 3601, 2003, with permission from Elsevier.)

$$\frac{d\varphi_I}{dt} = \frac{k_1}{(k_2 - k_1)}[-k_1 e^{-k_1 t_{max}} + k_2 e^{-k_2 t_{max}}] - k_2 \frac{(C_I)_0}{(C_A)_0} e^{-k_2 t_{max}} = 0$$

$$\Rightarrow \quad \frac{k_1}{(k_2 - k_1)}[-k_1 e^{(k_2 - k_1)t_{max}} + k_2] - k_2 \frac{(C_I)_0}{(C_A)_0} = 0$$

$$\Rightarrow \quad k_2 - k_1 e^{(k_2 - k_1)t_{max}} = \frac{k_2(k_2 - k_1)(C_I)_0}{k_1 (C_A)_0}$$

$$\Rightarrow \quad k_1 e^{(k_2 - k_1)t_{max}} = k_2 \left[1 - \frac{(k_2 - k_1)(C_I)_0}{k_1 (C_A)_0}\right]$$

$$\Rightarrow \quad e^{(k_2 - k_1)t_{max}} = \frac{k_2}{k_1}\left[1 - \frac{(k_2 - k_1)(C_I)_0}{k_1 (C_A)_0}\right]$$

$$\Rightarrow \quad (k_2 - k_1)t_{max} = \ln\left\{\frac{k_2}{k_1}\left[1 - \frac{(k_2 - k_1)(C_I)_0}{k_1 (C_A)_0}\right]\right\}$$

$$\Rightarrow \quad t_{max} = \frac{1}{(k_2 - k_1)} \ln\left\{\frac{k_2}{k_1}\left[1 - \frac{(k_2 - k_1)(C_I)_0}{k_1 (C_A)_0}\right]\right\} \tag{5.57}$$

From Equation 5.55,

$$(\varphi_I)_{max} = \frac{k_1}{(k_2 - k_1)}\left[\left\{\frac{k_2}{k_1}\left[1 - \frac{(k_2 - k_1)(C_I)_0}{k_1 (C_A)_0}\right]\right\}^{-k_1/(k_2 - k_1)} - \left\{\frac{k_2}{k_1}\left[1 - \frac{(k_2 - k_1)(C_I)_0}{k_1 (C_A)_0}\right]\right\}^{-k_2/(k_2 - k_1)}\right] + \frac{(C_I)_0}{(C_A)_0}\left\{\frac{k_2}{k_1}\left[1 - \frac{(k_2 - k_1)(C_I)_0}{k_1 (C_A)_0}\right]\right\}^{-k_2/(k_2 - k_1)} \tag{5.58}$$

In the special case when the species I and T are not present initially, $(C_I)_0 = (C_T)_0 = 0$, and Equations 5.55 and 5.56 reduce to

$$\varphi_I = \frac{k_1}{(k_2 - k_1)}[e^{-k_1 t} - e^{-k_2 t}] \tag{5.59}$$

$$\varphi_T = \left\{1 - e^{-k_1 t} - \frac{k_1}{(k_2 - k_1)}[e^{-k_1 t} - e^{-k_2 t}]\right\} \tag{5.60}$$

Similarly, Equations 5.57 and 5.58 reduce to

$$t_{max} = \frac{\ln(k_2/k_1)}{(k_2 - k_1)} \tag{5.61}$$

$$(\varphi_I)_{max} = \left(\frac{k_1}{k_2}\right)^{k_2/(k_2-k_1)} \tag{5.62}$$

It may be seen that Equations 5.49 through 5.62 become indeterminate when $k_1 = k_2$. In this case, from Equation 5.48, writing $k_1 = k_2 = k$

$$\frac{dC_I}{dt} + kC_I = k(C_A)_0 e^{-k_1 t} \tag{5.63}$$

The solution to this equation would be

$$C_I e^{\int kdt} = \int k(C_A)_0 e^{-kt} e^{\int kdt} dt + \text{constant}$$

$$\Rightarrow \quad C_I e^{kt} = \int k(C_A)_0 dt + \text{constant}$$

$$\Rightarrow \quad C_I e^{kt} = k(C_A)_0 t + \text{constant} \tag{5.64}$$

To evaluate the constant of integration we again use the boundary condition

$$\text{At } t = 0, \quad C_I = (C_I)_0$$

From Equation 5.64, at $t = 0$

$$\text{constant} = (C_I)_0$$

Substituting this value in Equation 5.64,

$$\begin{aligned} C_I e^{kt} &= k(C_A)_0 t + (C_I)_0 \\ \Rightarrow \quad C_I &= [k(C_A)_0 t + (C_I)_0] e^{-kt} \end{aligned} \tag{5.65}$$

$$\Rightarrow \quad \varphi_I = \left[kt + \frac{(C_I)_0}{(C_A)_0}\right] e^{-kt} \tag{5.66}$$

From the relation $(C_A)_0 + (C_I)_0 + (C_T)_0 = C_A + C_I + C_T$

$$C_T = (C_A)_0 + (C_I)_0 + (C_T)_0 - C_A - C_I$$

$$\Rightarrow \quad C_T = (C_A)_0 + (C_I)_0 + (C_T)_0 - (C_A)_0 e^{-kt} - [k(C_A)_0 t + (C_I)_0]e^{-kt}$$

$$\Rightarrow \quad C_T = (C_A)_0[1 - e^{-kt} - kte^{-kt}] + (C_I)_0[1 - e^{-kt}] + (C_T)_0 \tag{5.67}$$

$$\Rightarrow \quad \varphi_T = [1 - e^{-kt} - kte^{-kt}] + \frac{(C_I)_0}{(C_A)_0}[1 - e^{-kt}] + \frac{(C_T)_0}{(C_A)_0} \tag{5.68}$$

Again, for maximum φ_I from Equation 5.66,

$$\frac{d\varphi_I}{dt} == \left[kt_{max} + \frac{(C_I)_0}{(C_A)_0}\right](-ke^{-kt_{max}}) + ke^{-kt_{max}} = 0$$

$$\Rightarrow \quad kt_{max} + \frac{(C_I)_0}{(C_A)_0} = 1$$

$$\Rightarrow \quad t_{max} = \frac{1}{k}\left[1 - \frac{(C_I)_0}{(C_A)_0}\right] \tag{5.69}$$

From Equation 5.66,

$$(\varphi_I)_{max} = \frac{1}{e} \tag{5.70}$$

So, when the two specific reaction rate constants are equal, $(\varphi_I)_{max}$ is independent of the initial concentrations of the species. Further, in the special case when the species I and T are not present initially,

$$(C_I)_0 = (C_T)_0 = 0$$

Therefore, Equations 5.66, 5.68, and 5.69 reduce to

$$\varphi_I = kte^{-kt} \tag{5.71}$$

$$\varphi_T = [1 - e^{-kt} - kte^{-kt}] \tag{5.72}$$

$$t_{max} = \frac{1}{k} \tag{5.73}$$

EXAMPLE 5.6

The reaction

$$A \rightarrow I \rightarrow T$$

with $k_1 = 0.05\ \text{min}^{-1}$ and $k_2 = 0.02\ \text{min}^{-1}$, respectively, is carried out in a batch reactor with an initial concentration of A equal to 2.5 mol/L. What is the time required to get the maximum concentration of I? What is the composition of the reactor at this time?

Solution

From the problem statement

$$(C_A)_0 = 2.5\ \text{mol/L}, \quad (C_I)_0 = (C_T)_0 = 0$$

From Equation 5.61,

$$t_{max} = \frac{\ln(k_2/k_1)}{(k_2 - k_1)} = \frac{\ln(0.02/0.05)}{(0.02 - 0.05)} = 30.54\ \text{min}$$

At this time, from Equation 5.62,

$$(\varphi_I)_{max} = \left(\frac{k_1}{k_2}\right)^{k_2/(k_2-k_1)} = \left(\frac{0.05}{0.02}\right)^{0.02/(0.02-0.05)} = 0.543$$

$$(C_I)_{max} = (C_A)_0(\varphi_I)_{max} = 2.5 \times 0.543 = 1.358\ \text{mol/L}$$

At time t_{max}, from Equation 5.47,

$$\frac{C_A}{(C_A)_0} = e^{-k_1 t}$$

$$\Rightarrow \quad C_A = 2.5 \times e^{-0.05 \times 30.54} = 0.543\ \text{mol/L}$$

$$C_T = (C_A)_0 + (C_I)_0 + (C_T)_0 - C_A - C_I = 2.5 - 0.543 - 1.358 = 0.599\ \text{mol/L}$$

5.7.2 Parallel Reactions

Consider the parallel reaction represented by the balanced chemical Equation 3.9 (Chapter 3). For convenience, we reproduce the equation here

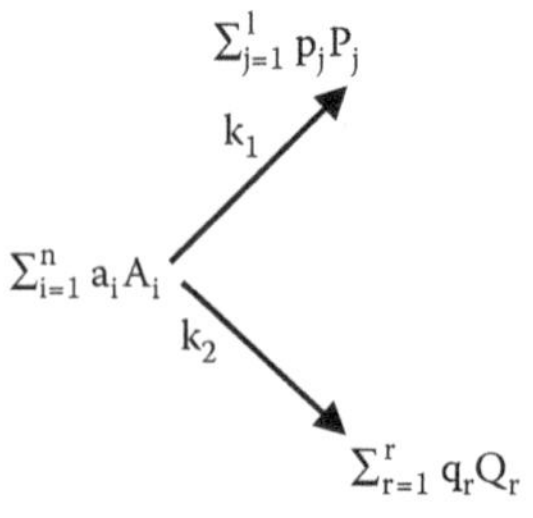

(3.38)

Further, we consider that each competing reaction step follows the law of mass action and $\left(C_{A_i}\right)_0 = a_i/a_{LR}\left(C_{A_{LR}}\right)_0$. Then, from Equation 3.19 (Chapter 3),

$$-r_{A_{LR}} = -\frac{d(C_{A_{LR}})}{dt} = (k_1 + k_2)\prod_{i=1}^{n} C_{A_i}^{a_i} \quad (5.74)$$

and, from Equation 5.32,

$$\frac{(a_{LR})^{\Sigma_{i=1}^{n} a_i}}{\left(1-\sum_{i=1}^{n} a_i\right)\prod_{i=1}^{n}(a_i)^{a_i}}\left[\left(C_{A_{LR}}\right)_0^{(1-\Sigma_{i=1}^{n} a_i)} - \left(C_{A_{LR}}\right)^{(1-\Sigma_{i=1}^{n} a_i)}\right] = (k_1 + k_2)t$$

$$\Rightarrow \left(C_{A_{LR}}\right) = \left[\left(C_{A_{LR}}\right)_0^{(1-\Sigma_{i=1}^{n} a_i)} - \frac{(k_1 + k_2)t\left(1-\sum_{i=1}^{n} a_i\right)\prod_{i=1}^{n}(a_i)^{a_i}}{(a_{LR})^{\Sigma_{i=1}^{n} a_i}}\right]^{1/(1-\Sigma_{i=1}^{n} a_i)} \quad (5.75)$$

For a product P_j,

$$r_{P_j} = \frac{d(C_{P_j})}{dt} = \left(\frac{p_j}{a_{LR}}\right)k_1\prod_{i=1}^{n} C_{A_i}^{a_i} \quad (5.76)$$

Similarly, for a product Q_r,

$$r_{Q_r} = \frac{d\left(C_{Q_r}\right)}{dt} = \left(\frac{q_r}{a_{LR}}\right)k_2\prod_{i=1}^{n} C_{A_i}^{a_i} \quad (5.77)$$

Dividing Equation 5.76 by Equation 5.77,

$$\frac{d\left(C_{P_j}\right)}{d\left(C_{Q_r}\right)} = \frac{p_j k_1 \prod_{i=1}^{n} C_{A_i}^{a_i}}{q_r k_2 \prod_{i=1}^{n} C_{A_i}^{a_i}} = \frac{p_j k_1 \prod_{i=1}^{n}\left[\left(C_{A_i}\right)_0 - a_i/a_{LR}\left\{\left(C_{A_{LR}}\right)_0 - C_{A_{LR}}\right\}\right]^{a_i}}{q_r k_2 \prod_{i=1}^{n}\left[\left(C_{A_i}\right)_0 - a_i/a_{LR}\left\{\left(C_{A_{LR}}\right)_0 - C_{A_{LR}}\right\}\right]^{a_i}}$$

$$= \left(\frac{p_j}{q_r}\right)\frac{k_1}{k_2} \quad (5.78)$$

Here, we define the instantaneous selectivity Φ_{inst} as (Carberry, 1966; Missen and Smith, 2000)

$$\Phi_{inst} = \frac{d(C_{P_j})}{d(C_{Q_r})} \quad (5.79)$$

$$\Rightarrow \int_{(C_{P_j})_0}^{C_{P_j}} d(C_{P_j}) = \int_{(C_{Q_r})_0}^{C_{Q_r}} \Phi_{inst} d(C_{Q_r}) \tag{5.80}$$

For the reacting system under consideration,

$$\Phi_{inst} = \frac{d(C_{P_j})}{d(C_{Q_r})} = \left(\frac{p_j}{q_r}\right)\frac{k_1}{k_2}$$

$$\Rightarrow \int_{(C_{P_j})_0}^{C_{P_j}} d(C_{P_j}) = \left(\frac{p_j}{q_r}\right)\frac{k_1}{k_2}\int_{(C_{Q_r})_0}^{C_{Q_r}} d(C_{Q_r})$$

$$\Rightarrow C_{P_j} - (C_{P_j})_0 = \left(\frac{p_j}{q_r}\right)\frac{k_1}{k_2}\left[C_{Q_r} - (C_{Q_r})_0\right]$$

$$\Rightarrow \frac{C_{P_j} - (C_{P_j})_0}{C_{Q_r} - (C_{Q_r})_0} = \left(\frac{p_j}{q_r}\right)\frac{k_1}{k_2} \tag{5.81}$$

The quantity $\left(C_{P_j} - (C_{P_j})_0\right)/\left(C_{Q_r} - (C_{Q_r})_0\right)$ is known as the overall selectivity Φ_{ovl}. For systems where Φ_{inst} is not constant, it may not be possible to compute the overall selectivity by integrating Equation 5.80.

Dividing Equation 5.76 by Equation 5.74,

$$\frac{d(C_{P_j})}{-d(C_{A_{LR}})} = \frac{(p_j / a_{LR})k_1 \prod_{i=1}^{n} C_{A_i}^{a_i}}{(k_1 + k_2)\prod_{i=1}^{n} C_{A_i}^{a_i}}$$

$$= \frac{(p_j / a_{LR})k_1 \prod_{i=1}^{n}\left[(C_{A_i})_0 - (a_i / a_{LR})\left\{(C_{A_{LR}})_0 - C_{A_{LR}}\right\}\right]^{a_i}}{(k_1 + k_2)\prod_{i=1}^{n}\left[(C_{A_i})_0 - (a_i / a_{LR})\left\{(C_{A_{LR}})_0 - C_{A_{LR}}\right\}\right]^{a_i}} = \frac{(p_j / a_{LR})k_1}{(k_1 + k_2)} \tag{5.82}$$

We define the instantaneous fractional yield Ψ_{inst} as (Carberry, 1966; Missen and Smith, 2000)

$$\Psi_{inst} = \frac{d(C_{P_j})}{-d(C_{A_{LR}})} \tag{5.83}$$

$$\Rightarrow \int_{(C_{P_j})_0}^{C_{P_j}} d(C_{P_j}) = -\int_{(C_{A_{LR}})_0}^{C_{A_{LR}}} \Psi_{inst} d(C_{A_{LR}})$$

$$\Rightarrow C_{P_j} - (C_{P_j})_0 = -\int_{(C_{A_{LR}})_0}^{C_{A_{LR}}} \Psi_{inst} d(C_{A_{LR}})$$

$$\Rightarrow \frac{C_{P_j} - (C_{P_j})_0}{(C_{A_{LR}})_0 - C_{A_{LR}}} = \frac{-1}{(C_{A_{LR}})_0 - C_{A_{LR}}} \int_{(C_{A_{LR}})_0}^{C_{A_{LR}}} \Psi_{inst} d(C_{A_{LR}}) \quad (5.84)$$

The quantity $C_{P_j} - (C_{P_j})_0 / (C_{A_{LR}})_0 - C_{A_{LR}}$ is known as the overall fractional yield Ψ_{ovl}.

For the reacting system under consideration,

$$\Psi_{inst} = \frac{d(C_{P_j})}{-d(C_{A_{LR}})} = \frac{(p_j / a_{LR})k_1}{(k_1 + k_2)}$$

$$\Rightarrow \int_{(C_{P_j})_0}^{C_{P_j}} d(C_{P_j}) = -\frac{(p_j / a_{LR})k_1}{(k_1 + k_2)} \int_{(C_{A_{LR}})_0}^{C_{A_{LR}}} d(C_{A_{LR}})$$

$$\Rightarrow C_{P_j} - (C_{P_j})_0 = \frac{(p_j / a_{LR})k_1}{(k_1 + k_2)} \left[(C_{A_{LR}})_0 - C_{A_{LR}} \right]$$

$$\Rightarrow \frac{C_{P_j} - (C_{P_j})_0}{(C_{A_{LR}})_0 - C_{A_{LR}}} = \frac{(p_j / a_{LR})k_1}{(k_1 + k_2)} \quad (5.85)$$

While Equation 5.85 is specific to the reacting system under consideration, Equations 5.83 and 5.84 have general validity. The equations can be used in engineering calculations as follows:

If the system composition is to be determined for a given reaction time:

Step 1

From Equation 5.75 determine $C_{A_{LR}}$ for the specified reaction time t. It may be noted that Equation 5.75 is applicable only when all the competing reactions follow the law of mass action. If the rate equations are other than that given by the law of mass action, the same must be used with Equation 5.10 to determine $C_{A_{LR}}$.

Step 2

Using Equation 5.83 along with Equation 5.84 determine $C_{P_j} - \left(C_{P_j}\right)_0$ and hence C_{P_j}.

Step 3

Determine C_{Q_r} by using the reaction stoichiometry, for example Equation 3.42 (Chapter 3).

If the time required for a given concentration of desired product is to be determined:

Step 1

Using Equation 5.83 along with Equation 5.84 determine $C_{A_{LR}}$ corresponding to the given concentration of the desired product C_{P_j}.

Step 2

From Equation 5.75 or the appropriate rate equations along with Equation 5.10 determine the reaction time "t" corresponding to $C_{A_{LR}}$ calculated in Step 1.

For most of the industrial applications involving parallel reactions, one of the competing reactions is desired while the others are undesired. Ideally, one would like to maximize both overall yield and overall selectivity. However, this may not be possible all the time and economic considerations often dictate the operation. It must be kept in mind that the reacting system at the end of the reaction contains the reactants as well as the products, both desired and undesired. If the cost of separation were less than the cost of operating the batch reactor, we would prefer low fractional yield. On the contrary, if the cost of separation were more than the cost of operating the batch reactor, we would prefer high fractional yield.

EXAMPLE 5.7

The parallel reaction

$$A \rightarrow P \quad (-r_A)_1 = r_P = 0.05C_A^2\ \text{mol/(L·min)}$$

$$A \rightarrow Q \ (-r_A)_2 = r_Q = 0.02C_A\ \text{mol/(L·min)}$$

is carried out in a batch reactor with 2 mol/L initial concentration of A and no P and Q present initially. Determine the composition of the reacting system after 10 min from the start.

Solution

$$-r_A = (-r_A)_1 + (-r_A)_2 = -\frac{dC_A}{dt} = 0.05C_A^2 + 0.02C_A \tag{5.7.1}$$

$$\Rightarrow \quad -\int_{2}^{C_A} \frac{dC_A}{0.05C_A^2 + 0.02C_A} = \int_0^t dt$$

$$\Rightarrow \quad -\int_{2}^{C_A} \left[\frac{1}{0.02C_A} - \frac{125}{(2.5C_A + 1)}\right] dC_A = \int_0^t dt$$

$$\Rightarrow \quad 50(\ln 2 - \ln C_A) + 50\left[\ln(2.5C_A + 1) - \ln 6\right] = t$$

$$\Rightarrow \quad 34.66 - 50\ln C_A + 50\ln(2.5C_A + 1) - 89.59 = 10$$

$$\Rightarrow \quad 50\left[\ln(2.5C_A + 1) - \ln C_A\right] = 64.93$$

$$\Rightarrow \quad \ln(2.5C_A + 1) - \ln C_A = 1.2986$$

$$\Rightarrow \quad \ln\frac{2.5C_A + 1}{C_A} = 1.2986$$

$$\frac{2.5C_A + 1}{C_A} = 3.664$$

$$\Rightarrow \quad C_A = 0.859 \text{ mol/L}$$

$$\frac{dC_Q}{dt} = 0.02C_A \tag{5.7.2}$$

From Equations 5.7.1 and 5.7.2,

$$\frac{dC_Q}{-dC_A} = \frac{0.02C_A}{0.05C_A^2 + 0.02C_A}$$

$$\Rightarrow \quad \int_0^{C_Q} dC_Q = -\int_2^{0.859} \frac{0.02C_A dC_A}{0.05C_A^2 + 0.02C_A}$$

$$\Rightarrow \quad C_Q = -\int_2^{0.859} \frac{dC_A}{2.5C_A + 1}$$

$$\Rightarrow \quad C_Q = 0.4\left[\ln 6 - \ln 3.1475\right]$$

$$\Rightarrow \quad C_Q = 0.258 \text{ mol/L}$$

From the relation

$$C_A + C_P + C_Q = (C_A)_0 + (C_P)_0 + (C_Q)_0$$

$$\Rightarrow \quad C_P = (C_A)_0 + (C_P)_0 + (C_Q)_0 - C_A - C_Q = 2 - 0.859 - 0.258 = 0.883 \text{ mol/L}$$

5.8 VARIABLE VOLUME BATCH REACTOR

As noted in Section 5.1, for gaseous-reacting systems the reactor can be operated either at constant volume or at constant pressure. If the reactor is operated under constant pressure, the reactor volume will vary with conversion as explained in Chapter 2, Section 2.3.1.1. Under these conditions, Equations 5.4 and 5.5 act as the governing performance equations for the batch reactor. Additionally, from Equation 2.43 (Chapter 2),

$$V_{BR} = (V_{BR})_0(1 + \varepsilon_{LR}X_{A_{LR}})$$

$$\Rightarrow \quad X_{A_{LR}} = \frac{1}{(V_{BR})_0\varepsilon_{LR}}[V_{BR} - (V_{BR})_0] \tag{5.86}$$

Therefore, measuring the volume of the reactor serves as a convenient way of measuring the conversion and hence the progress of the reaction.

From Equation 5.86,

$$\frac{dX_{A_{LR}}}{dV_{BR}} = \frac{1}{(V_{BR})_0\varepsilon_{LR}}$$

$$\Rightarrow \quad dX_{A_{LR}} = \frac{dV_{BR}}{(V_{BR})_0\varepsilon_{LR}} \tag{5.87}$$

Then, from Equation 5.5,

$$-(C_{A_{LR}})_0 \int_{(V_{BR})_0}^{V_{BR}} \frac{dV_{BR}}{-r_{A_{LR}}(V_{BR})_0\varepsilon_{LR}(1 + \varepsilon_{LR}X_{A_{LR}})} = \int_0^t dt = t$$

$$\Rightarrow \quad -\frac{(C_{A_{LR}})_0}{\varepsilon_{LR}} \int_{(V_{BR})_0}^{V_{BR}} \frac{dV_{BR}}{-r_{A_{LR}}V_{BR}} = \int_0^t dt = t \tag{5.88}$$

5.8.1 Single-Reactant nth-Order Reaction in Variable Volume Batch Reactor

Here, the rate of reaction can be expressed by Equation 5.11. Combining with Equations 2.45 (Chapter 2) and 5.86,

$$-r_A = kC_A^n$$

$$\Rightarrow \quad -r_A = kC_A^n$$

$$\Rightarrow \quad -r_A = k\left[\frac{N_A}{V}\right]^n = k\left[(C_A)_0(V_{BR})_0\right]^n\left[\frac{(1-(V_{BR}/(V_{BR})_0\varepsilon_A)+1/\varepsilon_A)}{V_{BR}}\right]^n \tag{5.89}$$

$$\Rightarrow \quad -r_A = k\left[(C_A)_0(V_{BR})_0\right]^n\left[\frac{(V_{BR})_0\varepsilon_A - V_{BR} + (V_{BR})_0}{(V_{BR})_0\varepsilon_A V_{BR}}\right]^n$$

$$\Rightarrow \quad -r_A = k\left[\frac{(C_A)_0}{\varepsilon_A}\right]^n\left[\frac{(V_{BR})_0\varepsilon_A - V_{BR} + (V_{BR})_0}{V_{BR}}\right]^n$$

From Equation 5.88,

$$\left[\frac{\varepsilon_A}{(C_A)_0}\right]^{n-1}\int_{(V_{BR})_0}^{V_{BR}}\left[\frac{1}{(V_{BR})_0\varepsilon_A - V_{BR} + (V_{BR})_0}\right]^n [V_{BR}]^{n-1} dV_{BR} = kt \quad (5.90)$$

Although analytical integration of Equation 5.90 is possible, it is avoided here owing to the mathematical complexity. For engineering calculations, numerical integration is more conducive and advisable. We deduce here integrated forms for reaction orders one and two.

First-order reaction

From Equation 5.90,

$$\int_{(V_{BR})_0}^{V_{BR}}\frac{dV_{BR}}{(V_{BR})_0\varepsilon_A - V_{BR} + (V_{BR})_0} = kt$$

$$\Rightarrow \quad \ln\left[\frac{(V_{BR})_0\varepsilon_A}{(V_{BR})_0\varepsilon_A - V_{BR} + (V_{BR})_0}\right] = kt \quad (5.91)$$

$$\Rightarrow \quad (V_{BR})_0\varepsilon_A - V_{BR} + (V_{BR})_0 = (V_{BR})_0\varepsilon_A e^{-kt}$$

$$\Rightarrow \quad V_{BR} = (V_{BR})_0(1 + \varepsilon_A - \varepsilon_A e^{-kt})$$

Equation 5.91 relates the reactor volume with the reaction time for a first-order reaction under variable volume conditions.

Second-order reaction

From Equation 5.90,

$$\frac{\varepsilon_A}{(C_A)_0}\int_{(V_{BR})_0}^{V_{BR}}\frac{V_{BR}}{[(V_{BR})_0\varepsilon_A - V_{BR} + (V_{BR})_0]^2} dV_{BR} = kt$$

$$\Rightarrow \frac{\varepsilon_A}{(C_A)_0}\left[\frac{\{(V_{BR})_0\varepsilon_A + (V_{BR})_0\}\{V_{BR} - (V_{BR})_0\}}{\{(V_{BR})_0\varepsilon_A - V_{BR} + (V_{BR})_0\}(V_{BR})_0\varepsilon_A} + \ln\left\{\frac{(V_{BR})_0\varepsilon_A - V_{BR} + (V_{BR})_0}{(V_{BR})_0\varepsilon_A}\right\}\right] = kt$$

(5.92)

5.8.2 Conversion in Variable Volume Batch Reactor

For a single-reactant nth-order reaction, from Equations 5.4 and 2.45 (Chapter 2),

$$(C_A)_0\int_0^{X_A}\frac{dX_A}{kC_A^n(1 + \varepsilon_A X_A)} = t$$

$$\Rightarrow \quad [(C_A)_0]^{1-n}\int_0^{X_A}\frac{[1 + \varepsilon_A X_A]^{n-1}}{[1 - X_A]^n} dX_A = kt \quad (5.93)$$

First-order reaction

Equation 5.93 reduces to

$$\int_0^{X_A} \frac{1}{1-X_A} dX_A = kt$$

$$\Rightarrow \quad \ln(1-X_A) = -kt$$

$$\Rightarrow \quad X_A = 1-e^{-kt} \tag{5.94}$$

We may see that Equation 5.94 is the same as Equation 5.22. Thus, conversion for a given reaction time is independent of reactor volume change for a first-order reaction. This can be used as another criterion to check if the reaction is first order. Once conversion is known, the reactant concentration can be calculated using Equation 2.45 (Chapter 2).

Second-order reaction

Equation 5.93 reduces to

$$\int_0^{X_A} \frac{1+\varepsilon_A X_A}{[1-X_A]^2} dX_A = (C_A)_0 kt$$

$$\Rightarrow \quad \varepsilon_A \ln(1-X_A) + \frac{(\varepsilon_A - 1)X_A}{1-X_A} = -(C_A)_0 kt \tag{5.95}$$

EXAMPLE 5.8

The second-order gaseous reaction

$$A \rightarrow B + C$$

is carried out in a variable volume batch reactor at 27°C and 1 atm pressure with pure A initially present. The specific reaction rate constant is 0.2 L/(mol · min). Calculate the conversion of A after 30 min.

SOLUTION

Since pure A is used initially, $(x_A)_0 = 1$, so from Equation 2.27 (Chapter 2) $(p_A)_0 = 1$ atm

From Equation 2.26 (Chapter 2),

$$(C_A)_0 = \frac{(p_A)_0}{RT} = \frac{1}{0.082 \times 300} = 0.041 \text{ mol/L}$$

From Equation 2.44 (Chapter 2),

$$\varepsilon_A = \frac{(N_A)_0 \left[\sum_{j=1}^{1} p_j/a_{LR} - \sum_{i=1}^{n} a_i/a_{LR}\right]}{(N_A)_0 + N_{inerts}} = 1+1-1 = 1$$

From Equation 5.95,

$$\varepsilon_A \ln(1-X_A)+\frac{(\varepsilon_A-1)X_A}{1-X_A}=-(C_A)_0 kt$$

$$\Rightarrow \quad \ln(1-X_A)=-0.041\times 0.2\times 30=-0.246$$

$$\Rightarrow \quad 1-X_A=0.78$$

$$\Rightarrow \quad X_A=0.22$$

5.9 HETEROGENEOUS REACTIONS IN BATCH REACTOR

Batch reactor for a heterogeneous reaction essentially means an enclosure in which the constituent phases of the reacting system interact with each other without inflow or outflow. For example, a batch digester in a pulp mill where a fixed amount of lignocellulosic raw material (solid phase) and a fixed amount of sodium hydroxide solution are made to react with each other to separate the cellulose from other constituents of the raw material. More common are the situations where one of the phases is present as a batch while the other flows past it. For example, a coal-fired stove in a rural household where a batch of coal burns in a stream of air going into the furnace while the product gases move out.

In Chapter 3, we discussed that the rate of a heterogeneous process is often expressed in terms of unit area of interface separating the phases, volume of one of the phases, or mass of one of the phases. Accordingly, the general mole balance equation for the batch reactor is to be adapted to match with the form of the rate equation at hand.

5.9.1 Fluid–Solid Noncatalytic Reactions

5.9.1.1 Solid Product Is Formed on the Surface of the Unreacted Core

Considering the shrinking core model to adequately represent the fluid–solid noncatalytic reaction represented by the balanced chemical Equation 3.51 (Chapter 3), Equation 3.68 relates the time of reaction with the conversion of the solid reactant A_2 for a single spherical particle of radius R. Consider a batch of solid A_2 consisting of spherical particles of n different radii, R_1, R_2, etc., with a given distribution of mole fractions x_1, x_2, etc. It is reacted in a stream of A_1 with concentration $(C_{A_1})_{fl}$. Suppose we start with a batch of $(N_{A_2})_0$ moles of solid A_2. It will consist of $x_1(N_{A_2})_0$, moles of A_2 of particle size R_1, $x_2(N_{A_2})_0$ moles of A_2 of particle size R_2, and so on. After a time t, the particles of size R_1 will be converted to the extent $(X_{A_2})_1$, the particles of size R_2 will be converted to the extent $(X_{A_2})_2$, and so on. Therefore,

Total moles of A_2 converted at time t = $\sum_{i=1}^{n} x_i (N_{A_2})_0 (X_{A_2})_i$

Therefore, total conversion of A_2 at time t is

$$X_{A_2}=\frac{\sum_{i=1}^{n} x_i (N_{A_2})_0 (X_{A_2})_i}{(N_{A_2})_0}=\sum_{i=1}^{n} x_i (X_{A_2})_i \qquad (5.96)$$

The numerical procedure can be conveniently performed by tabulating the data as follows:

Particle Radius	Mole Fraction	Conversion from Equation 3.68 (Chapter 3)	Weighted Conversion
R_1	x_1	$(X_{A_2})_1$	$x_1(X_{A_2})_1$
R_2	x_2	$(X_{A_2})_2$	$x_2(X_{A_2})_2$
R_3	x_3	$(X_{A_2})_3$	$x_3(X_{A_2})_3$
–	–	–	–
–	–	–	–
–	–	–	–
–	–	–	–
R_n	x_n	$(X_{A_2})_n$	$x_n(X_{A_2})_n$

Total conversion of A_2 = Sum of entries in the last column = $\sum_{i=1}^{n} x_i (X_{A_2})_i$

For the limiting cases where one of the resistances limits the rate of the process, the conversions $(X_{A_2})_1$, $(X_{A_2})_2$, etc., can be computed using the appropriate equations from Equations 3.71, 3.77, or 3.82 (Chapter 3) as the case may be.

5.9.1.2 No Solid Product Layer Is Formed on the Surface of the Unreacted Core

The method to calculate the total conversion of the solid reactant A_2 remains the same but the individual conversions $(X_{A_2})_1$, $(X_{A_2})_2$, etc., are to be calculated using Equation 3.96 or Equation 3.101 (Chapter 3). Here also, if one of the resistances is limiting the appropriate equations from Equations 3.105, 3.109, or 3.114, should be used as the case may be.

EXAMPLE 5.9

Spherical particles of graphite of uniform density with the size distribution

Particle radius (mm)	4	5	6	7	8
Mole fraction	0.2	0.55	0.15	0.05	0.05

are burnt in air at a temperature of 727°C and 1 atm pressure. The specific reaction rate constant based on unit surface area of graphite is 16 cm/s and the process is controlled by the chemical reaction proper. Calculate conversion after 5 min. The molar density of graphite particles is 0.18 mol/cm^3.

Solution

The reaction is

$$C(s) + O_2(g) \rightarrow CO_2(g)$$

For this reaction under chemical reaction control, the appropriate equation relating conversion and reaction time is Equation 3.114 (Chapter 3), which gives

$$t = \frac{\rho_C R[1-(1-X_C)^{1/3}]}{a_2 (C_{O_2})_{fl} k'}$$

$$\Rightarrow \quad 1-(1-X_C)^{\frac{1}{3}} = \frac{a_2 (C_{O_2})_{fl} k't}{\rho_C R}$$

$$\Rightarrow \quad 1-X_C = \left[1-\frac{a_2 (C_{O_2})_{fl} k't}{\rho_C R}\right]^3$$

$$\Rightarrow \quad X_C = 1-\left[1-\frac{a_2 (C_{O_2})_{fl} k't}{\rho_C R}\right]^3 \tag{5.9.1}$$

Air has 21 mole percent oxygen, so from Equation 2.26 (Chapter 2)

$$(C_{O_2})_{fl} = \frac{p_{O_2}}{RT} = \frac{0.21}{0.082 \times 1000} = 0.0026 \text{ mol/L} = 2.6 \text{ mol/m}^3$$

From the given data:

$$\rho_C = 0.18 \text{ mol/cm}^3 = 1.8 \times 10^5 \text{ mol/m}^3$$

$$a_2 = 1, \quad k' = 16 \text{ cm/s} = 0.16 \text{ m/s}, \quad t = 300 \text{ s}$$

Next, we make the following table:

Particle Radius (m)	Mole Fraction	Conversion from Equation 5.9.1	Weighted Conversion
0.004	0.2	0.44	0.088
0.005	0.55	0.36	0.198
0.006	0.15	0.31	0.047
0.007	0.05	0.27	0.014
0.008	0.05	0.24	0.012
Total conversion of C = Sum of entries in the last column = 0.36			

Therefore, conversion after 5 min is 36%.

5.9.2 Fluid–Fluid Noncatalytic Reactions

Such reactions are seldom carried out in batch reactors. For the irreversible reaction represented by the balanced chemical Equation 3.117 (Chapter 3), the rate is expressed by Equations 3.127 through 3.129 depending on the reactor characteristics taken as the basis for the rate. These rate expressions have also been deduced for steady state condition that prevails in a flow reactor and represent the local rate at any particular location in the reactor. Flow reactors will be considered in subsequent chapters.

5.9.3 Fluid–Solid Catalytic Reactions

For a constant volume batch reactor, Equation 5.10 is still applicable with $r_{A_{1}R}$ appropriately substituted by the rate expression for the fluid–solid catalytic reaction. For the single-reactant-type first-order reaction represented by reaction 3.148, from Equation 3.204 (Chapter 3),

$$\left(-\frac{dN_{A_1}}{dt}\right)=\frac{(C_{A_1})_{fl}}{1/(k_{fl}Ar_{ex})+1/(V_p k\xi)}$$

$$\Rightarrow \left(-\frac{d(C_{A_1})_{fl}}{dt}\right)=\frac{(C_{A_1})_{fl}}{(V_{fl}/k_{fl}Ar_{ex})+(V_{fl}/V_p k\xi)}=\left[\frac{1}{(V_{fl}/k_{fl}Ar_{ex})+(V_{fl}/V_p k\xi)}\right](C_{A_1})_{fl}$$

$$\Rightarrow \int_{(C_{A_1})_0}^{(C_{A_1})_{fl}}\frac{d(C_{A_1})_{fl}}{(C_{A_1})_{fl}}=-\left[\frac{1}{V_{fl}/k_{fl}Ar_{ex}+V_{fl}/V_p k\xi}\right]\int_0^t dt=-\left[\frac{1}{V_{fl}/k_{fl}Ar_{ex}+V_{fl}/V_p k\xi}\right]t$$

$$\Rightarrow \ln\frac{(C_{A_1})_{fl}}{(C_{A_1})_0}=-\left[\frac{1}{V_{fl}/k_{fl}Ar_{ex}+V_{fl}/V_p k\xi}\right]t \tag{5.97}$$

$$\Rightarrow (C_{A_1})_{fl}=(C_{A_1})_0 e^{-\left[\frac{1}{\frac{V_{fl}}{k_{fl}Ar_{ex}}+\frac{V_{fl}}{V_p k\xi}}\right]t} \tag{5.98}$$

$$\Rightarrow X_{A_1}=e^{-\left[\frac{1}{\frac{V_{fl}}{k_{fl}Ar_{ex}}+\frac{V_{fl}}{V_p k\xi}}\right]t} \tag{5.99}$$

For nth-order reaction limited by strong resistance to intrapore diffusion, the rate is expressed by Equation 3.188 (Chapter 3). Since the resistance to external mass transfer is negligible, $(C_{A_1})_{ES}=(C_{A_1})_{fl}$. Therefore, we have

$$(-r_{A_1})_{obs}=-\frac{d(C_{A_1})_{fl}}{dt}=\frac{Ar_{ex}}{V_p}\sqrt{\frac{2kD_{pore}}{n+1}}(C_{A_1})_{fl}^{(n+1)/2}$$

$$\Rightarrow \int_{(C_{A_1})_0}^{(C_{A_1})_{fl}}\frac{d(C_{A_1})_{fl}}{(C_{A_1})_{fl}^{(n+1)/2}}=-\frac{Ar_{ex}}{V_p}\sqrt{\frac{2kD_{pore}}{n+1}}\int_0^t dt=-\left[\frac{Ar_{ex}}{V_p}\sqrt{\frac{2kD_{pore}}{n+1}}\right]t$$

$$\Rightarrow \frac{2}{(1-n)}\left[(C_{A_1})_{fl}^{(1-n)/2}-(C_{A_1})_0^{(1-n)/2}\right]=-\left[\frac{Ar_{ex}}{V_p}\sqrt{\frac{2kD_{pore}}{n+1}}\right]t \tag{5.100}$$

$$\Rightarrow \quad \frac{2}{(1-n)}\left(C_{A_1}\right)_0^{(1-n)/2}\left[\left(1-X_{A_1}\right)^{(1-n)/2}-1\right]=-\left[\frac{Ar_{ex}}{V_p}\sqrt{\frac{2kD_{pore}}{n+1}}\right]t \qquad (5.101)$$

It is often the case that the catalyst particles are not all of the same size and shape. Consider a batch of catalyst particles in which a volume fraction (or weight fraction if the particle density remains the same) v_1 has effectiveness factor ξ_1, volume fraction v_2 has effectiveness factor ξ_2, and so on. For such composite catalyst samples, the mean effectiveness factor is given by (Aris, 1957)

$$\xi_{mean} = \sum_{i=1}^{n} \xi_i v_i \qquad (5.102)$$

In such cases ξ should be replaced with ξ_{mean} in Equations 5.97 through 5.101.

EXERCISES

5.1 Determine the order of reaction and the specific reaction rate constant from the concentration time data obtained in an isothermal constant volume batch reactor using the differential method of analysis. The homogeneous liquid-phase reaction is of the type

$$A \rightarrow B + C$$

a.

Time (min)	**0**	**2**	**5**	**10**	**15**	**30**
Concentration (mol/L)	2.5	1.123	0.338	0.046	0.006	1.54×10^{-5}

b.

Time (min)	**0**	**2**	**5**	**10**	**15**	**30**
Concentration (mol/L)	4	1.176	0.571	0.307	0.211	0.108

c.

Time (min)	**0**	**2**	**5**	**10**	**15**	**30**
Concentration (mol/L)	2	1.857	1.690	1.491	1.348	1.085

5.2 Determine the order of reaction and the specific reaction rate constant from the conversion time data obtained in an isothermal constant volume batch reactor using the differential method of analysis. The initial concentration of the reactant in each case is 1.5 mol/L. The homogeneous liquid-phase reaction is of the type

$$A \rightarrow B + C$$

a.

Time (min)	0	2	5	10	15	30
Conversion	0	0.6	0.789	0.882	0.918	0.957

b.

Time (min)	0	2	5	10	15	30
Conversion	0	0.058	0.139	0.259	0.362	0.593

c.

Time (min)	0	2	5	10	15	30
Conversion	0	0.113	0.227	0.348	0.425	0.555

5.3 Repeat Exercise 5.1 by the integral method of analysis.

5.4 Repeat Exercise 5.2 by the integral method of analysis.

5.5 In the batch reactor operation reported in Exercise 5.1, plot the concentration time graph for the product B.

5.6 In the batch reactor operation reported in Exercise 5.2, determine the concentration of the product B in the reactor 20 min from the start.

5.7 In the batch reactor operation reported in Exercise 5.1, some fresh reactant A is added to the reactor after 10 min from the start to make the concentration again equal to the initial concentration. What will be the concentration of A in the reactor after 30 min from the start?

5.8 In the batch reactor operation reported in Exercise 5.1, half of the reactor content is taken out after 10 min from the start and the remaining material is allowed to react further. What will be the concentration of A in the reactor after 30 min from the start?

5.9 The reaction reported in Exercise 5.2, part a, is carried out starting with a batch of 200 kmol A at an initial concentration of 2 mol/L. How much time is required to produce 150 kmol of B?

5.10 The homogeneous liquid-phase reaction

$$A + 2B \rightarrow C + D$$

is carried out in a constant volume batch reactor with initial concentrations of A and B of 2 and 3.5 mol/L, respectively. The reaction rate may be expressed as

$$-r_A = 0.2C_AC_B^2 \text{ mol/(L·min)}$$

Find the time required for the concentration of B to drop to 2 mol/L.

5.11 Repeat Exercise 5.10 for the case when the initial concentrations of A and B are 1.5 and 3.5 mol/L, respectively.

5.12 For the batch reactor operation in Exercise 5.10, determine the time required for 50% conversion of A.

5.13 For the batch reactor operation in Exercise 5.11, determine the time required for 50% conversion of A.

5.14 After solving Exercises 5.12 and 5.13, it will be clear to the students that the time required for a given conversion of A (or B) is dependent on the ratio $(C_A)_0/(C_B)_0$. Plot graphs between X_A and t for different values of $(C_A)_0/(C_B)_0$. Comment on the graphs so obtained.

5.15 Show that for nth-order reaction, ratio of time required for 50% completion to time required for 75% completion is independent of initial concentration and function of n alone.

5.16 The homogeneous liquid-phase reversible reaction

$$A + 2B \rightleftharpoons C + D$$

with $k_1 = 0.03\ L^2/(mol^2 \cdot min)$ and $k_2 = 0.02 L/(mol \cdot min)$, is carried out in a batch reactor with initial concentrations of A and B as 1 and 2.5 mol/L, respectively. C and D are not present initially. Determine the time required to achieve 80% of equilibrium conversion.

5.17 Repeat Exercise 5.16 when the initial concentration of C and D is 0.1 mol/L.

5.18 For the batch reactor operation in Exercise 5.16, determine the time required for 30% conversion of A.

5.19 The series reaction

$$A \rightarrow I \rightarrow T \rightarrow U$$

with the specific reaction rate constants for the consecutive reaction steps being k_1, k_2, and k_3, respectively, is carried out in a constant volume batch reactor. The initial concentration of A is $(C_A)_0$ and no I, T, or U is present initially. Derive an expression for $C_T/(C_A)_0$ as a function of reaction time in the batch reactor.

5.20 For Exercise 5.19, determine the reaction time at which $C_T/(C_A)_0$ will be maximum. Also, determine this maximum value of $C_T/(C_A)_0$.

5.21 For Exercise 5.19, derive an expression for $C_U/(C_A)_0$ as a function of reaction time in the batch reactor.

5.22 The multiple reaction

$$A \xrightarrow{k1} I \xrightarrow{k2} T \xrightarrow{k3} U$$
$$I \xrightarrow{k4} P$$

is carried out in a constant volume batch reactor. The initial concentration of A is $(C_A)_0$ and no I, T, U, or P is present initially. Derive an expression for $C_I/(C_A)_0$ as a function of reaction time in the batch reactor.

5.23 For Exercise 5.22, determine the reaction time at which $C_I/(C_A)_0$ will be maximum. Also determine this maximum value of $C_I/(C_A)_0$.

5.24 For Exercise 5.22, derive an expression for $C_T/(C_A)_0$ as a function of reaction time in the batch reactor.

5.25 For Exercise 5.22, determine the reaction time at which $C_T/(C_A)_0$ will be maximum. Also determine this maximum value of $C_T/(C_A)_0$.

5.26 For Exercise 5.22, derive an expression for $C_U/(C_A)_0$ as a function of reaction time in the batch reactor.

5.27 For Exercise 5.22, derive an expression for $C_P/(C_A)_0$ as a function of reaction time in the batch reactor.

5.28 The reaction

$$A \rightarrow I \rightarrow T$$

with $k_1 = 0.5\ min^{-1}$ and $k_2 = 0.05\ min^{-1}$, respectively, is carried out in a batch reactor with an initial concentrations of A, I, and T equal to 5, 0.2, and 0.1 mol/L, respectively. Find the composition of the reactor after 10 min from the start.

5.29 In Exercise 5.28, when will we get the maximum concentration of T? What will be this maximum concentration?

5.30 The liquid-phase parallel reactions

$$A \rightarrow I \quad r_I = 0.5C_A\ mol/(L \cdot min)$$

$$2A \rightarrow T + U \quad r_T = 0.1C_A^2\ mol/(L \cdot min)$$

are carried out in a constant volume batch reactor with 5 mol/L initial concentration of A. Species I, T, and U are not present initially.

a. At what time from the start the concentration of I will be 1 mol/L?
b. What is the instantaneous selectivity of I with respect to T at this time?
c. What is the instantaneous fractional yield of I at this time?

5.31 Using the data given in Exercise 5.30, plot the instantaneous fractional yields for the species I and T as a function of time on the same graph. What conclusion can you draw from these plots?

5.32 In a constant volume batch reactor, the following half-life data are obtained for the irreversible liquid-phase reaction:

$$A \rightarrow Products$$

when the reaction is carried out at different initial concentrations. Determine the order of reaction and the specific reaction rate constant.

a.

$(C_A)_0$ (mol/L)	1	1.2	1.4	1.6	1.8	2
$t_{1/2}$ (min)	7.5	5.2	3.8	2.9	2.3	1.9

b.

$(C_A)_0$ (mol/L)	1	1.2	1.4	1.6	1.8	2
$t_{1/2}$ (min)	14.6	13.4	12.4	11.6	10.9	10.4

5.33 In a constant volume batch reactor, the following half-life data are obtained for the irreversible liquid-phase reaction:

$$A + 2B \rightarrow C + D$$

$(C_A)_0$ (mol/L)	1	1.2	1.4	1.6	1.8	2
$(C_B)_0$ (mol/L)	2	2.4	2.8	3.2	3.6	4
$t_{1/2}$ (min)	37.5	26	19.1	14.6	11.6	9.4

when the reaction is carried out at different initial concentrations of A and B. Determine the specific reaction rate constant for the reaction.

5.34 The first-order irreversible gas-phase reaction

$$A \rightarrow B + C$$

is carried out in a variable volume batch reactor with pure A initially present. The specific reaction rate constant is 0.4 s^{-1}.

a. After how much time will the reactor volume be 30% more than the initial volume?
b. What is the conversion at this time?
c. What is the time required for the reaction volume to be double the initial volume?

5.35 Repeat Exercise 5.34 if the reaction is second order and the initial temperature and pressure are 500°C and 1 atm, respectively.

5.36 Repeat Example 5.8 if the initial gas mixture contains 80% A and 20% inerts by volume.

5.37 A gas solid noncatalytic reaction is carried out in a batch reactor for 15 min with 10 kg solid of varying size of the following size distribution:

Particle size (μm)	50	80	100	150	200
Mass (%)	17	22	28	13	20

The time required for complete reaction for these size fractions is 5, 9, 12, 17, and 20 min, respectively. Calculate the overall conversion if the chemical reaction is rate controlling and the shrinking core model is valid.

REFERENCES

Aris, R., On shape factors of irregular particles—I. *Chem. Eng. Sci.*, 6, 262, 1957.

Bardow, A. and Marquardt, W., Incremental and simultaneous identification of reaction kinetics: Methods and comparison. *Chem. Eng. Sci.*, 59, 2673, 2004.

Bateman, H., The solution of a system of differential equations occurring in the theory of radioactive transformations. *Proc. Camb. Phil. Soc.*, 15, 423, 1910.

Carberry, J.J., Yield in chemical reactor engineering. *Ind. Eng. Chem.*, 58, 40, 1966.

Chien, J.-Y., Kinetic analysis of irreversible consecutive reactions. *J. Am. Chem. Soc.*, 70, 2256, 1948.

Gurol, M.D. and Singer, P.C., Kinetics of ozone decomposition: A dynamic approach. *Environ. Sci. Technol.*, 16, 377, 1982.

Hill, T.L., The rate equations for consecutive reactions. *J. Am. Chem. Soc.*, 64, 465, 1942.

Kittrel, J.R. and Mezaki, R., Obtaining reaction rates from integral reactor data. *Brit. Chem. Eng. Process. Technol.*, 11, 1538, 1966.

Missen, R.W. and Smith, W.R., Yield, selectivity, and all that. *Chem. Eng. Edu.*, 34, 320, 2000.

Toledo-Velasquez, D., Gaud, H.T., and Connors, K.A., Misoprostol dehydration kinetics in aqueous solution in the presence of hydroxypropyl methylcellulose. *J. Pharm. Sci.*, 81, 145, 1992.

Yeow, Y.L., Wickramasinghe, S.R., Han, B., and Leong, Y.-K., A new method of processing the time-concentration data of reaction kinetics. *Chem. Eng. Sci.*, 58, 3601, 2003.

6 Ideal Isothermal Flow Reactors

A flow reactor, also known as a continuous reactor, is a vessel that holds an open reacting system within its confines while the reactants chemically transform into products. Since the reacting system is open, there is continuous entry of material into the vessel as well as continuous exit of material out of the vessel. To sum up, a flow reactor is a vessel in which reactants are continuously fed and products are continuously taken out. The reactants chemically transform into products while they flow through the reactor. Most industrial reactors are of the continuous type as they require lower operational cost compared to batch reactors. They are suited for large-scale operations owing to better economies of scale. An ideal flow reactor is one in which the flow of material (fluid) through the reactor can be modeled as one of the two ideal flow patterns (Levenspiel and Bischoff, 1963), namely plug flow or complete backmix flow. Further, an isothermal flow reactor has constant temperature both with respect to time as well as with respect to position within the reactor. These ideal conditions may be approximated but not rigorously met in a real flow reactor. Nevertheless, the study of the hypothetical idealized system provides vital understanding that can be used for the real systems as well.

6.1 IDEAL PLUG FLOW REACTOR

Ideal plug flow reactor (PFR), also known as piston flow reactor, is one where every fluid element entering the reactor spends the same amount of time inside the reactor. This is only possible when every fluid element is propelled by a unidirectional uniform flow field in parallel trajectories. There is no other phenomenon except the flow field, not even diffusion(!), causing the fluid to move. This is an idealized concept that cannot be rigorously realized in a real situation. However, there are real systems that closely approach this ideal behavior.

For the ideal PFR schematically represented in Figure 6.1, we write a material balance for the limiting reactant species over the differential volume element shown in the figure as

Input = Output + Accumulation + Disappearance due to chemical reaction

For the PFR at steady state, the accumulation within the volume element is zero. So we have

Input = Output + Disappearance due to chemical reaction

$$\Rightarrow \quad F_{A_{LR}} = (F_{A_{LR}} + dF_{A_{LR}}) + (-r_{A_{LR}})dV$$

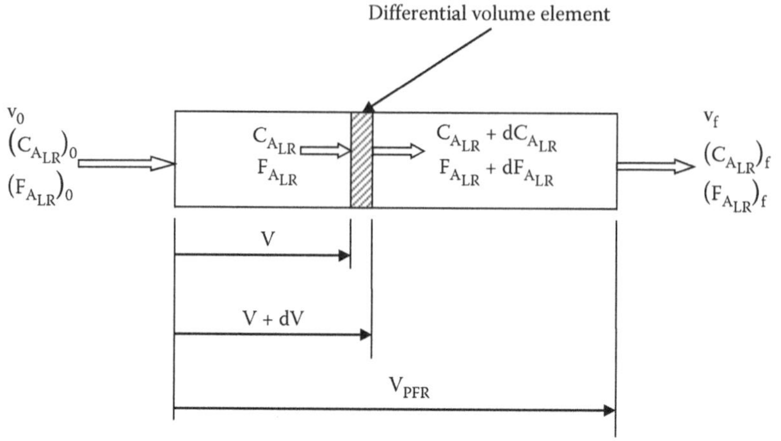

FIGURE 6.1 Schematic representation of an ideal PFR.

$$\Rightarrow \quad -dF_{A_{LR}} = (-r_{A_{LR}})dV$$

$$\Rightarrow \quad -d\left\{(F_{A_{LR}})_0(1 - X_{A_{LR}})\right\} = (-r_{A_{LR}})dV$$

$$\Rightarrow \quad (F_{A_{LR}})_0 dX_{A_{LR}} = (-r_{A_{LR}})dV$$

$$\Rightarrow \quad (F_{A_{LR}})_0 \int_0^{(X_{A_{LR}})_f} \frac{dX_{A_{LR}}}{-r_{A_{LR}}} = \int_0^{V_{PFR}} dV$$

$$\Rightarrow \quad V_{PFR} = (F_{A_{LR}})_0 \int_0^{(X_{A_{LR}})_f} \frac{dX_{A_{LR}}}{-r_{A_{LR}}} \tag{6.1}$$

This is the performance equation of an ideal isothermal PFR operating under the steady-state condition. It relates the volume of the PFR to the conversion achieved in the reactor for a given molar feed rate. The equation can be used for the design of PFR to know the volume of the reactor required if a specified molar feed rate of the reactant is to be converted to a desired extent. Alternatively, it can be used for the performance evaluation of an existing PFR of known volume to determine the conversion that will be achieved for a specified molar feed rate of the reactant. It is evident from Equation 6.1 that as long as the ratio $V_{PFR}/(F_{A_{LR}})_0$ is unaltered the conversion from a PFR remains unchanged. The higher this ratio, the higher is the conversion and vice versa. Non-steady-state reactor operation is important during start-up, shut down, or other process disturbances (Alopaeus et al., 2008).

From Equation 6.1, we have

$$\frac{V_{PFR}}{(F_{A_{LR}})_0} = \int_0^{(X_{A_{LR}})_f} \frac{dX_{A_{LR}}}{-r_{A_{LR}}}$$

$$\Rightarrow \frac{d}{dX_{A_{LR}}}\left[\frac{V_{PFR}}{(F_{A_{LR}})_0}\right] = \frac{1}{-r_{A_{LR}}} \text{ (from basic principles of calculus)} \tag{6.2}$$

If the volume of the PFR is constant, Equation 6.2 would reduce to

$$V_{PFR}\left\{\frac{d}{dX_{A_{LR}}}\left[\frac{1}{(F_{A_{LR}})_0}\right]\right\} = \frac{1}{-r_{A_{LR}}} \tag{6.3}$$

Equations 6.2 and 6.3 are very useful in the experimental study of reactions with unknown kinetics. The reaction for which the kinetics is to be studied is carried out in a PFR of known volume. Different experimental runs are carried out with different molar flow rate of the reactant and the resulting conversion is noted. From this a relation—often in the form of a graph—is obtained between $1/(F_{A_{LR}})_0$ and $X_{A_{LR}}$. From this relation the quantity $\left(d/dX_{A_{LR}}\right)\left[1/(F_{A_{LR}})_0\right]$ is determined for different values of $X_{A_{LR}}$. Multiplying the different values of $\{d/dX_{A_{LR}}[1/(F_{A_{LR}})_0]\}$ so obtained by V_{PFR} one can determine $1/-r_{A_{LR}}$ for different values of $X_{A_{LR}}$. This provides the rate of reaction at different conversion and, therefore, at different concentration.

Equation 6.1 can be rearranged into other convenient forms for ease of application in different situations. Referring to Figure 6.1, $(F_{A_{LR}})_0 = v_0(C_{A_{LR}})_0$; so from Equation 6.1,

$$V_{PFR} = v_0(C_{A_{LR}})_0 \int_0^{(X_{A_{LR}})_f} \frac{dX_{A_{LR}}}{-r_{A_{LR}}}$$

$$\Rightarrow \frac{V_{PFR}}{v_0} = (C_{A_{LR}})_0 \int_0^{(X_{A_{LR}})_f} \frac{dX_{A_{LR}}}{-r_{A_{LR}}} \tag{6.4}$$

Here we define a new quantity space time "τ" as

$$\tau_{PFR} = \frac{V_{PFR}}{v_0} \tag{6.5}$$

In a physical sense, it is the time required to process one reactor volume of feed through the reactor. It is a distinguishing characteristic of reactor operation. Equation 6.4 can be rewritten as

$$\tau_{PFR} = (C_{A_{LR}})_0 \int_0^{(X_{A_{LR}})_f} \frac{dX_{A_{LR}}}{-r_{A_{LR}}} \tag{6.6}$$

Equations 6.4 and 6.6 are alternative performance equations for the PFR. Equation 6.4 relates the volume of the PFR to the conversion achieved in the reactor for a given initial concentration of the reactant and the volumetric feed rate. The equation can be used for the design of PFR to know the volume of the reactor required if the reactant of a given initial concentration being fed to the reactor at a specified volumetric feed rate is to be converted to a desired extent. Alternatively, it can be used for the performance evaluation of an existing PFR of known volume to determine the conversion that will be achieved if the reactant of a given initial concentration is fed to the reactor at a specified volumetric feed rate. Similarly, Equation 6.6 relates the space time of reactor operation to the conversion achieved in the reactor for a given initial concentration of the reactant. It is evident from Equation 6.6 that as long as the ratio $\tau/(C_{A_{LR}})_0$ is unaltered, the conversion from a PFR remains unchanged. The higher this ratio, the higher is the conversion and vice versa.

From Equation 6.6,

$$\frac{\tau_{PFR}}{(C_{A_{LR}})_0} = \int_0^{(X_{A_{LR}})_f} \frac{dX_{A_{LR}}}{-r_{A_{LR}}}$$

$$\Rightarrow \quad \frac{d}{dX_{A_{LR}}}\left[\frac{\tau_{PFR}}{(C_{A_{LR}})_0}\right] = \frac{1}{-r_{A_{LR}}} \quad \text{(from basic principles of calculus)} \tag{6.7}$$

If the initial concentration of the reactant is constant, Equation 6.7 would reduce to

$$\frac{1}{(C_{A_{LR}})_0}\left\{\frac{d\tau_{PFR}}{dX_{A_{LR}}}\right\} = \frac{1}{-r_{A_{LR}}} \tag{6.8}$$

On the other hand, if the space time of the reactor operation is constant, Equation 6.7 would reduce to

$$\tau_{PFR}\left\{\frac{d}{dX_{A_{LR}}}\left[\frac{1}{(C_{A_{LR}})_0}\right]\right\} = \frac{1}{-r_{A_{LR}}} \tag{6.9}$$

Equations 6.7 through 6.9 can be used in a similar fashion as Equations 6.2 and 6.3 for the study of reaction kinetics.

The use of Equations 6.1, 6.4, and 6.6 requires the integral $\int_0^{(X_{A_{LR}})_f} \left(dX_{A_{LR}}/-r_{A_{LR}}\right)$ to be evaluated. If the expression for $-r_{A_{LR}}$ is not so complicated, we may analytically evaluate the integral. If the expression is complicated, numerical or graphical integration is recommended.

The performance equation for the PFR can also be written in terms of the concentration of the limiting reactant. From Equation 2.45 (Chapter 2),

$$C_{A_{LR}} = \frac{(C_{A_{LR}})_0(1-X_{A_{LR}})}{1+\varepsilon_{LR}X_{A_{LR}}}$$

$$\Rightarrow \quad X_{A_{LR}} = \frac{(C_{A_{LR}})_0 - C_{A_{LR}}}{(C_{A_{LR}})_0 + \varepsilon_{LR}C_{A_{LR}}}$$

$$\Rightarrow \quad dX_{A_{LR}} = -\frac{(1+\varepsilon_{LR})(C_{A_{LR}})_0}{\left[(C_{A_{LR}})_0 + \varepsilon_{LR}C_{A_{LR}}\right]^2} dC_{A_{LR}} \tag{6.10}$$

From Equations 6.1 and 6.10,

$$V_{PFR} = -(F_{A_{LR}})_0 \int_{(C_{A_{LR}})_0}^{(C_{A_{LR}})_f} \frac{(1+\varepsilon_{LR})(C_{A_{LR}})_0}{(-r_{A_{LR}})[(C_{A_{LR}})_0 + \varepsilon_{LR}C_{A_{LR}}]^2} dC_{A_{LR}} \tag{6.11}$$

And for reacting systems with no volume change,

$$V_{PFR} = -\frac{(F_{A_{LR}})_0}{(C_{A_{LR}})_0} \int_{(C_{A_{LR}})_0}^{(C_{A_{LR}})_f} \frac{1}{(-r_{A_{LR}})} dC_{A_{LR}} \tag{6.12}$$

Similarly,

$$\tau_{PFR} = -(C_{A_{LR}})_0 \int_{(C_{A_{LR}})_0}^{(C_{A_{LR}})_f} \frac{(1+\varepsilon_{LR})(C_{A_{LR}})_0}{(-r_{A_{LR}})[(C_{A_{LR}})_0 + \varepsilon_{LR}C_{A_{LR}}]^2} dC_{A_{LR}} \tag{6.13}$$

And for reacting systems with no volume change,

$$\tau_{PFR} = -\int_{(C_{A_{LR}})_0}^{(C_{A_{LR}})_f} \frac{1}{(-r_{A_{LR}})} dC_{A_{LR}} \tag{6.14}$$

6.1.1 Single-Reactant nth-Order Reaction in a PFR

From Equations 5.11 (Chapter 5) and 2.45 (Chapter 2),

$$-r_A = kC_A^n = k\frac{(C_A)_0^n(1-X_A)^n}{(1+\varepsilon_A X_A)^n} \tag{6.15}$$

From Equations 6.1 and 6.15,

$$V_{PFR} = (F_A)_0 \int_0^{(X_A)_f} \frac{(1+\varepsilon_A X_A)^n}{k(C_A)_0^n(1-X_A)^n} dX_A$$

$$\Rightarrow \quad V_{PFR} = \frac{(F_A)_0}{k(C_A)_0^n} \int_0^{(X_A)_f} \frac{(1+\varepsilon_A X_A)^n}{(1-X_A)^n} dX_A \tag{6.16}$$

For liquid-phase reactions, and the gas-phase reactions where the stoichiometry is such that the reacting system neither expands nor contracts, $\varepsilon_A = 0$. Equation 6.16 then reduces to

$$V_{PFR} = \frac{(F_A)_0}{k(C_A)_0^n} \int_0^{(X_A)_f} \frac{dX_A}{(1-X_A)^n}$$

$$\Rightarrow \quad V_{PFR} = \frac{(F_A)_0}{k(C_A)_0^n(n-1)}[(1-(X_A)_f)^{1-n} - 1] \quad \text{for } n \neq 1 \tag{6.17}$$

For a first-order reaction, we have

$$V_{PFR} = \frac{(F_A)_0}{k(C_A)_0} \int_0^{(X_A)_f} \frac{dX_A}{(1-X_A)}$$

$$\Rightarrow \quad V_{PFR} = -\frac{(F_A)_0}{k(C_A)_0} \ln(1-(X_A)_f) \tag{6.18}$$

Similarly, from Equations 6.6 and 6.15,

$$\tau_{PFR} = (C_A)_0 \int_0^{(X_A)_f} \frac{(1+\varepsilon_A X_A)^n}{k(C_A)_0^n(1-X_A)^n} dX_A$$

$$\Rightarrow \quad \tau_{PFR} = \frac{1}{k(C_A)_0^{n-1}} \int_0^{(X_A)_f} \frac{(1+\varepsilon_A X_A)^n}{(1-X_A)^n} dX_A \tag{6.19}$$

$$\Rightarrow \quad k\tau_{PFR} = \frac{1}{(C_A)_0^{n-1}} \int_0^{(X_A)_f} \frac{(1+\varepsilon_A X_A)^n}{(1-X_A)^n} dX_A \tag{6.20}$$

For liquid-phase reactions, and the gas-phase reactions where the stoichiometry is such that the reacting system neither expands nor contracts, $\varepsilon_A = 0$. Equation 6.19 then reduces to

$$\tau_{PFR} = \frac{1}{k(C_A)_0^{n-1}} \int_0^{(X_A)_f} \frac{dX_A}{(1-X_A)^n}$$

$$\Rightarrow \quad \tau_{PFR} = \frac{1}{k(C_A)_0^{n-1}(n-1)}[(1-(X_A)_f)^{1-n} - 1] \quad \text{for } n \neq 1 \tag{6.21}$$

$$\Rightarrow \quad k\tau_{PFR} = \frac{1}{(C_A)_0^{n-1}(n-1)}[(1-(X_A)_f)^{1-n} - 1] \tag{6.22}$$

For a first-order reaction, we have

$$\tau_{PFR} = \frac{1}{k} \int_0^{(X_A)_f} \frac{dX_A}{(1-X_A)}$$

$$\Rightarrow \quad \tau_{PFR} = -\frac{1}{k}\ln(1-(X_A)_f) \tag{6.23}$$

$$\Rightarrow k\tau_{PFR} = -\ln(1-(X_A)_f) \tag{6.24}$$

We can see that the quantity $k\tau_{PFR}$ for a first-order reaction is uniquely related to the conversion in a PFR.

EXAMPLE 6.1

For a reaction, the reaction rate at different concentrations is measured in a batch reactor and reported below.

Concentration (mol/L)	52	41	34	29	25	22	20	19	18.5
Reaction rate (mol/L/min)	6.5	3.64	2.92	2.18	1.56	0.89	0.58	0.52	0.5

The reaction is to be carried out in an isothermal PFR to process 1000 mol/min of feed of A to a conversion of 50%. Calculate the volume of PFR needed for the purpose.

Solution

From the given data we prepare the Table EXE6.1

From Equation 6.1,

TABLE EXE6.1
Calculation Table for Graphical Integration

Concentration (mol/L)	Reaction Rate (mol/(L · min))	Conversion	1/Rate
52	6.5	0	0.153846
41	3.64	0.21	0.274725
34	2.92	0.35	0.342466
29	2.18	0.44	0.458716
25	1.56	0.52	0.641026
22	0.89	0.58	1.123596
20	0.58	0.62	1.724138
19	0.52	0.64	1.923077
18.5	0.5	0.644	2

$$V_{PFR} = (F_A)_0 \int_0^{(X_A)_f} \frac{dX_A}{-r_A}$$

$$\Rightarrow \quad V_{PFR} = 1000 \int_0^{0.5} \frac{dX_A}{-r_A} \tag{6.1.1}$$

In order to estimate the integral $\int_0^{0.5} (dX_A/-r_A)$ we plot a graph between conversion "X_A" and the inverse of reaction rate "$1/-r_A$" as shown in Figure EXE6.1.

From the properties of integral the value of the desired integral is the area ABCD. From the graph this area is $76 \times 0.002 = 0.152$ L/(min · mol). So, from Equation 6.1.1

$$V_{PFR} = 1000 \times 0.152 = 152 \text{ L}$$

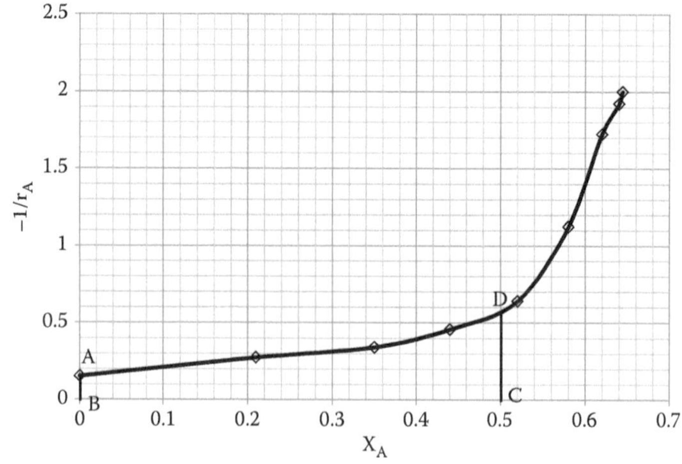

FIGURE EXE6.1 Graphical integration for PFR volume.

EXAMPLE 6.2

The second-order gaseous reaction

$$A \rightarrow B + C$$

is to be carried out in an isothermal PFR at 27°C and 1 atm pressure with a feed of pure A. The specific reaction rate constant is 0.2 L/(mol · min). Calculate the space time of the reactor required to get 90% conversion.

Solution

Since pure A is used as feed, $(x_A)_0 = 1$, so from Equation 2.27 (Chapter 2) $(p_A)_0 = 1$ atm

From Equation 2.26 (Chapter 2),

$$(C_A)_0 = \frac{(p_A)_0}{RT} = \frac{1}{0.082 \times 300} = 0.041 \text{ mol/L}$$

From Equation 2.44 (Chapter 2),

$$\varepsilon_A = \frac{(N_A)_0\left[\sum_{j=1}^{l}(p_j/a_{LR}) - \sum_{i=1}^{n}(a_i/a_{LR})\right]}{(N_A)_0 + N_{inerts}} = 1 + 1 - 1 = 1$$

From Equation 6.19,

$$\tau = \frac{1}{k(C_A)_0}\int_0^{0.9}\frac{(1+X_A)^2}{(1-X_A)^2}dX_A$$

$$\Rightarrow \quad \tau = \frac{1}{0.2 \times 0.041}\left[\frac{4}{1-X_A} + 4\ln(1-X_A) - (1-X_A)\right]_0^{0.9}$$

$$\Rightarrow \quad \tau = \frac{1}{0.2 \times 0.041}\left[\frac{4}{0.1} + 4\ln 0.1 - 0.1 - 4 - 0 + 1\right]$$

$$\Rightarrow \quad \tau = \frac{1}{0.082}[40 - 9.21 - 0.1 - 4 - 0 + 1] = 337.7 \text{ min}$$

In terms of reactant concentration, from Equations 6.11 and 5.11 (Chapter 5),

$$V_{PFR} = -(F_A)_0 \int_{(C_A)_0}^{(C_A)_f} \frac{(1+\varepsilon_A)(C_A)_0}{kC_A^n[(C_A)_0 + \varepsilon_A C_A]^2} dC_A \tag{6.25}$$

and from Equations 6.13 and 5.11 (Chapter 5),

$$\tau_{PFR} = -(C_A)_0 \int_{(C_A)_0}^{(C_A)_f} \frac{(1+\varepsilon_A)(C_A)_0}{kC_A^n[(C_A)_0 + \varepsilon_A C_A]^2} dC_A \tag{6.26}$$

For liquid-phase reactions, and the gas-phase reactions where the stoichiometry is such that the reacting system neither expands nor contracts, $\varepsilon_A = 0$. Equation 6.25 then reduces to

$$V_{PFR} = -(F_A)_0 \int_{(C_A)_0}^{(C_A)_f} \frac{1}{k(C_A)_0 C_A^n} dC_A$$

$$\Rightarrow \quad V_{PFR} = \frac{(F_A)_0}{k(C_A)_0(n-1)}[(C_A)_f^{1-n} - (C_A)_0^{1-n}] \quad \text{for } n \neq 1 \tag{6.27}$$

It may be noted that Equation 6.26 can be directly deduced from Equation 6.17 by combining with Equation 2.25 (Chapter 2). Similarly, from Equation 6.18 in combination with Equation 2.25 (Chapter 2) for a first-order reaction

$$V_{PFR} = \frac{(F_A)_0}{k(C_A)_0} \ln \frac{(C_A)_0}{C_A} \tag{6.28}$$

From Equations 6.21 and 2.25 (Chapter 2),

$$\tau_{PFR} = \frac{1}{k(n-1)}[(C_A)_f^{1-n} - (C_A)_0^{1-n}] \quad \text{for } n \neq 1 \tag{6.29}$$

$$\Rightarrow \quad k\tau_{PFR} = \frac{1}{(n-1)}[(C_A)_f^{1-n} - (C_A)_0^{1-n}] \tag{6.30}$$

And, for a first-order reaction from Equation 6.28,

$$\tau_{PFR} = \frac{1}{k} \ln \frac{(C_A)_0}{C_A} \tag{6.31}$$

$$\Rightarrow \quad k\tau_{PFR} = \ln \frac{(C_A)_0}{C_A} \tag{6.32}$$

$$\Rightarrow \quad C_A = (C_A)_0 e^{-k\tau_{PFR}} \tag{6.33}$$

Here, we can see that Equations 6.29 and 6.30 for a PFR for constant volume-reacting system are identical to Equation 5.16 (Chapter 5) for a constant volume batch reactor with reaction time "t" for a batch reactor being replaced by space time "τ" for PFR. In the same way, Equations 6.32 and 6.33 for PFR are identical to Equations 5.19 and 5.21 (Chapter 5) for batch reactor with "t" replaced with "τ." Thus, the space time for a PFR has the same significance as the reaction time for a batch reactor. For constant volume-reacting systems, the performance equations for a batch reactor can also be used for PFR if "t" is replaced with "τ" in the corresponding equation.

6.1.2 General Reaction Involving Several Reactants

For the general reaction represented by the balanced chemical Equation 5.24 (Chapter 5), we may combine Equation 6.11 with Equation 5.28 (Chapter 5) to have

$$V_{PFR} = -\left(F_{A_{LR}}\right)_0 \int_{(C_{A_{LR}})_0}^{(C_{A_{LR}})_f} \frac{(1+\varepsilon_{LR})\left(C_{A_{LR}}\right)_0 \left[\left(C_{A_{LR}}\right)_0 + \varepsilon_{LR} C_{A_{LR}}\right]^2}{k\prod_{i=1}^{n}\left[\left(C_{A_i}\right)_0 - (a_i/a_{LR})\left\{\left(C_{A_{LR}}\right)_0 - C_{A_{LR}}\right\}\right]^{a_i}} dC_{A_{LR}} \quad (6.34)$$

And, combining Equation 6.13 with Equation 5.28 (Chapter 5), we have

$$\tau_{PFR} = -\left(C_{A_{LR}}\right)_0 \int_{(C_{A_{LR}})_0}^{(C_{A_{LR}})_f} \frac{(1+\varepsilon_{LR})\left(C_{A_{LR}}\right)_0 \left[\left(C_{A_{LR}}\right)_0 + \varepsilon_{LR} C_{A_{LR}}\right]^2}{k\prod_{i=1}^{n}\left[\left(C_{A_i}\right)_0 - (a_i/a_{LR})\left\{\left(C_{A_{LR}}\right)_0 - C_{A_{LR}}\right\}\right]^{a_i}} dC_{A_{LR}} \quad (6.35)$$

For the special case when $(C_{A_i})_0 = (a_i/a_{LR})(C_{A_{LR}})_0$ Equations 6.34 and 6.35 reduce to

$$V_{PFR} = -\left(F_{A_{LR}}\right)_0 \int_{(C_{A_{LR}})_0}^{(C_{A_{LR}})_f} \frac{(1+\varepsilon_{LR})\left(C_{A_{LR}}\right)_0}{k\prod_{i=1}^{n}(a_i/a_{LR})^{a_i}\left(C_{A_{LR}}\right)^{a_i}\left[\left(C_{A_{LR}}\right)_0 + \varepsilon_{LR} C_{A_{LR}}\right]^2} dC_{A_{LR}} \quad (6.36)$$

and

$$\tau_{PFR} = -\left(C_{A_{LR}}\right)_0 \int_{(C_{A_{LR}})_0}^{(C_{A_{LR}})_f} \frac{(1+\varepsilon_{LR})\left(C_{A_{LR}}\right)_0}{k\prod_{i=1}^{n}(a_i/a_{LR})^{a_i}\left(C_{A_{LR}}\right)^{a_i}\left[\left(C_{A_{LR}}\right)_0 + \varepsilon_{LR} C_{A_{LR}}\right]^2} dC_{A_{LR}} \quad (6.37)$$

For liquid-phase reactions, and the gas-phase reactions where the stoichiometry is such that the reacting system neither expands nor contracts, $\varepsilon = 0$. Equations 6.34 and 6.35 then reduce to

$$V_{PFR} = -(F_{A_{LR}})_0 \int_{(C_{A_{LR}})_0}^{(C_{A_{LR}})_f} \frac{1}{k(C_{A_{LR}})_0 \prod_{i=1}^{n} \left[(C_{A_i})_0 - (a_i/a_{LR})\{(C_{A_{LR}})_0 - C_{A_{LR}}\}\right]^{a_i}} dC_{A_{LR}} \tag{6.38}$$

and

$$\tau_{PFR} = -\frac{1}{k} \int_{(C_{A_{LR}})_0}^{(C_{A_{LR}})_f} \frac{1}{\prod_{i=1}^{n} \left[(C_{A_i})_0 - (a_i/a_{LR})\{(C_{A_{LR}})_0 - C_{A_{LR}}\}\right]^{a_i}} dC_{A_{LR}} \tag{6.39}$$

$$\Rightarrow \quad k\tau_{PFR} = -\int_{(C_{A_{LR}})_0}^{(C_{A_{LR}})_f} \frac{1}{\prod_{i=1}^{n} \left[(C_{A_i})_0 - (a_i/a_{LR})\{(C_{A_{LR}})_0 - C_{A_{LR}}\}\right]^{a_i}} dC_{A_{LR}} \tag{6.40}$$

Similarly, Equations 6.36 and 6.37 reduce to

$$V_{PFR} = -(F_{A_{LR}})_0 \int_{(C_{A_{LR}})_0}^{(C_{A_{LR}})_f} \frac{1}{k(C_{A_{LR}})_0 \prod_{i=1}^{n} (a_i/a_{LR})^{a_i} (C_{A_{LR}})^{a_i}} dC_{A_{LR}} \tag{6.41}$$

and

$$\tau_{PFR} = -\frac{1}{k} \int_{(C_{A_{LR}})_0}^{(C_{A_{LR}})_f} \frac{1}{\prod_{i=1}^{n} (a_i/a_{LR})^{a_i} (C_{A_{LR}})^{a_i}} dC_{A_{LR}} \tag{6.42}$$

$$\Rightarrow \quad k\tau_{PFR} = -\int_{(C_{A_{LR}})_0}^{(C_{A_{LR}})_f} \frac{1}{\prod_{i=1}^{n} (a_i/a_{LR})^{a_i} (C_{A_{LR}})^{a_i}} dC_{A_{LR}} \tag{6.43}$$

The general algorithm for engineering calculations can be summarized as follows:

Step 1

Write down the rate equation in terms of the concentrations of the reactants. The rate equation can be of the form of Equation 5.25 (Chapter 5) or any other form as determined from the experiment.

Step 2

From reaction stoichiometry express the concentration of each reactant in terms of the concentration of the limiting reactant.

Step 3

Substitute the concentration terms of each reactant in the rate equation written in Step 1 with the equivalent expression in terms of the limiting reactant as formulated in Step 2. Now, we have the rate equation in terms of the concentration of the limiting reactant alone.

Step 4

Use the rate equation obtained in Step 3 in the performance equation of PFR to determine the desired quantities.

EXAMPLE 6.3

The gas-phase irreversible reaction

$$A + B \rightarrow C$$

follows the law of mass action with a specific reaction rate constant of 0.3 L/(min · mol). The reaction is to be carried out in an isothermal PFR at 500°C and 2 atm pressure. The feed to the reactor consists of a gas mixture consisting of 1 volume A and 1 volume B at a total feed rate of 20 L/min. What should be the volume of the PFR to achieve a conversion of 80%?

Solution

Since the feed consists of 1 volume A and 1 volume B, we have $(C_B)_0 = (a_B/a_A)(C_A)_0$ and Equation 6.36 is applicable.

Mole fraction of A in feed, $(x_A)_0 = (1/(1+1)) = 0.5$

From Equations 2.26 and 2.27 (Chapter 2),

$$(C_A)_0 = \frac{(p_A)_0}{RT} = \frac{(x_A)_0(p)_0}{RT} = \frac{1 \times 2}{2 \times 0.082 \times 773} = 0.0158 \frac{\text{mol}}{\text{L}}$$

From Equation 2.44 (Chapter 2),

$$\varepsilon_A = \frac{1-(1+1)}{1+1} = -0.5$$

From Equation 2.45 (Chapter 2),

$$(C_A)_f = \frac{(C_A)_0(1-X_A)}{1+\varepsilon_A X_A} = \frac{0.0158 \times (1-0.8)}{1-(0.8/2)} = 0.0053 \frac{\text{mol}}{\text{L}}$$

$$(F_A)_0 = (C_A)_0 \times v_0 = 0.0158 \times 20 = 0.316 \text{ mol/min}$$

From Equation 6.36,

$$V_{PFR} = -(F_A)_0 \int_{(C_A)_0}^{(C_A)_f} \frac{(1+\varepsilon_A)(C_A)_0}{k\prod_{i=1}^{n}(a_i/a_A)^{a_i}(C_A)^{a_i}[(C_A)_0+\varepsilon_A C_A]^2} dC_A$$

$$\Rightarrow \quad V_{PFR} = -0.316 \int_{0.0158}^{0.0053} \frac{0.5\times 0.0158}{0.3\times C_A^2[0.0158-(C_A/2)]^2} dC_A$$

$$\Rightarrow \quad V_{PFR} = 0.033 \int_{0.0053}^{0.0158} \frac{1}{C_A^2[0.0316-C_A]^2} dC_A \qquad (6.3.1)$$

Next, we numerically evaluate the integral $\int_{0.0053}^{0.0158}(1/(C_A^2[0.0316-C_A]^2))dC_A$. For this purpose, we prepare the following table (Table EXE6.3):

By Simpson's one-third rule,

$$\int_{0.0053}^{0.0158} \frac{1}{C_A^2[0.0316-C_A]^2} dC_A = \frac{0.0015}{3}[51{,}467{,}937.37+16{,}046{,}179.72$$

$$+4(35{,}162{,}424.52+21{,}909{,}620.82+17{,}268{,}887.58)+$$

$$2(26{,}738{,}188.04+19{,}004{,}263.23+16{,}339{,}385.48)]$$

$$\Rightarrow \quad \int_{0.0053}^{0.0158} \frac{1}{C_A^2[0.0316-C_A]^2} dC_A = 244{,}520.76$$

From Equation 6.3.1,

$$V_{PFR} = 0.033\times 244{,}520.76 = 8069.2 \text{ L}$$

TABLE EXE6.3
Calculation for Numerical Integration

C_A	$\frac{1}{C_A^2[0.0316-C_A]^2}$
0.0053	51,467,937.37
0.0068	35,162,424.52
0.0083	26,738,188.04
0.0098	21,909,620.82
0.0113	19,004,263.23
0.0128	17,268,887.58
0.0143	16,339,385.48
0.0158	16,046,179.72

6.1.3 Reversible Reactions

For constant volume-reacting systems, consider the general reversible reaction represented by the balanced chemical Equation 5.33 (Chapter 5). For this reaction, from Equations 5.34 (Chapter 5) and 6.1,

$$V_{PFR} = \left(F_{A_{LR}}\right)_0 \int_0^{(X_{A_{LR}})_f} \frac{dX_{A_{LR}}}{k_1\left(C_{A_{LR}}\right)_0^{\sum_{i=1}^{n} a_i} \prod_{i=1}^{n}\left[\left(\frac{\left(C_{A_i}\right)_0}{\left(C_{A_{LR}}\right)_0}\right)-\left(\frac{a_i}{a_{LR}}\right)X_{A_{LR}}\right]^{a_i} - k_2\left(C_{A_{LR}}\right)_0^{\sum_{j=1}^{1} p_j} \prod_{j=1}^{1}\left[\left(\frac{\left(C_{P_j}\right)_0}{\left(C_{A_{LR}}\right)_0}\right)+\left(\frac{p_j}{a_{LR}}\right)X_{A_{LR}}\right]^{p_j}} \tag{6.44}$$

If we integrate the RHS of Equation 6.44 between the limits 0 and $\left(X_{A_{LR}}\right)_{eq}$ we can get the volume of the PFR required to achieve equilibrium. $\left(X_{A_{LR}}\right)_{eq}$ is related to the equilibrium constant by Equations 4.21 and 4.22 (Chapter 4) for constant volume systems. Here, numerical or graphical integration is recommended for systems consisting of many components with large stoichiometric coefficients.

From Equations 5.34 (Chapter 5) and 6.6,

$$\tau_{PFR} = \left(C_{A_{LR}}\right)_0 \int_0^{(X_{A_{LR}})_f} \frac{dX_{A_{LR}}}{k_1\left(C_{A_{LR}}\right)_0^{\sum_{i=1}^{n} a_i} \prod_{i=1}^{n}\left[\left(\frac{\left(C_{A_i}\right)_0}{\left(C_{A_{LR}}\right)_0}\right)-\left(\frac{a_i}{a_{LR}}\right)X_{A_{LR}}\right]^{a_i} - k_2\left(C_{A_{LR}}\right)_0^{\sum_{j=1}^{1} p_j} \prod_{j=1}^{1}\left[\left(\frac{\left(C_{P_j}\right)_0}{\left(C_{A_{LR}}\right)_0}\right)+\left(\frac{p_j}{a_{LR}}\right)X_{A_{LR}}\right]^{p_j}} \tag{6.45}$$

If we integrate the RHS of Equation 6.45 between the limits 0 and $\left(X_{A_{LR}}\right)_{eq}$ we can get the space time of the PFR required to achieve equilibrium.

EXAMPLE 6.4

The liquid-phase reversible reaction

$$A \rightleftharpoons B$$

with $k_1 = 0.05\ \text{min}^{-1}$ and $k_2 = 0.02\ \text{min}^{-1}$, is carried out in a PFR with a total volumetric feed rate of 100 L/min with initial concentrations of A and B as 1 and 0.1 mol/L, respectively. Determine the volume of PFR required to achieve equilibrium.

Solution

For this reaction from Example 5.4 (Chapter 5),

$$(X_A)_{eq} = 0.68$$

$$(F_A)_0 = v_0(C_A)_0 = 100 \times 1 = 100 \text{ mol/min}$$

From Equation 6.44,

$$V_{PFR} = 100 \int_0^{(X_A)_{eq}} \frac{dX_A}{k_1(C_A)_0(1-X_A) - k_2(C_A)_0[((C_B)_0/(C_A)_0) + X_A]}$$

$$\Rightarrow \quad V_{PFR} = 100 \int_0^{0.68} \frac{dX_A}{0.05(1-X_A) - 0.02[0.1+X_A]}$$

$$\Rightarrow \quad V_{PFR} = \frac{100}{(0.02-0.05)}[\ln\{0.05 \times 0.32 - 0.02 \times 0.78\} - \ln\{0.05 - 0.002\}]$$

$$\Rightarrow \quad V_{PFR} = 159.33 \text{ L}$$

6.1.4 Multiple Reactions

6.1.4.1 Series Reactions

Consider the series reaction

$$A \rightarrow I \rightarrow T$$

with the specific reaction rate constants for the two consecutive reaction steps being k_1 and k_2, respectively. If this reaction is carried out in a PFR, then from Equations 5.43 (Chapter 5) and 6.14, for species A

$$\tau_{PFR} = -\int_{(C_A)_0}^{(C_A)_f} \frac{1}{k_1 C_A} dC_A$$

$$\Rightarrow \quad C_A = (C_A)_0 e^{-k_1 \tau_{PFR}} \tag{6.46}$$

$$\Rightarrow \quad \frac{C_A}{(C_A)_0} = e^{-k_1 \tau_{PFR}} \tag{6.47}$$

For species I, from Equations 5.44 (Chapter 5), 6.46, and 6.14

$$\tau_{PFR} = \int_{(C_I)_0}^{(C_I)_f} \frac{1}{k_1(C_A)_0 e^{-k_1 \tau_{PFR}} - k_2 C_I} dC_I$$

$$\Rightarrow \quad d\tau_{PFR} = \frac{1}{k_1(C_A)_0 e^{-k_1\tau_{PFR}} - k_2C_I} dC_I$$

$$\Rightarrow \quad \frac{dC_I}{d\tau_{PFR}} = k_1(C_A)_0 e^{-k_1\tau_{PFR}} - k_2C_I$$

$$\Rightarrow \quad \frac{dC_I}{d\tau_{PFR}} + k_2C_I = k_1(C_A)_0 e^{-k_1\tau_{PFR}} \tag{6.48}$$

This equation is similar to Equation 5.48 (Chapter 5) with the variable "t" replaced with the variable "τ_{PFR}." The solution to this equation will also be similar to that for Equation 5.48 (Chapter 5). Therefore, we have

$$C_I = \frac{k_1(C_A)_0}{(k_2 - k_1)}[e^{-k_1\tau_{PFR}} - e^{-k_2\tau_{PFR}}] + (C_I)_0 e^{-k_2\tau_{PFR}} \tag{6.49}$$

$$\Rightarrow \quad \frac{C_I}{(C_A)_0} = \frac{k_1}{(k_2 - k_1)}[e^{-k_1\tau_{PFR}} - e^{-k_2\tau_{PFR}}] + \frac{(C_I)_0}{(C_A)_0} e^{-k_2\tau_{PFR}} \tag{6.50}$$

In analogy with Equation 5.57 (Chapter 5) the dimensionless concentration $C_I/(C_A)_0$ will be maximum when the space time is

$$(\tau_{PFR})_{max} = \frac{1}{(k_2 - k_1)} \ln\left\{\frac{k_2}{k_1}\left[1 - \frac{(k_2 - k_1)(C_I)_0}{k_1(C_A)_0}\right]\right\} \tag{6.51}$$

And the maximum value of the dimensionless concentration $C_I/(C_A)_0$ will be

$$\left(\frac{C_I}{(C_A)_0}\right)_{max} = \frac{k_1}{(k_2 - k_1)}\left[\left\{\frac{k_2}{k_1}\left[1 - \frac{(k_2 - k_1)(C_I)_0}{k_1(C_A)_0}\right]\right\}^{-k_1/(k_2-k_1)} - \left\{\frac{k_2}{k_1}\left[1 - \frac{(k_2 - k_1)(C_I)_0}{k_1(C_A)_0}\right]\right\}^{-k_2/(k_2-k_1)}\right] + \frac{(C_I)_0}{(C_A)_0}\left\{\frac{k_2}{k_1}\left[1 - \frac{(k_2 - k_1)(C_I)_0}{k_1(C_A)_0}\right]\right\}^{-k_2/(k_2-k_1)} \tag{6.52}$$

Here, we can clearly see the similarity between Equations 6.51 and 5.57 (Chapter 5) and that between Equations 6.52 and 5.58 (Chapter 5) with t_{max} being replaced by $(\tau_{PFR})_{max}$. This similarity holds good for the special cases when the species I and T are not present initially and when $k_1 = k_2 = k$.

The outgoing concentration of the species T is related to the space time of PFR as

$$C_T = (C_A)_0\left\{1 - e^{-k_1\tau_{PFR}} - \frac{k_1}{(k_2 - k_1)}[e^{-k_1\tau_{PFR}} - e^{-k_2\tau_{PFR}}]\right\} + (C_I)_0(1 - e^{-k_1\tau_{PFR}}) + (C_T)_0 \quad (6.53)$$

$$\Rightarrow \frac{C_T}{(C_A)_0} = \left\{1 - e^{-k_1\tau_{PFR}} - \frac{k_1}{(k_2 - k_1)}[e^{-k_1\tau_{PFR}} - e^{-k_2\tau_{PFR}}]\right\} + \frac{(C_I)_0}{(C_A)_0}(1 - e^{-k_1\tau_{PFR}}) + \frac{(C_T)_0}{(C_A)_0} \quad (6.54)$$

In terms of the volume of the PFR, Equation 6.47 changes to

$$\frac{C_A}{(C_A)_0} = e^{-k_1[V_{PFR}(C_A)_0/(F_A)_0]} \quad (6.55)$$

Equation 6.50 changes to

$$\frac{C_I}{(C_A)_0} = \frac{k_1}{(k_2 - k_1)}\left[e^{-k_1[V_{PFR}(C_A)_0/(F_A)_0]} - e^{-k_2[V_{PFR}(C_A)_0/(F_A)_0]}\right] + \frac{(C_I)_0}{(C_A)_0}e^{-k_2[V_{PFR}(C_A)_0/(F_A)_0]} \quad (6.56)$$

and Equation 6.54 changes to

$$\frac{C_T}{(C_A)_0} = \left\{1 - e^{-k_1[V_{PFR}(C_A)_0/(F_A)_0]} - \frac{k_1}{(k_2 - k_1)}\left[e^{-k_1[V_{PFR}(C_A)_0/(F_A)_0]} - e^{-k_2[V_{PFR}(C_A)_0/(F_A)_0]}\right]\right\} + \frac{(C_I)_0}{(C_A)_0}(1 - e^{-k_1[V_{PFR}(C_A)_0/(F_A)_0]}) + \frac{(C_T)_0}{(C_A)_0} \quad (6.57)$$

The volume of PFR for which the concentration of I is maximum is given by

$$(V_{PFR})_{max} = \frac{(F_A)_0}{(C_A)_0(k_2 - k_1)}\ln\left\{\frac{k_2}{k_1}\left[1 - \frac{(k_2 - k_1)(C_I)_0}{k_1(C_A)_0}\right]\right\} \quad (6.58)$$

In the special case when the species I and T are not present initially, $(C_I)_0 = (C_T)_0 = 0$, and Equations 6.50 and 6.54 reduce to

$$\frac{C_I}{(C_A)_0} = \frac{k_1}{(k_2 - k_1)}[e^{-k_1\tau_{PFR}} - e^{-k_2\tau_{PFR}}] \quad (6.59)$$

$$\Rightarrow \quad \frac{C_I}{(C_A)_0} = \frac{k_1}{(k_2 - k_1)}\left[e^{-k_1[V_{PFR}(C_A)_0/(F_A)_0]} - e^{-k_2[V_{PFR}(C_A)_0/(F_A)_0]}\right] \tag{6.60}$$

and

$$\frac{C_T}{(C_A)_0} = \left\{1 - e^{-k_1\tau_{PFR}} - \frac{k_1}{(k_2 - k_1)}[e^{-k_1\tau_{PFR}} - e^{-k_2\tau_{PFR}}]\right\} \tag{6.61}$$

$$\Rightarrow \quad \frac{C_T}{(C_A)_0} = \left\{1 - e^{-k_1[V_{PFR}(C_A)_0/(F_A)_0]} - \frac{k_1}{(k_2 - k_1)}\left[e^{-k_1[V_{PFR}(C_A)_0/(F_A)_0]} - e^{-k_2[V_{PFR}(C_A)_0/(F_A)_0]}\right]\right\} \tag{6.62}$$

EXAMPLE 6.5

The reaction

$$A \rightarrow I \rightarrow T$$

with $k_1 = 0.05\ \text{min}^{-1}$ and $k_2 = 0.02\ \text{min}^{-1}$, respectively, is carried out in an isothermal PFR with an initial concentration of A equal to 2.5 mol/L and volumetric feed rate of 100 L/min. What is the volume of PFR required to get the maximum concentration of I? What is the composition of the outlet stream from the reactor under this condition?

Solution

From the problem statement,

$$(C_A)_0 = 2.5\ \text{mol/L}, \quad (C_I)_0 = (C_T)_0 = 0, \quad v_0 = 100\ \text{L/min}$$

From Equation 6.58,

$$(V_{PFR})_{max} = \frac{(F_A)_0 \ln(k_2/k_1)}{(C_A)_0(k_2 - k_1)}$$

$$\Rightarrow \quad (V_{PFR})_{max} = \frac{v_0 \ln(k_2/k_1)}{(k_2 - k_1)} = \frac{100 \times \ln(0.02/0.05)}{(0.02 - 0.05)} = 30.54\ \text{L}$$

From Equation 6.52,

$$\left(\frac{C_I}{(C_A)_0}\right)_{max} = \left(\frac{k_1}{k_2}\right)^{k_2/(k_2-k_1)} = \left(\frac{0.05}{0.02}\right)^{0.02/(0.02-0.05)} = 0.543$$

$$\Rightarrow \quad (C_I)_{max} = (C_A)_0 \times 0.543 = 1.358\ \text{mol/L}$$

From Equation 6.55,

$$\frac{C_A}{(C_A)_0} = e^{-k_1[(V_{PFR})_{max}(C_A)_0/(F_A)_0]} = e^{-k_1[(V_{PFR})_{max}/v_0]}$$

$$\Rightarrow \quad C_A = 2.5 \times e^{-0.05\times(3054/100)} = 0.543 \text{ mol/L}$$

$$C_T = (C_A)_0 + (C_I)_0 + (C_T)_0 - C_A - C_I = 2.5 - 0.543 - 1.358 = 0.599 \text{ mol/L}$$

6.1.4.2 Parallel Reactions

Consider the parallel reactions

$$\sum_{i=1}^{n} a_i A_i \rightarrow \sum_{j=1}^{l} p_j P_j \tag{6.63}$$

$$\sum_{i=1}^{n} a_i A_i \rightarrow \sum_{r=1}^{r} q_r Q_r \tag{6.64}$$

If, for the first reaction the reaction rate is $\left(-r_{A_{LR}}\right)_1$, and for the second reaction the rate is $\left(-r_{A_{LR}}\right)_2$, then the overall rate of reaction with respect to the limiting reactant is

$$-r_{A_{LR}} = \left(-r_{A_{LR}}\right)_1 + \left(-r_{A_{LR}}\right)_2 \tag{6.65}$$

From Equation 6.1,

$$V_{PFR} = \left(F_{A_{LR}}\right)_0 \int_0^{(X_{A_{LR}})_f} \frac{dX_{A_{LR}}}{-r_{A_{LR}}}$$

$$\Rightarrow \quad V_{PFR} = \left(F_{A_{LR}}\right)_0 \int_0^{(X_{A_{LR}})_f} \frac{dX_{A_{LR}}}{\left(-r_{A_{LR}}\right)_1 + \left(-r_{A_{LR}}\right)_2} \tag{6.66}$$

Alternatively, from Equation 6.6,

$$\tau_{PFR} = \left(C_{A_{LR}}\right)_0 \int_0^{(X_{A_{LR}})_f} \frac{dX_{A_{LR}}}{-r_{A_{LR}}}$$

$$\Rightarrow \quad \tau_{PFR} = \left(C_{A_{LR}}\right)_0 \int_0^{(X_{A_{LR}})_f} \frac{dX_{A_{LR}}}{\left(-r_{A_{LR}}\right)_1 + \left(-r_{A_{LR}}\right)_2} \tag{6.67}$$

Equations 6.66 and 6.67 relate the total conversion of the limiting reactant to the volume of the PFR and the space time, respectively. Once the conversion of the limiting reactant is known, its outlet concentration can be determined from Equation 2.45 (Chapter 2).

For the determination of the outlet concentration of any of the product chemical species, we make use of the concept of fractional yield. For the product species P_j from Equation 5.84 (Chapter 5),

$$\frac{\left(C_{P_j}\right)_f-\left(C_{P_j}\right)_0}{\left(C_{A_{LR}}\right)_0-\left(C_{A_{LR}}\right)_f}=\frac{-1}{\left(C_{A_{LR}}\right)_0-\left(C_{A_{LR}}\right)_f}\int_{(C_{A_{LR}})_0}^{(C_{A_{LR}})_f}\Psi_{inst}d\left(C_{A_{LR}}\right) \quad (6.68)$$

And, for the reacting system under consideration

$$\frac{C_{P_j}-\left(C_{P_j}\right)_0}{\left(C_{A_{LR}}\right)_0-C_{A_{LR}}}=\frac{(p_j/a_{LR})k_1}{(k_1+k_2)} \quad (6.69)$$

The steps used in engineering calculations are explained next.

If the outlet composition is to be determined for a given reactor volume or space time:

Step 1
From either Equation 6.66 or Equation 6.67, determine the total conversion of the limiting reactant achieved in the reactor.

Step 2
Determine the outlet concentration of the limiting reactant using Equation 2.45 (Chapter 2).

Step 3
Determine the outlet concentration of the different product species using Equation 6.68. For each product species, the corresponding Ψ_{inst} should be calculated using Equation 5.83 (Chapter 5).

If the reactor volume or space time required for a given outlet concentration of desired product is to be determined:

Step 1
Using Equation 5.83 (Chapter 5) along with Equation 6.68, determine the outlet concentration of the limiting reactant $\left(C_{A_{LR}}\right)_f$ corresponding to the given concentration of the desired product $\left(C_{P_j}\right)_f$.

Step 2
Using Equation 2.45 (Chapter 2), determine the total conversion of the limiting reactant $\left(X_{A_{LR}}\right)_f$ corresponding to the concentration $\left(C_{A_{LR}}\right)_f$.

Step 3

Using Equation 6.66, determine the volume of the PFR required corresponding to the conversion $(X_{A_{LR}})_f$. Alternatively, use Equation 6.67 to determine the space time.

EXAMPLE 6.6

The liquid-phase parallel reaction

$$A \rightarrow P \; (-r_A)_1 = r_P = 0.05C_A \text{ mol/(L·min)}$$

$$A \rightarrow Q \; (-r_A)_2 = r_Q = 0.02C_A \text{ mol/(L·min)}$$

is carried out in an isothermal PFR of 100-L volume with 2 mol/L initial concentration of A and no P and Q present initially. The volumetric flow rate into the reactor is 10 L/min. Determine the outlet composition from the reactor.

SOLUTION

$$-r_A = (-r_A)_1 + (-r_A)_2 = 0.05C_A + 0.02C_A = 0.07C_A$$

$$\Rightarrow \quad -r_A = 0.07 \times 2 \times (1 - X_A) \tag{6.6.1}$$

$$\tau_{PFR} = \frac{V_{PFR}}{v_0} = \frac{100}{10} = 10 \text{ min}$$

From Equation 6.67,

$$\tau_{PFR} = (C_A)_0 \int_0^{(X_A)_f} \frac{dX_A}{(-r_{A_{LR}})_1 + (-r_{A_{LR}})_2}$$

$$\Rightarrow \quad \tau_{PFR} = (C_A)_0 \int_0^{(X_A)_f} \frac{dX_A}{0.14(1 - X_A)}$$

$$\Rightarrow \quad -\ln[1 - (X_A)_f] = \frac{10 \times 0.14}{2}$$

$$\Rightarrow \quad 1 - (X_A)_f = 0.496$$

$$\Rightarrow \quad (X_A)_f = 0.505$$

From Equation 2.45 (Chapter 2), for liquid-phase reaction

$$(C_A)_f = (C_A)_0[1 - (X_A)_f] = 2 \times 0.496 = 0.992 \text{ mol/L}$$

$$(\Psi_{inst})_P = \frac{r_P}{-r_A} = \frac{0.05C_A}{0.07C_A} = \frac{5}{7}$$

From Equation 6.68,

$$(C_P)_f - (C_P)_0 = -\int_{(C_A)_0}^{(C_A)_f} (\Psi_{inst})_P d(C_A)$$

$$\Rightarrow \quad (C_P)_f = \frac{5}{7} \times [(C_A)_0 - (C_A)_f] = \frac{5}{7} \times [2 - 0.992] = 0.72 \text{ mol/L}$$

$$C_Q = (C_A)_0 + (C_P)_0 + (C_Q)_0 - (C_A)_f - (C_P)_f = 2 - 0.992 - 0.72 = 0.288 \text{ mol/L}$$

6.1.5 Combination of PFRs

6.1.5.1 Series Combination

Consider the schematic shown in Figure 6.2a, which shows three PFRs of volumes $(V_{PFR})_1$, $(V_{PFR})_2$, and $(V_{PFR})_3$, respectively, joined in series. The output from the first PFR is the feed for the second PFR and the output from the second PFR is the feed for the third PFR. For each of these PFRs, from Equation 6.1

$$(V_{PFR})_1 = (F_{A_{LR}})_0 \int_0^{(X_{A_{LR}})_1} \frac{dX_{A_{LR}}}{-r_{A_{LR}}} \quad (6.70)$$

$$(V_{PFR})_2 = (F_{A_{LR}})_0 \int_{(X_{A_{LR}})_1}^{(X_{A_{LR}})_2} \frac{dX_{A_{LR}}}{-r_{A_{LR}}} \quad (6.71)$$

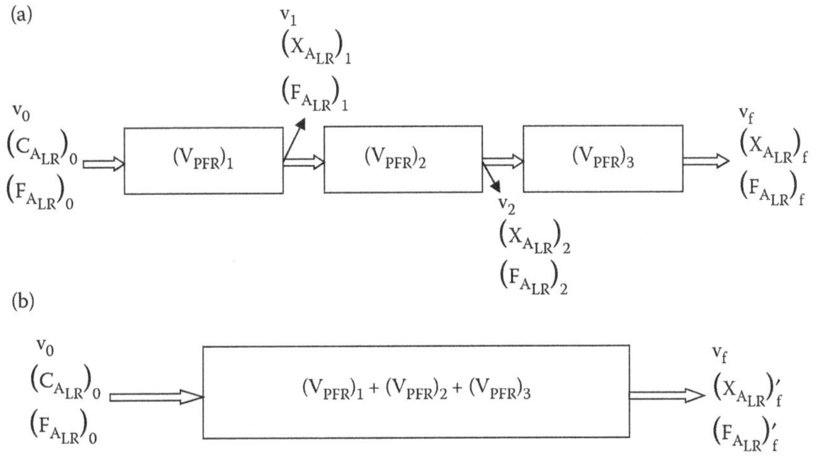

FIGURE 6.2 Schematic representation of ideal PFRs in series.

$$(V_{PFR})_3 = (F_{ALR})_0 \int_{(X_{ALR})_2}^{(X_{ALR})_f} \frac{dX_{ALR}}{-r_{ALR}} \tag{6.72}$$

Adding Equations 6.70 through 6.72,

$$(V_{PFR})_1 + (V_{PFR})_2 + (V_{PFR})_3 = (F_{ALR})_0 \left[\int_0^{(X_{ALR})_1} \frac{dX_{ALR}}{-r_{ALR}} + \int_{(X_{ALR})_1}^{(X_{ALR})_2} \frac{dX_{ALR}}{-r_{ALR}} + \int_{(X_{ALR})_2}^{(X_{ALR})_f} \frac{dX_{ALR}}{-r_{ALR}} \right]$$

$$\Rightarrow (V_{PFR})_1 + (V_{PFR})_2 + (V_{PFR})_3 = (F_{ALR})_0 \int_0^{(X_{ALR})_f} \frac{dX_{ALR}}{-r_{ALR}} \tag{6.73}$$

Next consider the schematic shown in Figure 6.2b, which shows a single PFR of volume $(V_{PFR})_1 + (V_{PFR})_2 + (V_{PFR})_3$. For this PFR, from Equation 6.1

$$(V_{PFR})_1 + (V_{PFR})_2 + (V_{PFR})_3 = (F_{ALR})_0 \int_0^{(X_{ALR})'_f} \frac{dX_{ALR}}{-r_{ALR}} \tag{6.74}$$

Comparing Equations 6.73 and 6.74 it is clear that $(X_{ALR})_f = (X_{ALR})'_f$. Thus, several PFRs of different volumes joined in series give the same conversion as a single PFR of volume equal to the sum of volumes of the PFRs joined in series.

6.1.5.2 Parallel Combination

Consider the schematic shown in Figure 6.3a, which shows two PFRs of volumes $(V_{PFR})_1$, and $(V_{PFR})_2$, respectively, joined in parallel. The volumetric feed rate v_0 is split into two streams and fed into the two PFRs so that

$$v_1 + v_2 = v_0 \tag{6.75}$$

$$(F_{ALR})_1 + (F_{ALR})_2 = (F_{ALR})_0 \tag{6.76}$$

For the first PFR, from Equation 6.1

$$(V_{PFR})_1 = (F_{ALR})_1 \int_0^{(X_{ALR})_{1f}} \frac{dX_{ALR}}{-r_{ALR}}$$

$$\Rightarrow \frac{(V_{PFR})_1}{(F_{ALR})_1} = \int_0^{(X_{ALR})_{1f}} \frac{dX_{ALR}}{-r_{ALR}} \tag{6.77}$$

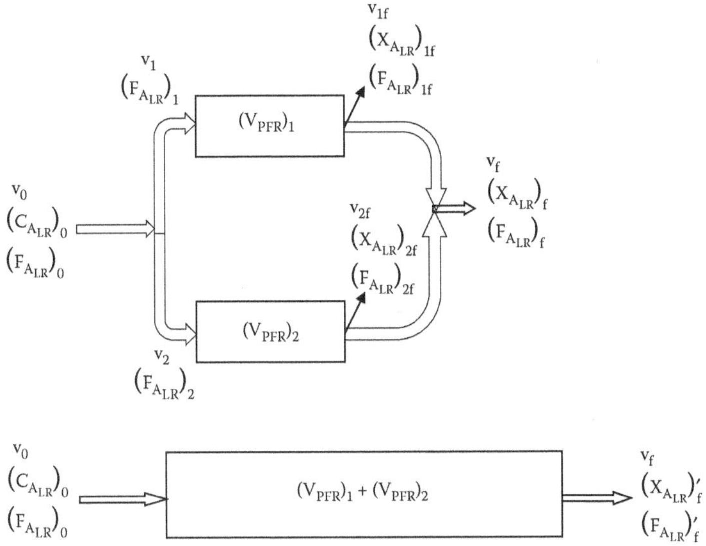

FIGURE 6.3 Schematic representation of ideal PFRs in parallel.

And, for the second PFR, from Equation 6.1

$$(V_{PFR})_2 = \left(F_{A_{LR}}\right)_2 \int_0^{(X_{A_{LR}})_{2f}} \frac{dX_{A_{LR}}}{-r_{A_{LR}}}$$

$$\Rightarrow \frac{(V_{PFR})_2}{\left(F_{A_{LR}}\right)_2} = \int_0^{(X_{A_{LR}})_{2f}} \frac{dX_{A_{LR}}}{-r_{A_{LR}}} \tag{6.78}$$

If the feed $\left(F_{A_{LR}}\right)_0$ is split in such a way that

$$\frac{\left(F_{A_{LR}}\right)_1}{\left(F_{A_{LR}}\right)_2} = \frac{(V_{PFR})_1}{(V_{PFR})_2}$$

$$\Rightarrow \frac{\left(F_{A_{LR}}\right)_1 + \left(F_{A_{LR}}\right)_2}{\left(F_{A_{LR}}\right)_2} = \frac{(V_{PFR})_1 + (V_{PFR})_2}{(V_{PFR})_2}$$

$$\Rightarrow \frac{\left(F_{A_{LR}}\right)_0}{\left(F_{A_{LR}}\right)_2} = \frac{(V_{PFR})_1 + (V_{PFR})_2}{(V_{PFR})_2}$$

$$\Rightarrow \left(F_{A_{LR}}\right)_0 = \frac{(V_{PFR})_1 + (V_{PFR})_2}{(V_{PFR})_2} \times \left(F_{A_{LR}}\right)_2 \tag{6.79}$$

then the left-hand sides of Equations 6.77 and 6.78 are equal, so

$$\int_0^{(X_{ALR})_{1f}} \frac{dX_{ALR}}{-r_{ALR}} = \int_0^{(X_{ALR})_{2f}} \frac{dX_{ALR}}{-r_{ALR}}$$

$$\Rightarrow \quad (X_{ALR})_{1f} = (X_{ALR})_{2f} \tag{6.80}$$

From the schematic shown in Figure 6.3a and Equation 6.80,

$$\begin{aligned}
&(F_{ALR})_f = (F_{ALR})_{1f} + (F_{ALR})_{2f} \\
\Rightarrow\quad &(F_{ALR})_0\left[1-(X_{ALR})_f\right] = (F_{ALR})_1\left[1-(X_{ALR})_{1f}\right] + (F_{ALR})_2\left[1-(X_{ALR})_{2f}\right] \\
\Rightarrow\quad &(F_{ALR})_0\left[1-(X_{ALR})_f\right] = (F_{ALR})_1\left[1-(X_{ALR})_{1f}\right] + (F_{ALR})_2\left[1-(X_{ALR})_{1f}\right] \\
\Rightarrow\quad &(F_{ALR})_0\left[1-(X_{ALR})_f\right] = \left[(F_{ALR})_1 + (F_{ALR})_2\right]\left[1-(X_{ALR})_{1f}\right] \\
\Rightarrow\quad &(F_{ALR})_0\left[1-(X_{ALR})_f\right] = (F_{ALR})_0\left[1-(X_{ALR})_{1f}\right]
\end{aligned} \tag{6.81}$$

$$\Rightarrow \quad (X_{ALR})_f = (X_{ALR})_{1f} = (X_{ALR})_{2f} \tag{6.82}$$

Now consider the schematic shown in Figure 6.3b, which shows a single PFR of volume $(V_{PFR})_1 + (V_{PFR})_2$. For this PFR, from Equations 6.1 and 6.79

$$\begin{aligned}
&(V_{PFR})_1 + (V_{PFR})_2 = (F_{ALR})_0 \int_0^{(X_{ALR})'_f} \frac{dX_{ALR}}{-r_{ALR}} \\
\Rightarrow\quad &(V_{PFR})_1 + (V_{PFR})_2 = \frac{(V_{PFR})_1 + (V_{PFR})_2}{(V_{PFR})_2} \times (F_{ALR})_2 \int_0^{(X_{ALR})'_f} \frac{dX_{ALR}}{-r_{ALR}}
\end{aligned} \tag{6.83}$$

$$\Rightarrow \quad \frac{(V_{PFR})_2}{(F_{ALR})_2} = \int_0^{(X_{ALR})'_f} \frac{dX_{ALR}}{-r_{ALR}} \tag{6.84}$$

Comparing Equations 6.78, 6.80, 6.82, and 6.84, we have

$$\left(X_{A_{LR}}\right)_{1f} = \left(X_{A_{LR}}\right)_{2f} = \left(X_{A_{LR}}\right)_f' = \left(X_{A_{LR}}\right)_f \tag{6.85}$$

The result of Equation 6.85 can be extended to n PFRs joined in parallel provided the feed is split between them in such a way that

$$\frac{\left(F_{A_{LR}}\right)_1}{(V_{PFR})_1} = \frac{\left(F_{A_{LR}}\right)_2}{(V_{PFR})_2} = \frac{\left(F_{A_{LR}}\right)_3}{(V_{PFR})_3} = \cdots = \frac{\left(F_{A_{LR}}\right)_n}{(V_{PFR})_n}$$

Thus, PFRs joined in parallel give the same combined conversion as a single PFR of volume equal to the sum of volumes of the PFRs joined in parallel provided the total feed is split into the feed to individual reactors in the ratio of the volumes of the individual reactors. If the feed is split in any other ratio, the combined conversion will be less than that obtained in a single PFR of volume equal to the sum of volumes of the PFRs joined in parallel.

Considering the complexity of plant layout required for combining PFRs, it is better to have a single large PFR rather than several small PFRs joined either in series or in parallel.

EXAMPLE 6.7

A second-order liquid-phase reaction

$$A \rightarrow \text{Products}$$

with reaction rate $-r_A = C_A^2$ mol/(L·min) is carried out in two PFRs, volumes 30 and 70 L, respectively, joined in parallel. A feed of 100 mol/min at a concentration of 1 mol/L is equally split between the two reactors. Find the conversion in the combined stream when the outlet from the two reactors is mixed together. Also, compare this conversion with the conversion that would be achieved if the entire feed of 100 mol/min were treated in a single PFR of volume 100 L.

Solution

Here,

$$(V_{PFR})_1 = 30 \text{ L}, \quad (V_{PFR})_2 = 70 \text{ L},$$

$$(F_A)_1 = 50 \text{ mol/min}, \quad (F_A)_2 = 50 \text{ mol/min}, \quad (F_A)_0 = 100 \text{ mol/min},$$

$$(C_A)_0 = 1 \text{ mol/L}$$

For the first reactor, from Equation 6.77

$$(V_{PFR})_1 = (F_A)_1 \int_0^{(X_A)_{1f}} \frac{dX_A}{-r_A}$$

$$\Rightarrow \quad 30 = 50 \int_0^{(X_A)_{1f}} \frac{dX_A}{(C_A)_0^2(1-X_A)^2}$$

$$\Rightarrow \quad \frac{1}{1-(X_A)_{1f}} - 1 = 0.6$$

$$\Rightarrow \quad (X_A)_{1f} = 0.375$$

For the second reactor, from Equation 6.77

$$(V_{PFR})_2 = (F_{A_{LR}})_2 \int_0^{(X_{A_{LR}})_{2f}} \frac{dX_{A_{LR}}}{-r_{A_{LR}}}$$

$$\Rightarrow \quad 70 = 50 \int_0^{(X_A)_{2f}} \frac{dX_A}{(C_A)_0^2(1-X_A)^2}$$

$$\Rightarrow \quad \frac{1}{1-(X_A)_{2f}} - 1 = 1.4$$

$$\Rightarrow \quad (X_A)_{2f} = 0.583$$

From Equation 6.81

$$(F_A)_0\left[1-(X_A)_f\right] = (F_A)_1\left[1-(X_A)_{1f}\right] + (F_A)_2\left[1-(X_A)_{2f}\right]$$

$$\Rightarrow \quad 100\left[1-(X_A)_f\right] = 50\left[1-0.375\right] + 50\left[1-0.583\right]$$

$$\Rightarrow \quad (X_A)_f = 0.479$$

For the single reactor, from Equation 6.83

$$(V_{PFR})_1 + (V_{PFR})_2 = (F_A)_0 \int_0^{(X_A)'_f} \frac{dX_A}{-r_A}$$

$$\Rightarrow \quad 100 = 100 \int_0^{(X_A)'_f} \frac{dX_{A_{LR}}}{(C_A)_0^2(1-X_A)^2}$$

$$\Rightarrow \quad \frac{1}{1-(X_A)'_f} - 1 = 1$$

$$\Rightarrow \quad (X_A)'_f = 0.5$$

We can see that $(X_A)_f < (X_A)'_f$

6.2 IDEAL BACKMIX FLOW REACTOR

Ideal backmix flow reactor (BFR) is one where the fluid inside the reactor is ideally mixed to the extent that at every point inside the reactor, the concentration of each chemical species is the same. Owing to this ideal mixing, the outgoing stream from the reactor has the same concentration of the chemical species as everywhere inside the reactor. On the contrary, a fluid element that enters the reactor instantly mixes with the entire fluid present within the reactor and loses its separate identity. Again, this is an idealized concept that cannot be rigorously realized in a real situation. However, there are real systems that closely approach this ideal behavior. These reactors are also known as continuously stirred tank reactor (CSTR).

For the ideal BFR schematically represented in Figure 6.4, we write a material balance for the limiting reactant species over the reactor shown in the figure as

$$\text{Input} = \text{Output} + \text{Accumulation} + \text{Disappearance due to chemical reaction}$$

For the BFR at steady state, the accumulation within the reactor is zero. So we have

$$\text{Input} = \text{Output} + \text{Disappearance due to chemical reaction}$$

$$\Rightarrow \quad \left(F_{A_{LR}}\right)_0 = \left(F_{A_{LR}}\right)_f + \left(-r_{A_{LR}}\right)_f V_{BFR}$$

$$\Rightarrow \quad \left(F_{A_{LR}}\right)_0 = \left(F_{A_{LR}}\right)_0\left[1-\left(X_{A_{LR}}\right)_f\right] + \left(-r_{A_{LR}}\right)_f V_{BFR}$$

$$\Rightarrow \quad V_{BFR} = \frac{\left(F_{A_{LR}}\right)_0\left(X_{A_{LR}}\right)_f}{\left(-r_{A_{LR}}\right)_f} \tag{6.86}$$

This is the performance equation of an ideal isothermal BFR operating under steady-state condition. It relates the volume of the BFR to the conversion achieved in the reactor for a given molar feed rate. The equation can be used for the design of BFR to know the volume of the reactor required if a specified molar feed rate of the reactant is to be converted to a desired extent. Alternatively, it can be used for the performance evaluation of an existing BFR of known volume to determine the conversion that will be achieved for a specified molar feed rate of the reactant. It is

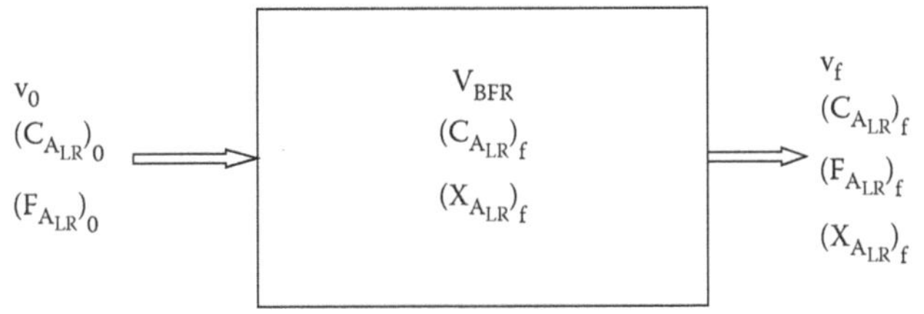

FIGURE 6.4 Schematic representation of an ideal BFR.

evident from Equation 6.86 that as long as the ratio $V_{BFR}/(F_{A_{LR}})_0$ is unaltered, the conversion from a BFR remains unchanged. The higher this ratio, the higher is the conversion and vice versa.

The equation can be rearranged into other convenient forms for ease of application in different situations. Referring to Figure 6.4, $(F_{A_{LR}})_0 = v_0(C_{A_{LR}})_0$; so from Equation 6.86

$$V_{BFR} = \frac{v_0(C_{A_{LR}})_0(X_{A_{LR}})_f}{(-r_{A_{LR}})_f}$$

$$\Rightarrow \quad \frac{V_{BFR}}{v_0} = \frac{(C_{A_{LR}})_0(X_{A_{LR}})_f}{(-r_{A_{LR}})_f} \tag{6.87}$$

$$\Rightarrow \tau_{BFR} = \frac{(C_{A_{LR}})_0(X_{A_{LR}})_f}{(-r_{A_{LR}})_f} \tag{6.88}$$

Equations 6.87 and 6.88 are alternative performance equations for the BFR. Equation 6.87 relates the volume of the BFR to the conversion achieved in the reactor for a given initial concentration of the reactant and the volumetric feed rate. The equation can be used for the design of BFR to know the volume of the reactor required if the reactant of a given initial concentration being fed to the reactor at a specified volumetric feed rate is to be converted to a desired extent. Alternatively, it can be used for the performance evaluation of an existing BFR of known volume to determine the conversion that will be achieved if the reactant of a given initial concentration is fed to the reactor at a specified volumetric feed rate. Similarly, Equation 6.88 relates the space time of reactor operation to the conversion achieved in the reactor for a given initial concentration of the reactant. It is evident from Equation 6.88 that as long as the ratio $\tau/(C_{A_{LR}})_0$ is unaltered, the conversion from a BFR remains unchanged. The higher this ratio, the higher is the conversion and vice versa.

The performance equation for the BFR can also be written in terms of the concentration of the limiting reactant. From Equation 2.45 (Chapter 2)

$$(C_{A_{LR}})_f = \frac{(C_{A_{LR}})_0\left[1-(X_{A_{LR}})_f\right]}{1+\varepsilon_{LR}(X_{A_{LR}})_f}$$

$$\Rightarrow \quad (X_{A_{LR}})_f = \frac{(C_{A_{LR}})_0-(C_{A_{LR}})_f}{(C_{A_{LR}})_0+\varepsilon_{LR}(C_{A_{LR}})_f}$$

From Equation 6.86,

$$V_{BFR} = \frac{(F_{A_{LR}})_0\left[(C_{A_{LR}})_0 - (C_{A_{LR}})_f\right]}{(-r_{A_{LR}})_f\left[(C_{A_{LR}})_0 + \varepsilon_{LR}(C_{A_{LR}})_f\right]} \tag{6.89}$$

And, from Equation 6.88,

$$\tau_{BFR} = \frac{(C_{A_{LR}})_0\left[(C_{A_{LR}})_0 - (C_{A_{LR}})_f\right]}{(-r_{A_{LR}})_f\left[(C_{A_{LR}})_0 + \varepsilon_{LR}(C_{A_{LR}})_f\right]} \tag{6.90}$$

6.2.1 Single-Reactant nth-Order Reaction in a BFR

From Equations 6.15 and 6.86,

$$V_{BFR} = \frac{(F_A)_0(X_A)_f[1+\varepsilon_A(X_A)_f]^n}{k(C_A)_0^n[1-(X_A)_f]^n} \tag{6.91}$$

For liquid-phase reactions, and the gas-phase reactions where the stoichiometry is such that the reacting system neither expands nor contracts, $\varepsilon_A = 0$. Equation 6.91 then reduces to

$$V_{BFR} = \frac{(F_A)_0(X_A)_f}{k(C_A)_0^n[1-(X_A)_f]^n} \tag{6.92}$$

Similarly, from Equations 6.15 and 6.88,

$$\tau_{BFR} = \frac{(C_A)_0(X_A)_f[1+\varepsilon_A(X_A)_f]^n}{k(C_A)_0^n[1-(X_A)_f]^n}$$

$$\Rightarrow \quad \tau_{BFR} = \frac{(X_A)_f[1+\varepsilon_A(X_A)_f]^n}{k(C_A)_0^{(n-1)}[1-(X_A)_f]^n} \tag{6.93}$$

For liquid-phase reactions, and the gas-phase reactions where the stoichiometry is such that the reacting system neither expands nor contracts, $\varepsilon_A = 0$. Equation 6.93 then reduces to

$$\tau_{BFR} = \frac{(X_A)_f}{k(C_A)_0^{(n-1)}[1-(X_A)_f]^n} \tag{6.94}$$

For first-order reactions, Equation 6.94 gives

$$\tau_{BFR} = \frac{(X_A)_f}{k[1-(X_A)_f]} \tag{6.95}$$

$$\Rightarrow \quad k\tau_{BFR} = \frac{(X_A)_f}{[1-(X_A)_f]} \tag{6.96}$$

We can see that the quantity $k\tau_{BFR}$ for a first-order reaction is uniquely related to the conversion in a BFR.

EXAMPLE 6.8

Repeat Example 6.1 for an ideal isothermal BFR.

SOLUTION

From the data given in Example 6.1,

$$(F_A)_0 = 1000 \text{ mol/min}, \quad (X_A)_f = 0.5$$

From Figure EXE6.1,
At 0.5 conversion, the value of $1/-r_A$ is 0.58.
From Equation 6.86,

$$V_{BFR} = \frac{(F_A)_0(X_A)_f}{(-r_A)} = 1000 \times 0.5 \times 0.58 = 290 \text{ L}$$

EXAMPLE 6.9

Repeat Example 6.2 for an ideal isothermal BFR.

SOLUTION

From Solution to Example 6.2,

$$(C_A)_0 = 0.041 \text{ mol/L}, \quad k = 0.2 \text{ L/(mol}\cdot\text{min)}$$

and

$$\varepsilon_A = 1, \quad (X_A)_f = 0.9$$

Order of reaction $n = 2$
From Equation 6.93,

$$\tau_{BFR} = \frac{(X_A)_f[1+\varepsilon_A(X_A)_f]^n}{k(C_A)_0^{(n-1)}[1-(X_A)_f]^n}$$

$$\tau_{BFR} = \frac{0.9 \times [1+0.9]^2}{0.2 \times 0.041[1-0.9]^2} = 39{,}622 \text{ min}$$

6.2.2 General Reaction Involving Several Reactants

For the general reaction represented by the balanced chemical Equation 5.24 (Chapter 5), we may combine Equation 6.89 with Equation 5.28 (Chapter 5) to have

$$V_{BFR} = \frac{(F_{A_{LR}})_0\left[(C_{A_{LR}})_0 - (C_{A_{LR}})_f\right]}{k\prod_{i=1}^{n}\left[(C_{A_i})_0 - (a_i/a_{LR})\left\{(C_{A_{LR}})_0 - C_{A_{LR}}\right\}\right]^{a_i}\left[(C_{A_{LR}})_0 + \varepsilon_{LR}(C_{A_{LR}})_f\right]} \tag{6.97}$$

And, from Equations 6.90 and 5.28 (Chapter 5)

$$\tau_{BFR} = \frac{(C_{A_{LR}})_0\left[(C_{A_{LR}})_0 - (C_{A_{LR}})_f\right]}{k\prod_{i=1}^{n}\left[(C_{A_i})_0 - (a_i/a_{LR})\left\{(C_{A_{LR}})_0 - C_{A_{LR}}\right\}\right]^{a_i}\left[(C_{A_{LR}})_0 + \varepsilon_{LR}(C_{A_{LR}})_f\right]} \tag{6.98}$$

$$\Rightarrow k\tau_{BFR} = \frac{(C_{A_{LR}})_0\left[(C_{A_{LR}})_0 - (C_{A_{LR}})_f\right]}{\prod_{i=1}^{n}\left[(C_{A_i})_0 - (a_i/a_{LR})\left\{(C_{A_{LR}})_0 - C_{A_{LR}}\right\}\right]^{a_i}\left[(C_{A_{LR}})_0 + \varepsilon_{LR}(C_{A_{LR}})_f\right]} \tag{6.99}$$

For the special case when $(C_{A_i})_0 = (a_i/a_{LR})(C_{A_{LR}})_0$ Equations 6.97 and 6.98 reduce to

$$V_{BFR} = \frac{(F_{A_{LR}})_0\left[(C_{A_{LR}})_0 - (C_{A_{LR}})_f\right]}{k\prod_{i=1}^{n}(a_i/a_{LR})^{a_i}(C_{A_{LR}})^{a_i}\left[(C_{A_{LR}})_0 + \varepsilon_{LR}(C_{A_{LR}})_f\right]} \tag{6.100}$$

and

$$\tau_{BFR} = \frac{(C_{A_{LR}})_0\left[(C_{A_{LR}})_0 - (C_{A_{LR}})_f\right]}{k\prod_{i=1}^{n}(a_i/a_{LR})^{a_i}(C_{A_{LR}})^{a_i}\left[(C_{A_{LR}})_0 + \varepsilon_{LR}(C_{A_{LR}})_f\right]} \tag{6.101}$$

$$\Rightarrow \quad k\tau_{BFR} = \frac{(C_{A_{LR}})_0\left[(C_{A_{LR}})_0 - (C_{A_{LR}})_f\right]}{\prod_{i=1}^{n}(a_i/a_{LR})^{a_i}(C_{A_{LR}})^{a_i}\left[(C_{A_{LR}})_0 + \varepsilon_{LR}(C_{A_{LR}})_f\right]} \tag{6.102}$$

For liquid-phase reactions, and the gas-phase reactions where the stoichiometry is such that the reacting system neither expands nor contracts, $\varepsilon_A = 0$. Equations 6.97 and 6.98 then reduce to

$$V_{BFR} = \frac{(F_{A_{LR}})_0\left[(C_{A_{LR}})_0 - (C_{A_{LR}})_f\right]}{k(C_{A_{LR}})_0\prod_{i=1}^{n}\left[(C_{A_i})_0 - (a_i/a_{LR})\left\{(C_{A_{LR}})_0 - C_{A_{LR}}\right\}\right]^{a_i}} \tag{6.103}$$

and

$$\tau_{BFR} = \frac{(C_{A_{LR}})_0\left[(C_{A_{LR}})_0 - (C_{A_{LR}})_f\right]}{k(C_{A_{LR}})_0 \prod_{i=1}^{n}\left[(C_{A_i})_0 - (a_i/a_{LR})\left\{(C_{A_{LR}})_0 - C_{A_{LR}}\right\}\right]^{a_i}} \quad (6.104)$$

$$\Rightarrow k\tau_{BFR} = \frac{(C_{A_{LR}})_0\left[(C_{A_{LR}})_0 - (C_{A_{LR}})_f\right]}{(C_{A_{LR}})_0 \prod_{i=1}^{n}\left[(C_{A_i})_0 - (a_i/a_{LR})\left\{(C_{A_{LR}})_0 - C_{A_{LR}}\right\}\right]^{a_i}} \quad (6.105)$$

Similarly, Equations 6.100 and 6.101 reduce to

$$V_{BFR} = \frac{(F_{A_{LR}})_0\left[(C_{A_{LR}})_0 - (C_{A_{LR}})_f\right]}{k(C_{A_{LR}})_0 \prod_{i=1}^{n}(a_i/a_{LR})^{a_i}\left(C_{A_{LR}}\right)^{a_i}} \quad (6.106)$$

and

$$\tau_{BFR} = \frac{(C_{A_{LR}})_0\left[(C_{A_{LR}})_0 - (C_{A_{LR}})_f\right]}{k(C_{A_{LR}})_0 \prod_{i=1}^{n}(a_i/a_{LR})^{a_i}\left(C_{A_{LR}}\right)^{a_i}} \quad (6.107)$$

$$\Rightarrow k\tau_{BFR} = \frac{(C_{A_{LR}})_0\left[(C_{A_{LR}})_0 - (C_{A_{LR}})_f\right]}{(C_{A_{LR}})_0 \prod_{i=1}^{n}(a_i/a_{LR})^{a_i}\left(C_{A_{LR}}\right)^{a_i}} \quad (6.108)$$

The general algorithm for engineering calculations can be summarized as follows:

Step 1
Write down the rate equation in terms of the concentrations of the reactants. The rate equation can be of the form of Equation 5.28 (Chapter 5) or any other form as determined from the experiment.

Step 2
From reaction stoichiometry express the concentration of each reactant in terms of the concentration of the limiting reactant.

Step 3
Substitute the concentration terms of each reactant in the rate equation written in Step 1 with the equivalent expression in terms of the limiting reactant as formulated in Step 2. Now, we have the rate equation in terms of the concentration of the limiting reactant alone.

Step 4
Use the rate equation obtained in Step 3 in the performance equation of BFR to determine the desired quantities.

EXAMPLE 6.10

Repeat Example 6.3 for an isothermal BFR.

Solution

From Solution to Example 6.3,

$$(C_A)_0 = 0.0158 \frac{\text{mol}}{\text{L}}$$

$$\varepsilon_A = -0.5$$

$$k = 0.3 \text{ L/(min}\cdot\text{mol)}$$

From Equation 2.45 (Chapter 2),

$$(C_A)_f = \frac{(C_A)_0(1-X_A)}{1+\varepsilon_A X_A} = \frac{0.0158\times(1-0.8)}{1-(0.8/2)} = 0.0053 \frac{\text{mol}}{\text{L}}$$

$$(F_A)_0 = (C_A)_0 \times v_0 = 0.0158\times 20 = 0.316 \text{ mol/min}$$

Since the feed consists of 1 volume A and 1 volume B, we have $(C_B)_0 = (a_B/a_A)(C_A)_0$ and Equation 6.100 is applicable. So,

$$V_{BFR} = \frac{(F_A)_0[(C_A)_0 - (C_A)_f]}{k(C_A)_f^2[(C_A)_0 + \varepsilon_A(C_A)_f]}$$

$$\Rightarrow \quad V_{BFR} = \frac{0.316\times[0.0158-0.0053]}{0.3\times 0.0053\times 0.0053\times[0.0158-0.5\times 0.0053]} = 299,41 \text{ L}$$

6.2.3 Reversible Reactions

For constant volume-reacting systems consider the general reversible reaction represented by the balanced chemical Equation 5.33 (Chapter 5). For this reaction, from Equations 5.34 (Chapter 5) and 6.86,

$$V_{BFR} = \frac{\left(F_{A_{LR}}\right)_0 \left(X_{A_{LR}}\right)_f}{k_1\left(C_{A_{LR}}\right)_0^{\Sigma_{i=1}^{n} a_i} \prod_{i=1}^{n}\left[\frac{\left(C_{A_i}\right)_0}{\left(C_{A_{LR}}\right)_0} - \frac{a_i}{a_{LR}} X_{A_{LR}}\right]^{a_i} - k_2\left(C_{A_{LR}}\right)_0^{\Sigma_{j=1}^{l} p_j} \prod_{j=1}^{l}\left[\frac{\left(C_{P_j}\right)_0}{\left(C_{A_{LR}}\right)_0} + \frac{p_j}{a_{LR}} X_{A_{LR}}\right]^{p_j}} \quad (6.109)$$

If we substitute $(X_{A_{LR}})_f$ in Equation 6.109 with $(X_{A_{LR}})_{eq}$ we can get the volume of the BFR required to achieve equilibrium. $(X_{A_{LR}})_{eq}$ is related to the equilibrium constant by Equations 4.21 and 4.22 (Chapter 4) for constant volume systems.

From Equations 5.34 (Chapter 5) and 6.88,

$$\tau_{BFR} = \frac{(C_{A_{LR}})_0 (X_{A_{LR}})_f}{k_1 (C_{A_{LR}})_0^{\sum_{i=1}^{n} a_i} \prod_{i=1}^{n} \left[\left(\frac{(C_{A_i})_0}{(C_{A_{LR}})_0} \right) - \left(\frac{a_i}{a_{LR}} X_{A_{LR}} \right) \right]^{a_i} - k_2 (C_{A_{LR}})_0^{\sum_{j=1}^{1} p_j} \prod_{j=1}^{1} \left[\left(\frac{(C_{P_j})_0}{(C_{A_{LR}})_0} \right) + \left(\frac{p_j}{a_{LR}} \right) X_{A_{LR}} \right]^{p_j}} \quad (6.110)$$

If we substitute $(X_{A_{LR}})_f$ in Equation 6.110 with $(X_{A_{LR}})_{eq}$ we can get the space time of the BFR required to achieve equilibrium.

EXAMPLE 6.11

Repeat Example 6.4 for an ideal isothermal BFR.

SOLUTION

From the data given in Example 6.4 and Solution to Example 6.4,
$k_1 = 0.05\ \text{min}^{-1}$ and $k_2 = 0.02\ \text{min}^{-1}$

$$(X_A)_{eq} = 0.68, \quad (F_A)_0 = 100\ \text{mol/min}$$

$$(C_A)_0 = 1\ \text{mol/L} \quad (C_B)_0 = 0.1\ \text{mol/L}$$

From Equation 6.109,

$$V_{BFR} = \frac{(F_A)_0 (X_A)_{eq}}{k_1 (C_A)_0 [1-(X_A)_{eq}] - k_2 (C_A)_0 [((C_B)_0/(C_A)_0) + (X_A)_{eq}]}$$

$$\Rightarrow \quad V_{BFR} = \frac{100 \times 0.68}{0.05 \times 1 \times (1-0.68) - 0.02 \times 1 \times [0.1+0.68]} = 1{,}700{,}00\ \text{L} = 170\ \text{m}^3$$

6.2.4 Multiple Reactions

6.2.4.1 Series Reactions

Consider the series reaction

$$A \rightarrow I \rightarrow T$$

with the specific reaction rate constants for the two consecutive reaction steps being k_1 and k_2, respectively. If this reaction is carried out in a BFR, then from Equations 5.43 (Chapter 5) and 6.90, for species A

$$\tau_{BFR} = \frac{[(C_A)_0 - (C_A)_f]}{k_1(C_A)_f}$$

$$\Rightarrow \quad (C_A)_f = \frac{(C_A)_0}{1+k_1\tau_{BFR}} \tag{6.111}$$

$$\Rightarrow \quad \frac{(C_A)_f}{(C_A)_0} = \frac{1}{1+k_1\tau_{BFR}} \tag{6.112}$$

From Equations 5.44 (Chapter 5), 6.90, and 6.111 for species I,

$$\tau_{BFR} = \frac{[(C_I)_0 - (C_I)_f]}{k_2(C_I)_f - k_1(C_A)_f}$$

$$\Rightarrow \quad \tau_{BFR} = \frac{[(C_I)_0 - (C_I)_f]}{k_2(C_I)_f - k_1(C_A)_f}$$

$$\Rightarrow \quad (1+k_2\tau_{BFR})(C_I)_f = (C_I)_0 + \frac{k_1\tau_{BFR}(C_A)_0}{1+k_1\tau_{BFR}}$$

$$\Rightarrow \quad (C_I)_f = \frac{(C_I)_0}{(1+k_2\tau_{BFR})} + \frac{k_1\tau_{BFR}(C_A)_0}{(1+k_1\tau_{BFR})(1+k_2\tau_{BFR})} \tag{6.113}$$

$$\Rightarrow \quad \frac{(C_I)_f}{(C_A)_0} = \frac{(C_I)_0}{(1+k_2\tau_{BFR})(C_A)_0} + \frac{k_1\tau_{BFR}}{(1+k_1\tau_{BFR})(1+k_2\tau_{BFR})} \tag{6.114}$$

Let us now determine the space time of the BFR when we get the maximum outlet concentration of I. From Equation 6.113, for maximum $(C_I)_f$,

$$\frac{d(C_I)_f}{d\tau_{BFR}} = 0 = \frac{-k_2(C_I)_0}{[1+k_2(\tau_{BFR})_{max}]^2} +$$

$$\frac{k_1(C_A)_0[1+k_1(\tau_{BFR})_{max}][1+k_2(\tau_{BFR})_{max}] - k_1(C_A)_0(\tau_{BFR})_{max}[k_2[1+k_1(\tau_{BFR})_{max}]+k_1[1+k_2(\tau_{BFR})_{max}]]}{[1+k_1(\tau_{BFR})_{max}]^2[1+k_2(\tau_{BFR})_{max}]^2}$$

$$\Rightarrow \quad k_1^2 k_2\left[1+\frac{(C_I)_0}{(C_A)_0}\right](\tau_{BFR})_{max}^2 + 2k_1k_2\frac{(C_I)_0}{(C_A)_0}(\tau_{BFR})_{max} + k_2\frac{(C_I)_0}{(C_A)_0} - k_1 = 0$$

$$\Rightarrow \quad (\tau_{BFR})_{max} = \frac{-k_2((C_I)_0/(C_A)_0) + \sqrt{[k_1k_2 + k_1k_2((C_I)_0/(C_A)_0) - k_2^2((C_I)_0/(C_A)_0)]}}{k_1k_2[1+((C_I)_0/(C_A)_0)]} \tag{6.115}$$

From Equation 6.113 the maximum outlet concentration of I is obtained by substituting the value of $(\tau_{BFR})_{max}$ from Equation 6.115.

The outgoing concentration of the species T can be determined from Equation 3.37 (Chapter 3) as

$$(C_T)_f = (C_A)_0 + (C_I)_0 + (C_T)_0 - (C_A)_f - (C_I)_f$$

$$\Rightarrow \quad (C_T)_f = (C_A)_0 + (C_I)_0 + (C_T)_0 - \frac{(C_A)_0}{(1+k_1\tau_{BFR})} - \frac{(C_I)_0}{(1+k_2\tau_{BFR})}$$

$$- \frac{k_1\tau_{BFR}(C_A)_0}{(1+k_1\tau_{BFR})(1+k_2\tau_{BFR})}$$

$$\Rightarrow \quad (C_T)_f = (C_A)_0\left[1 - \frac{1}{(1+k_1\tau_{BFR})} - \frac{k_1\tau_{BFR}}{(1+k_1\tau_{BFR})(1+k_2\tau_{BFR})}\right]$$

$$+ (C_I)_0\left[1 - \frac{1}{(1+k_2\tau_{BFR})}\right] + (C_T)_0$$

$$\Rightarrow \quad (C_T)_f = (C_A)_0\left[\frac{k_1k_2\tau_{BFR}^2}{(1+k_1\tau_{BFR})(1+k_2\tau_{BFR})}\right] + (C_I)_0\left[\frac{k_2\tau_{BFR}}{(1+k_2\tau_{BFR})}\right] + (C_T)_0 \tag{6.116}$$

$$\Rightarrow \quad \frac{(C_T)_f}{(C_A)_0} = \left[\frac{k_1k_2\tau_{BFR}^2}{(1+k_1\tau_{BFR})(1+k_2\tau_{BFR})}\right] + \frac{(C_I)_0}{(C_A)_0}\left[\frac{k_2\tau_{BFR}}{(1+k_2\tau_{BFR})}\right] + \frac{(C_T)_0}{(C_A)_0} \tag{6.117}$$

In the special case when the species I and T are not present initially, $(C_I)_0 = (C_T)_0 = 0$, and Equations 6.114 and 6.117 reduce to

$$\frac{(C_I)_f}{(C_A)_0} = \frac{k_1\tau_{BFR}}{(1+k_1\tau_{BFR})(1+k_2\tau_{BFR})} \tag{6.118}$$

and

$$\frac{(C_T)_f}{(C_A)_0} = \frac{k_1k_2\tau_{BFR}^2}{(1+k_1\tau_{BFR})(1+k_2\tau_{BFR})} \tag{6.119}$$

Similarly, Equation 6.115 reduces to

$$(\tau_{BFR})_{max} = \frac{1}{\sqrt{k_1 k_2}} \tag{6.120}$$

Substituting this value in Equation 6.118, for this special case

$$\left(\frac{(C_I)_f}{(C_A)_0}\right)_{max} = \frac{1}{[1+\sqrt{k_2/k_1}]^2} \tag{6.121}$$

In terms of the volume of the BFR using the relation

$$\tau_{BFR} = \frac{(C_A)_0 V_{BFR}}{(F_A)_0} = \frac{(C_I)_0 V_{BFR}}{(F_I)_0} = \frac{(C_T)_0 V_{BFR}}{(F_T)_0},$$

Equation 6.111 changes to

$$(C_A)_f = \frac{(F_A)_0 (C_A)_0}{[(F_A)_0 + k_1 V_{BFR}(C_A)_0]} \tag{6.122}$$

Equation 6.113 changes to

$$(C_I)_f = \frac{(F_I)_0 (C_I)_0}{[(F_I)_0 + k_2 V_{BFR}(C_I)_0]} + \frac{k_1 (C_A)_0^2 V_{BFR} (F_A)_0}{[(F_A)_0 + k_1 V_{BFR}(C_A)_0][(F_A)_0 + k_2 V_{BFR}(C_A)_0]} \tag{6.123}$$

and Equation 6.116 changes to

$$(C_T)_f = \frac{k_1 k_2 (C_A)_0^3 V_{BFR}^2}{[(F_A)_0 + k_1 V_{BFR}(C_A)_0][(F_A)_0 + k_2 V_{BFR}(C_A)_0]} + \frac{k_2 (C_I)_0^2 V_{BFR}}{[(F_I)_0 + k_2 V_{BFR}(C_I)_0]} + (C_T)_0 \tag{6.124}$$

EXAMPLE 6.12

Repeat Example 6.5 for an ideal isothermal BFR.

Solution

From the data given in Example 6.5 and Solution to Example 6.5,

$$(C_A)_0 = 2.5 \text{ mol/L}, \quad (C_I)_0 = (C_T)_0 = 0, \quad v_0 = 100 \text{ L/min}$$

$k_1 = 0.05\ \text{min}^{-1}$ and $k_2 = 0.02\ \text{min}^{-1}$

$$(F_A)_0 = (C_A)_0 \times v_0 = 2.5 \times 100 = 250\ \text{mol/min}$$

From Equation 6.120,

$$(\tau_{BFR})_{max} = \frac{(V_{BFR})_{max}}{v_0} = \frac{1}{\sqrt{k_1 k_2}}$$

$$\Rightarrow \quad (V_{BFR})_{max} = \frac{v_0}{\sqrt{k_1 k_2}} = \frac{100}{\sqrt{0.05 \times 0.02}} = 3162.28\ \text{L}$$

From Equation 6.122,

$$(C_A)_f = \frac{(F_A)_0 (C_A)_0}{[(F_A)_0 + k_1 (V_{BFR})_{max} (C_A)_0]} = \frac{250 \times 2.5}{[250 + 0.05 \times 3162.28 \times 2.5]} = 0.96\ \text{mol/L}$$

From Equation 6.121,

$$\left(\frac{(C_I)_f}{(C_A)_0}\right)_{max} = \frac{1}{[1 + \sqrt{k_2/k_1}]^2} = \frac{1}{[1 + \sqrt{0.02/0.05}]^2} = 0.61$$

$$\Rightarrow \quad ((C_I)_f)_{max} = 0.61 \times 2.5 = 1.525\ \text{mol/L}$$

$$(C_T)_f = (C_A)_0 - (C_A)_f - ((C_I)_f)_{max} - 2.5 - 0.96 - 1.525 = 0.015\ \text{mol/L}$$

6.2.4.2 Parallel Reactions

Consider the parallel reactions represented by the balanced chemical Equations 6.63 and 6.64. For this reaction carried out in a BFR, from Equations 5.74 (Chapter 5) and 6.89 considering $\varepsilon_{LR} = 0$

$$V_{BFR} = \frac{(F_{A_{LR}})_0 \left[(C_{A_{LR}})_0 - (C_{A_{LR}})_f\right]}{(-r_{A_{LR}})_f (C_{A_{LR}})_0}$$

$$\Rightarrow \quad V_{BFR} = \frac{(F_{A_{LR}})_0 \left[(C_{A_{LR}})_0 - (C_{A_{LR}})_f\right]}{\left[(-r_{A_{LR}})_{1,f} + (-r_{A_{LR}})_{2,f}\right](C_{A_{LR}})_0} \tag{6.125}$$

$$\Rightarrow \quad V_{BFR} = \frac{(F_{A_{LR}})_0 \left[(C_{A_{LR}})_0 - (C_{A_{LR}})_f\right]}{(k_1 + k_2)(C_{A_{LR}})_0 \prod_{i=1}^{n} C_{A_i,f}^{a_i}} \tag{6.126}$$

This equation relates the outgoing concentration of the limiting reactant from the BFR with the volume of the BFR. Alternatively, from the relation $V_{BFR} = (F_{A_{LR}})_0 \tau_{BFR} / (C_{A_{LR}})_0$

$$\tau_{BFR} = \frac{\left[\left(C_{A_{LR}}\right)_0 - \left(C_{A_{LR}}\right)_f\right]}{\left[\left(-r_{A_{LR}}\right)_{1,f} + \left(-r_{A_{LR}}\right)_{2,f}\right]} \tag{6.127}$$

$$\tau_{BFR} = \frac{\left[\left(C_{A_{LR}}\right)_0 - \left(C_{A_{LR}}\right)_f\right]}{(k_1 + k_2)\prod_{i=1}^{n} C_{A_{i,f}}^{a_i}} \tag{6.128}$$

which relates the outgoing concentration of the limiting reactant from the BFR with the space time of the BFR.

For the product species P_j, a material balance over the reactor gives

$$\text{Input} + \text{Generation due to chemical reaction} = \text{Output}$$

$$\left(F_{P_j}\right)_0 + r_{P_j} V_{BFR} = \left(F_{P_j}\right)_f$$

$$\Rightarrow \left(C_{P_j}\right)_0 v_0 + r_{P_j} V_{BFR} = \left(C_{P_j}\right)_f v_0$$

$$\Rightarrow V_{BFR} = \frac{v_0\left[\left(C_{P_j}\right)_f - \left(C_{P_j}\right)_0\right]}{r_{P_j}} \tag{6.129}$$

$$\Rightarrow \tau_{BFR} = \frac{\left[\left(C_{P_j}\right)_f - \left(C_{P_j}\right)_0\right]}{r_{P_j}} \tag{6.130}$$

In combination with Equation 5.76 (Chapter 5),

$$V_{BFR} = \frac{v_0\left[\left(C_{P_j}\right)_f - \left(C_{P_j}\right)_0\right]}{(p_j/a_{LR})k_1\prod_{i=1}^{n} C_{A_{i,f}}^{a_i}} \tag{6.131}$$

From Equations 6.125 and 6.129,

$$\frac{v_0\left[\left(C_{P_j}\right)_f - \left(C_{P_j}\right)_0\right]}{r_{P_j}} = \frac{\left(F_{A_{LR}}\right)_0\left[\left(C_{A_{LR}}\right)_0 - \left(C_{A_{LR}}\right)_f\right]}{\left[\left(-r_{A_{LR}}\right)_1 + \left(-r_{A_{LR}}\right)_2\right]\left(C_{A_{LR}}\right)_0}$$

$$\Rightarrow \frac{\left[\left(C_{P_j}\right)_f - \left(C_{P_j}\right)_0\right]}{\left[\left(C_{A_{LR}}\right)_0 - \left(C_{A_{LR}}\right)_f\right]} = \frac{r_{P_j}}{\left[\left(-r_{A_{LR}}\right)_1 + \left(-r_{A_{LR}}\right)_2\right]} \tag{6.132}$$

and from Equations 6.126 and 6.129,

$$\frac{\left[\left(C_{P_j}\right)_f-\left(C_{P_j}\right)_0\right]}{\left[\left(C_{A_{LR}}\right)_0-\left(C_{A_{LR}}\right)_f\right]}=\frac{(p_j/a_{LR})k_1\prod_{i=1}^{n}C_{A_{i,f}}^{a_i}}{(k_1+k_2)\prod_{i=1}^{n}C_{A_{i,f}}^{a_i}}$$

$$\Rightarrow\quad\frac{\left[\left(C_{P_j}\right)_f-\left(C_{P_j}\right)_0\right]}{\left[\left(C_{A_{LR}}\right)_0-\left(C_{A_{LR}}\right)_f\right]}=\frac{(p_j/a_{LR})k_1}{(k_1+k_2)} \tag{6.133}$$

While Equations 6.126, 6.128, 6.131, and 6.133 are specific to the reacting system under consideration, Equations 6.125, 6.127, 6.129, 6.130, and 6.132 have general validity.

The steps used in engineering calculations are explained next.

If the outlet composition is to be determined for a given reactor volume or space time:

Step 1
From either Equation 6.125 or Equation 6.127 determine the outlet concentration of the limiting reactant.

Step 2
Determine the outlet concentration of the different product species using Equation 6.132.

If the reactor volume or space time required for a given outlet concentration of a desired product is to be determined:

Step 1
Using Equation 6.132 determine the outlet concentration of the limiting reactant $\left(C_{A_{LR}}\right)_f$ corresponding to the given concentration of the desired product $\left(C_{P_j}\right)_f$.

Step 2
Using Equation 6.125 determine the volume of the BFR required corresponding to the concentration $\left(C_{A_{LR}}\right)_f$. Alternatively, use Equation 6.127 to determine the space time.

EXAMPLE 6.13

Repeat Example 6.6 for an ideal isothermal BFR.

SOLUTION

$$-r_A=(-r_A)_1+(-r_A)_2=0.05C_A+0.02C_A=0.07C_A\ \text{mol/(L}\cdot\text{min)}$$

$$r_P = 0.05C_A \text{ mol/(L·min)}$$

$$(C_A)_0 = 2 \text{ mol/L}$$

$$V_{BFR} = 100 \text{ L} \quad v_0 = 10 \text{ L/min}$$

$$(F_A)_0 = v_0 \times (C_A)_0 = 10 \times 2 = 20 \text{ mol/min}$$

From Equation 6.125,

$$V_{BFR} = \frac{(F_A)_0\left[(C_A)_0 - (C_A)_f\right]}{\left[(-r_A)_1 + (-r_A)_2\right](C_A)_0} = \frac{(F_A)_0[(C_A)_0 - (C_A)_f]}{0.07(C_A)_f(C_A)_0}$$

$$\Rightarrow \quad (C_A)_f = \frac{(F_A)_0(C_A)_0}{0.07V_{BFR}(C_A)_0 + (F_A)_0} = \frac{20 \times 2}{0.07 \times 100 \times 2 + 20} = 1.17 \text{ mol/L}$$

From Equation 6.129,

$$V_{BFR} = \frac{v_0[(C_P)_f - (C_P)_0]}{r_P} = \frac{v_0(C_P)_f}{0.05(C_A)_f}$$

$$\Rightarrow \quad (C_P)_f = \frac{0.05(C_A)_f V_{BFR}}{v_0} = \frac{0.05 \times 1.17 \times 100}{10} = 0.59 \text{ mol/L}$$

$$C_Q = (C_A)_0 + (C_P)_0 + (C_Q)_0 - (C_A)_f - (C_P)_f = 2 - 1.17 - 0.59 = 0.24 \text{ mol/L}$$

6.2.5 Combination of BFRs

6.2.5.1 Series Combination

Consider the schematic shown in Figure 6.5a, which shows three BFRs of volumes $(V_{BFR})_1$, $(V_{BFR})_2$, and $(V_{BFR})_3$, respectively, joined in series. The output from the first BFR is the feed for the second BFR and the output from the second BFR is the feed for the third BFR.

For the first BFR, from Equation 6.86,

$$(V_{BFR})_1 = \frac{\left(F_{A_{LR}}\right)_0\left(X_{A_{LR}}\right)_1}{\left(-r_{A_{LR}}\right)_1} \tag{6.134}$$

For the second BFR, writing the material balance for the limiting reactant

$$\text{Input} = \text{Output} + \text{Disappearance due to chemical reaction}$$

$$\Rightarrow \quad \left(F_{A_{LR}}\right)_1 = \left(F_{A_{LR}}\right)_2 + \left(-r_{A_{LR}}\right)_2(V_{BFR})_2$$

$$\Rightarrow \quad \left(F_{A_{LR}}\right)_0\left[1 - \left(X_{A_{LR}}\right)_1\right] = \left(F_{A_{LR}}\right)_0\left[1 - \left(X_{A_{LR}}\right)_2\right] + \left(-r_{A_{LR}}\right)_2(V_{BFR})_2$$

$$\Rightarrow \quad (V_{BFR})_2 = \frac{\left(F_{A_{LR}}\right)_0\left[\left(X_{A_{LR}}\right)_2 - \left(X_{A_{LR}}\right)_1\right]}{\left(-r_{A_{LR}}\right)_2} \tag{6.135}$$

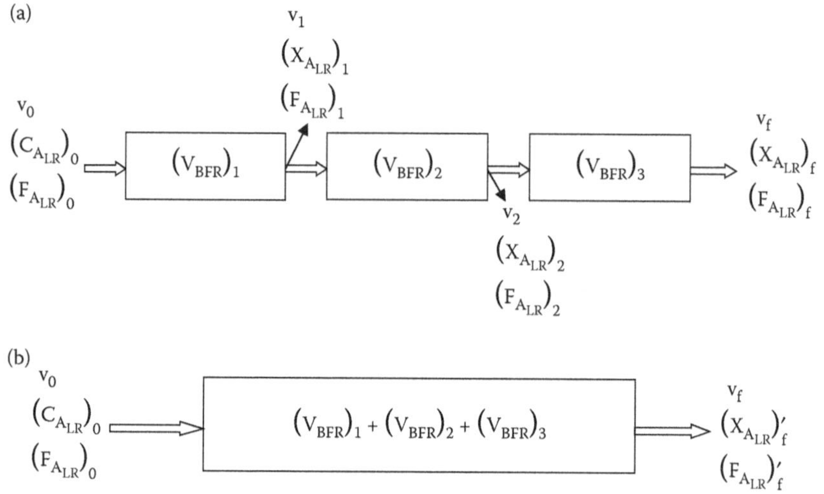

FIGURE 6.5 Schematic representation of ideal BFRs in series.

Similarly, for the third BFR

$$(V_{BFR})_3 = \frac{(F_{A_{LR}})_0\left[(X_{A_{LR}})_f - (X_{A_{LR}})_2\right]}{(-r_{A_{LR}})_3} \tag{6.136}$$

Adding Equations 6.134 through 6.136

$$(V_{BFR})_1 + (V_{BFR})_2 + (V_{BFR})_3 = (F_{A_{LR}})_0\left[\frac{(X_{A_{LR}})_1}{(-r_{A_{LR}})_1} + \frac{\left[(X_{A_{LR}})_2 - (X_{A_{LR}})_1\right]}{(-r_{A_{LR}})_2} + \frac{\left[(X_{A_{LR}})_f - (X_{A_{LR}})_2\right]}{(-r_{A_{LR}})_3}\right]$$

$$\Rightarrow \frac{(V_{BFR})_1 + (V_{BFR})_2 + (V_{BFR})_3}{(F_{A_{LR}})_0} = \left[\frac{(X_{A_{LR}})_1}{(-r_{A_{LR}})_1} + \frac{\left[(X_{A_{LR}})_2 - (X_{A_{LR}})_1\right]}{(-r_{A_{LR}})_2} + \frac{\left[(X_{A_{LR}})_f - (X_{A_{LR}})_2\right]}{(-r_{A_{LR}})_3}\right] \tag{6.137}$$

$$\Rightarrow \frac{(V_{BFR})_1 + (V_{BFR})_2 + (V_{BFR})_3}{(F_{A_{LR}})_0}$$

$$= \left[(X_{A_{LR}})_1 \left\{ \frac{1}{(-r_{A_{LR}})_1} - \frac{1}{(-r_{A_{LR}})_2} \right\} + (X_{A_{LR}})_2 \left\{ \frac{1}{(-r_{A_{LR}})_2} - \frac{1}{(-r_{A_{LR}})_3} \right\} + \frac{(X_{A_{LR}})_f}{(-r_{A_{LR}})_3} \right] \tag{6.138}$$

For the single BFR shown in Figure 6.5b, from Equation 6.86

$$(V_{BFR})_1 + (V_{BFR})_2 + (V_{BFR})_3 = \frac{(F_{A_{LR}})_0 (X_{A_{LR}})'_f}{(-r_{A_{LR}})_f}$$

$$\Rightarrow \frac{(V_{BFR})_1 + (V_{BFR})_2 + (V_{BFR})_3}{(F_{A_{LR}})_0} = \frac{(X_{A_{LR}})'_f}{(-r_{A_{LR}})_f} \tag{6.139}$$

Comparing Equations 6.138 and 6.139

$$\left[(X_{A_{LR}})_1 \left\{ \frac{1}{(-r_{A_{LR}})_1} - \frac{1}{(-r_{A_{LR}})_2} \right\} + (X_{A_{LR}})_2 \left\{ \frac{1}{(-r_{A_{LR}})_2} - \frac{1}{(-r_{A_{LR}})_3} \right\} + \frac{(X_{A_{LR}})_f}{(-r_{A_{LR}})_3} \right]$$

$$= \frac{(X_{A_{LR}})'_f}{(-r_{A_{LR}})_f} \tag{6.140}$$

From the schematic shown in Figure 6.5a,

$$(X_{A_{LR}})_1 < (X_{A_{LR}})_2 < (X_{A_{LR}})_f$$

So,

$$(-r_{A_{LR}})_1 > (-r_{A_{LR}})_2 > (-r_{A_{LR}})_3$$

and

$$\frac{1}{(-r_{A_{LR}})_1} < \frac{1}{(-r_{A_{LR}})_2} < \frac{1}{(-r_{A_{LR}})_3}$$

Therefore,

$$\left\{ \frac{1}{(-r_{A_{LR}})_1} - \frac{1}{(-r_{A_{LR}})_2} \right\} < 0$$

and

$$\left\{\frac{1}{\left(-r_{A_{LR}}\right)_2}-\frac{1}{\left(-r_{A_{LR}}\right)_3}\right\}<0$$

Thus, for Equation 6.140 to be valid,

$$\frac{\left(X_{A_{LR}}\right)_f}{\left(-r_{A_{LR}}\right)_3}>\frac{\left(X_{A_{LR}}\right)'_f}{\left(-r_{A_{LR}}\right)_f} \tag{6.141}$$

If $\left(X_{A_{LR}}\right)_f=\left(X_{A_{LR}}\right)'_f$, $\left(-r_{A_{LR}}\right)_3=\left(-r_{A_{LR}}\right)_f$ and the inequality 6.141 will be violated. So,

$$\left(X_{A_{LR}}\right)_f \neq \left(X_{A_{LR}}\right)'_f$$

If $\left(X_{A_{LR}}\right)_f<\left(X_{A_{LR}}\right)'_f$, $\left(-r_{A_{LR}}\right)_3>\left(-r_{A_{LR}}\right)_f$, and

$$\frac{\left(X_{A_{LR}}\right)_f}{\left(-r_{A_{LR}}\right)_3}<\frac{\left(X_{A_{LR}}\right)'_f}{\left(-r_{A_{LR}}\right)_f}$$

again violating inequality 6.141.

If $\left(X_{A_{LR}}\right)_f>\left(X_{A_{LR}}\right)'_f$, $\left(-r_{A_{LR}}\right)_3<\left(-r_{A_{LR}}\right)_f$, and

$$\frac{\left(X_{A_{LR}}\right)_f}{\left(-r_{A_{LR}}\right)_3}>\frac{\left(X_{A_{LR}}\right)'_f}{\left(-r_{A_{LR}}\right)_f}$$

conforming to inequality 6.141. So,

$$\left(X_{A_{LR}}\right)_f>\left(X_{A_{LR}}\right)'_f \tag{6.142}$$

This reasoning can be extended to any number of BFRs joined in series so that we can conclude that several BFRs of different volumes joined in series give higher conversion compared to a single BFR of volume equal to the sum of volumes of the BFRs joined in series.

Similarly, Equations 6.134 through 6.136 can be further extended for any number of BFRs joined in series. This combination of equations can be used for the performance evaluation of such series of BFRs. The steps are outlined next for three BFRs and can be extended further for any number of BFRs:

Step 1

From Equation 6.134 compute the conversion in the outgoing stream from the first BFR, $\left(X_{A_{LR}}\right)_1$.

Step 2

Substitute the value of $(X_{A_{LR}})_1$ calculated in Step 1 into Equation 6.135 and calculate the conversion in the outgoing stream from the second BFR, $(X_{A_{LR}})_2$.

Step 3

Substitute the value of $(X_{A_{LR}})_2$ calculated in Step 2 into Equation 6.136 and calculate the conversion in the outgoing stream from the third BFR, $(X_{A_{LR}})_f$.

If n BFRs of equal volume are joined in series, we have, from Equations 6.134 through 6.136

$$\frac{(X_{A_{LR}})_1}{(-r_{A_{LR}})_1} = \frac{\left[(X_{A_{LR}})_2 - (X_{A_{LR}})_1\right]}{(-r_{A_{LR}})_2} = \cdots = \frac{\left[(X_{A_{LR}})_n - (X_{A_{LR}})_{n-1}\right]}{(-r_{A_{LR}})_n} \tag{6.143}$$

And, from Equation 6.137,

$$\frac{nV_{BFR}}{(F_{A_{LR}})_0} = \left[\frac{(X_{A_{LR}})_1}{(-r_{A_{LR}})_1} + \frac{\left[(X_{A_{LR}})_2 - (X_{A_{LR}})_1\right]}{(-r_{A_{LR}})_2} + \cdots + \frac{\left[(X_{A_{LR}})_n - (X_{A_{LR}})_{n-1}\right]}{(-r_{A_{LR}})_n}\right] \tag{6.144}$$

Consider a single reactant first-order reaction with no volume change with reaction being carried out in n BFRs of equal volume joined in series. For this case, from Equation 6.134, for the first BFR,

$$\frac{V_{BFR}}{(F_A)_0} = \frac{(X_A)_1}{k(C_A)_0[1-(X_A)_1]}$$

$$\Rightarrow \quad k\tau_{BFR} = \frac{(X_A)_1}{[1-(X_A)_1]}$$

$$\Rightarrow \quad (X_A)_1 = \frac{k\tau_{BFR}}{[1+k\tau_{BFR}]} = 1 - \frac{1}{[1+k\tau_{BFR}]} \tag{6.145}$$

From Equations 6.135 and 6.145,

$$\frac{V_{BFR}}{(F_A)_0} = \frac{(X_A)_2 - (X_A)_1}{k(C_A)_0[1-(X_A)_2]}$$

$$\Rightarrow \quad k\tau_{BFR} = \frac{(X_A)_2 - \dfrac{k\tau_{BFR}}{[1+k\tau_{BFR}]}}{[1-(X_A)_2]}$$

$$\Rightarrow \quad (X_A)_2 = 1 - \frac{1}{[1+k\tau_{BFR}]^2} \tag{6.146}$$

From Equations 6.136 and 6.146,

$$\frac{V_{BFR}}{(F_A)_0} = \frac{(X_A)_3 - (X_A)_2}{k(C_A)_0[1-(X_A)_3]}$$

$$\Rightarrow \quad k\tau_{BFR} = \frac{(X_A)_3 - \left\{1 - \dfrac{k\tau_{BFR}}{[1+k\tau_{BFR}]^2}\right\}}{[1-(X_A)_3]} \tag{6.147}$$

$$\Rightarrow \quad (X_A)_2 = 1 - \frac{1}{[1+k\tau_{BFR}]^3}$$

And, for the nth BFR,

$$(X_A)_n = 1 - \frac{1}{[1+k\tau_{BFR}]^n} \tag{6.148}$$

For reacting systems with no volume change with reaction, Equation 6.144 can also be written in terms of the concentration of the limiting reactant. From Equations 2.24 (Chapter 2) and 6.143,

$$\frac{(C_A)_0 - (C_A)_1}{\left(-r_{A_{LR}}\right)_1} = \frac{[(C_A)_1 - (C_A)_2]}{\left(-r_{A_{LR}}\right)_2} = \cdots = \frac{[(C_A)_{n-1} - (C_A)_n]}{\left(-r_{A_{LR}}\right)_n} \tag{6.149}$$

And, from Equations 2.24 (Chapter 2) and 6.144,

$$\frac{nV_{BFR}}{\left(F_{A_{LR}}\right)_0} = \left[\frac{(C_A)_0 - (C_A)_1}{(C_A)_0\left(-r_{A_{LR}}\right)_1} + \frac{[(C_A)_1 - (C_A)_2]}{(C_A)_0\left(-r_{A_{LR}}\right)_2} + \cdots + \frac{[(C_A)_{n-1} - (C_A)_n]}{(C_A)_0\left(-r_{A_{LR}}\right)_n}\right]$$

$$\Rightarrow \quad n\tau_{BFR} = \frac{(C_A)_0 - (C_A)_1}{\left(-r_{A_{LR}}\right)_1} + \frac{[(C_A)_1 - (C_A)_2]}{\left(-r_{A_{LR}}\right)_2} + \cdots + \frac{[(C_A)_{n-1} - (C_A)_n]}{\left(-r_{A_{LR}}\right)_n} \tag{6.150}$$

Similarly, Equation 6.148 can be written in terms of the concentration of the limiting reactant. From Equations 2.24 (Chapter 2) and 6.148,

$$\frac{(C_A)_0 - (C_A)_n}{(C_A)_0} = 1 - \frac{1}{[1+k\tau_{BFR}]^n}$$

$$\Rightarrow \quad \frac{(C_A)_n}{(C_A)_0} = \frac{1}{[1+k\tau_{BFR}]^n} \tag{6.151}$$

EXAMPLE 6.14

The gas-phase irreversible reaction

$$A \rightarrow 2B$$

has a specific reaction rate constant 0.3 L/(min · mol). The reaction is to be carried out in two isothermal BFRs, each of volume 500 L, joined in series and operating at 500°C and 2 atm pressure. The feed to the first reactor consists of pure A at a total feed rate of 20 L/min. What is the conversion achieved after the second reactor?

SOLUTION

From Equations 2.26 and 2.27 (Chapter 2),

$$(C_A)_0 = \frac{(p_A)_0}{RT} = \frac{2}{0.082 \times 773} = 0.0316\ \frac{\text{mol}}{\text{L}}$$

From Equation 2.44 (Chapter 2),

$$\varepsilon_A = \frac{1+1-1}{1} = 1$$

$$(F_A)_0 = (C_A)_0 \times v_0 = 0.0316 \times 20 = 0.632\ \text{mol/min}$$

For the first reactor, from Equation 6.134,

$$(V_{BFR})_1 = \frac{(F_A)_0 (X_A)_1}{(-r_A)_1}$$

$$\Rightarrow\ 500 = \frac{0.632(X_A)_1}{k(C_A)_1} = \frac{0.632(X_A)_1}{k(C_A)_0[1-(X_A)_1]} = \frac{0.632(X_A)_1}{0.3 \times 0.0316[1-(X_A)_1]}$$

$$\Rightarrow\ (X_A)_1 = 0.88$$

For the second reactor, from Equation 6.135,

$$(V_{BFR})_2 = \frac{(F_A)_0[(X_A)_2 - (X_A)_1]}{(-r_A)_2}$$

$$\Rightarrow\ 500 = \frac{0.632[(X_A)_2 - 0.88]}{k(C_A)_2} = \frac{0.632[(X_A)_2 - 0.88]}{k(C_A)_0[1-(X_A)_2]} = \frac{0.632[(X_A)_2 - 0.88]}{0.3 \times 0.0316[1-(X_A)_2]}$$

$$\Rightarrow\ (X_A)_2 = 0.98$$

6.2.5.2 Parallel Combination

Consider the schematic shown in Figure 6.6a, which shows two BFRs of volumes $(V_{BFR})_1$, and $(V_{BFR})_2$, respectively, joined in parallel.

The volumetric feed rate v_0 is split into two streams and fed into the two BFRs so that

$$v_1 + v_2 = v_0 \tag{6.152}$$

$$(F_{A_{LR}})_1 + (F_{A_{LR}})_2 = (F_{A_{LR}})_0 \tag{6.153}$$

For the first BFR, from Equation 6.86,

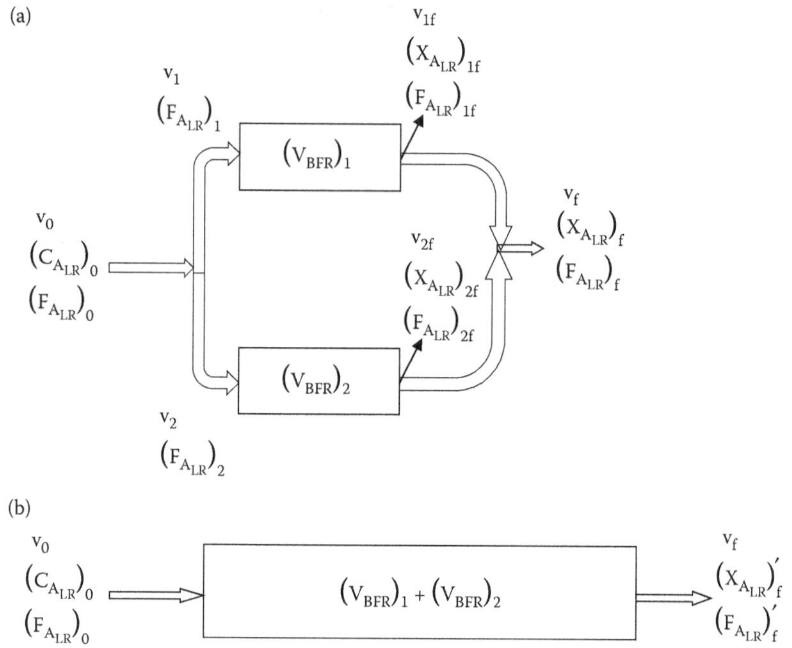

FIGURE 6.6 Schematic representation of ideal BFRs in parallel.

$$(V_{BFR})_1 = \frac{(F_{A_{LR}})_1 (X_{A_{LR}})_{1f}}{(-r_{A_{LR}})_{1f}} \tag{6.154}$$

$$\Rightarrow \frac{(V_{BFR})_1}{(F_{A_{LR}})_1} = \frac{(X_{A_{LR}})_{1f}}{(-r_{A_{LR}})_{1f}} \tag{6.155}$$

For the second BFR, from Equation 6.86,

$$(V_{BFR})_2 = \frac{(F_{A_{LR}})_2 (X_{A_{LR}})_{2f}}{(-r_{A_{LR}})_{2f}} \tag{6.156}$$

$$\Rightarrow \frac{(V_{BFR})_2}{(F_{A_{LR}})_2} = \frac{(X_{A_{LR}})_{2f}}{(-r_{A_{LR}})_{2f}} \tag{6.157}$$

If the feed $(F_{A_{LR}})_0$ is split in such a way that

$$\frac{(F_{A_{LR}})_1}{(F_{A_{LR}})_2} = \frac{(V_{BFR})_1}{(V_{BFR})_2} \tag{6.158}$$

$$\Rightarrow \frac{(F_{A_{LR}})_1 + (F_{A_{LR}})_2}{(F_{A_{LR}})_2} = \frac{(V_{BFR})_1 + (V_{BFR})_2}{(V_{BFR})_2}$$

$$\Rightarrow \frac{(F_{A_{LR}})_0}{(F_{A_{LR}})_2} = \frac{(V_{BFR})_1 + (V_{BFR})_2}{(V_{BFR})_2}$$

$$\Rightarrow (F_{A_{LR}})_0 = \frac{(V_{BFR})_1 + (V_{BFR})_2}{(V_{BFR})_2} \times (F_{A_{LR}})_2 \tag{6.159}$$

then the left-hand sides of Equations 6.155 and 6.157 are equal, so

$$\frac{(X_{A_{LR}})_{1f}}{(-r_{A_{LR}})_{1f}} = \frac{(X_{A_{LR}})_{2f}}{(-r_{A_{LR}})_{2f}}$$

$$\Rightarrow (X_{A_{LR}})_{1f} = (X_{A_{LR}})_{2f} \tag{6.160}$$

From the schematic shown in Figure 6.6a and Equation 6.160,

$$(F_{A_{LR}})_f = (F_{A_{LR}})_{1f} + (F_{A_{LR}})_{2f}$$

$$\Rightarrow (F_{A_{LR}})_0\left[1-(X_{A_{LR}})_f\right] = (F_{A_{LR}})_1\left[1-(X_{A_{LR}})_{1f}\right] + (F_{A_{LR}})_2\left[1-(X_{A_{LR}})_{2f}\right]$$

$$\Rightarrow (F_{A_{LR}})_0\left[1-(X_{A_{LR}})_f\right] = (F_{A_{LR}})_1\left[1-(X_{A_{LR}})_{1f}\right] + (F_{A_{LR}})_2\left[1-(X_{A_{LR}})_{1f}\right] \tag{6.161}$$

$$\Rightarrow (F_{A_{LR}})_0\left[1-(X_{A_{LR}})_f\right] = \left[(F_{A_{LR}})_1 + (F_{A_{LR}})_2\right]\left[1-(X_{A_{LR}})_{1f}\right]$$

$$\Rightarrow (F_{A_{LR}})_0\left[1-(X_{A_{LR}})_f\right] = (F_{A_{LR}})_0\left[1-(X_{A_{LR}})_{1f}\right]$$

$$\Rightarrow (X_{A_{LR}})_f = (X_{A_{LR}})_{1f} = (X_{A_{LR}})_{2f} \tag{6.162}$$

Now consider the schematic shown in Figure 6.6b, which shows a single BFR of volume $(V_{BFR})_1 + (V_{BFR})_2$. For this BFR, from Equations 6.86 and 6.159,

$$(V_{BFR})_1 + (V_{BFR})_2 = \frac{(F_{A_{LR}})_0 (X_{A_{LR}})_f'}{(-r_{A_{LR}})_f'} \tag{6.163}$$

$$\Rightarrow (V_{BFR})_1 + (V_{BFR})_2 = \frac{(V_{BFR})_1 + (V_{BFR})_2}{(V_{BFR})_2} \times (F_{A_{LR}})_2 \times \frac{(X_{A_{LR}})_f'}{(-r_{A_{LR}})_f'}$$

$$\Rightarrow \quad \frac{(V_{BFR})_2}{(F_{A_{LR}})_2} = \frac{(X_{A_{LR}})'_f}{(-r_{A_{LR}})'_f} \tag{6.164}$$

Comparing Equations 6.157, 6.160, 6.162, and 6.164, we have

$$(X_{A_{LR}})_{1f} = (X_{A_{LR}})_{2f} = (X_{A_{LR}})'_f = (X_{A_{LR}})_f \tag{6.165}$$

The result of Equation 6.165 can be extended to n BFRs joined in parallel provided the feed is split between them in such a way that

$$\frac{(F_{A_{LR}})_1}{(V_{BFR})_1} = \frac{(F_{A_{LR}})_2}{(V_{BFR})_2} = \frac{(F_{A_{LR}})_3}{(V_{BFR})_3} = \cdots = \frac{(F_{A_{LR}})_n}{(V_{BFR})_n}$$

Thus, BFRs joined in parallel give the same combined conversion as a single BFR of volume equal to the sum of volumes of the BFRs joined in parallel provided the total feed is split into the feed to individual reactors in the ratio of the volumes of the individual reactors. If the feed is split in any other ratio, the combined conversion will be less than that obtained in a single BFR of volume equal to the sum of volumes of the BFRs joined in parallel.

Considering the complexity of plant layout required for combining BFRs, it is better to have a single large BFR rather than several small BFRs joined in parallel.

EXAMPLE 6.15

Repeat Example 6.7 if the PFRs are replaced with BFRs.

SOLUTION

From the data given in Example 6.7 and Solution to Example 6.7,

$$(V_{BFR})_1 = 30 \text{ L}, \quad (V_{BFR})_2 = 70 \text{ L},$$

$$(F_A)_1 = 50 \text{ mol/min}, \quad (F_A)_2 = 50 \text{ mol/min}, \quad (F_A)_0 = 100 \text{ mol/min}$$

$$(C_A)_0 = 1 \text{ mol/L}$$

For the first reactor, from Equation 6.155,

$$\frac{(V_{BFR})_1}{(F_A)_1} = \frac{(X_A)_{1f}}{(-r_A)_{1f}} = \frac{(X_A)_{1f}}{(C_A)_0^2[1-(X_A)_{1f}]^2}$$

$$\Rightarrow \quad \frac{30}{50} = \frac{(X_A)_{1f}}{[1-(X_A)_{1f}]^2}$$

$$\Rightarrow \quad (X_A)_{1f} = 0.29$$

For the second reactor, from Equation 6.157

$$\frac{(V_{BFR})_2}{(F_A)_2} = \frac{(X_A)_{2f}}{(-r_A)_{2f}} = \frac{(X_A)_{2f}}{(C_A)_0^2[1-(X_A)_{2f}]^2}$$

$$\Rightarrow \quad \frac{70}{50} = \frac{(X_A)_{2f}}{[1-(X_A)_{2f}]^2}$$

$$\Rightarrow \quad (X_A)_{2f} = 0.44$$

From Equation 6.161,

$$(F_A)_0[1-(X_A)_f] = (F_A)_1[1-(X_A)_{1f}] + (F_A)_2[1-(X_A)_{2f}]$$

$$\Rightarrow \quad 100[1-(X_A)_f] = 50[1-0.29] + 50[1-0.44]$$

$$\Rightarrow \quad (X_A)_f = 0.365$$

For the single reactor from Equation 6.162,

$$(V_{BFR})_1 + (V_{BFR})_2 = \frac{(F_A)_0(X_A)'_f}{(-r_A)'_f}$$

$$\Rightarrow \quad \frac{(V_{BFR})_1 + (V_{BFR})_2}{(F_A)_0} = \frac{(X_A)'_f}{(C_A)_0^2[1-(X_A)'_f]^2}$$

$$\Rightarrow \quad \frac{30+70}{100} = \frac{(X_A)'_f}{[1-(X_A)'_f]^2}$$

$$\Rightarrow \quad (X_A)'_f = 0.38$$

We can see that $(X_A)_f < (X_A)'_f$

6.3 COMPARISON OF PFR AND BFR

6.3.1 Single Irreversible Reactions

If we compare the results of Examples 6.1 and 6.8, we can see that for the same conversion and feed conditions, the volume required for BFR is more than the volume required for PFR. Similarly, from the results of Examples 6.2 and 6.9 we find that for the same conversion and feed condition the space time required for BFR is more than the space time required for PFR. Let us find out if we can generalize these results for all single irreversible reactions. For a PFR, from Equation 6.1

$$V_{PFR} = (F_{ALR})_0 \int_0^{(X_{ALR})_f} \frac{dX_{ALR}}{-r_{ALR}}$$

$$\Rightarrow \quad \frac{V_{PFR}}{(F_{ALR})_0} = \int_0^{(X_{ALR})_f} \frac{dX_{ALR}}{-r_{ALR}} \tag{6.166}$$

From the properties of integral, as $dX_{A_{LR}} \rightarrow 0$,

$$\int_0^{(X_{A_{LR}})_f} \frac{dX_{A_{LR}}}{-r_{A_{LR}}} = \left[\left(\frac{1}{-r_{A_{LR}}}\right)_{X_{A_{LR}}=0} + \left(\frac{1}{-r_{A_{LR}}}\right)_{X_{A_{LR}}=dX_{A_{LR}}} + \left(\frac{1}{-r_{A_{LR}}}\right)_{X_{A_{LR}}=2dX_{A_{LR}}} + \ldots\infty \text{ terms} + \left(\frac{1}{-r_{A_{LR}}}\right)_f\right] dX_{A_{LR}}$$

$$\Rightarrow \frac{V_{PFR}}{(F_{A_{LR}})_0} = \left[\left(\frac{1}{-r_{A_{LR}}}\right)_{X_{A_{LR}}=0} + \left(\frac{1}{-r_{A_{LR}}}\right)_{X_{A_{LR}}=dX_{A_{LR}}} + \left(\frac{1}{-r_{A_{LR}}}\right)_{X_{A_{LR}}=2dX_{A_{LR}}} + \ldots\infty \text{ terms} + \left(\frac{1}{-r_{A_{LR}}}\right)_{X_{A_{LR}}=(X_{A_{LR}})_f}\right] dX_{A_{LR}} \quad (6.167)$$

For a BFR, from Equation 6.86,

$$V_{BFR} = \frac{(F_{A_{LR}})_0 (X_{A_{LR}})_f}{(-r_{A_{LR}})_f}$$

$$\Rightarrow \frac{V_{BFR}}{(F_{A_{LR}})_0} = \frac{(X_{A_{LR}})_f}{(-r_{A_{LR}})_f} = \frac{1}{(-r_{A_{LR}})_f} \int_0^{(X_{A_{LR}})_f} dX_{A_{LR}}$$

$$\Rightarrow \frac{V_{BFR}}{(F_{A_{LR}})_0} = \left[\left(\frac{1}{-r_{A_{LR}}}\right)_f + \left(\frac{1}{-r_{A_{LR}}}\right)_f + \left(\frac{1}{-r_{A_{LR}}}\right)_f + \ldots\infty \text{ terms} + \left(\frac{1}{-r_{A_{LR}}}\right)_f\right] dX_{A_{LR}} \quad (6.168)$$

For a single irreversible reaction, the rate of reaction decreases with increasing conversion; therefore,

$$(-r_{A_{LR}})_{X_{A_{LR}}=0} > (-r_{A_{LR}})_{X_{A_{LR}}=dX_{A_{LR}}} > (-r_{A_{LR}})_{X_{A_{LR}}=dX_{A_{LR}}} > \ldots > (-r_{A_{LR}})_f$$

$$\Rightarrow \left(\frac{1}{-r_{A_{LR}}}\right)_{X_{A_{LR}}=0} < \left(\frac{1}{-r_{A_{LR}}}\right)_{X_{A_{LR}}=dX_{A_{LR}}} < \left(\frac{1}{-r_{A_{LR}}}\right)_{X_{A_{LR}}=2dX_{A_{LR}}} < \ldots < \left(\frac{1}{-r_{A_{LR}}}\right)_f \quad (6.169)$$

Thus, the RHS of Equation 6.167 is less than the RHS of Equation 6.168, so that

$$\frac{V_{PFR}}{(F_{ALR})_0} < \frac{V_{BFR}}{(F_{ALR})_0}$$

$$\Rightarrow \quad V_{PFR} < V_{BFR} \tag{6.170}$$

Therefore, we can conclude that to achieve the same conversion for a single irreversible reaction, a PFR requires less volume than a BFR. The following information is important while comparing the performance of a PFR and a BFR:

1. The higher the reaction order, the lower is the ratio V_{PFR}/V_{BFR} for the same final conversion. So, the advantage of using a PFR is more pronounced for higher-order reactions.
2. For any reaction order the higher the final conversion required, the lower is the ratio V_{PFR}/V_{BFR}. So, the advantage of using a PFR is more pronounced for higher conversions.
3. The larger the value of ε_{LR}, the lower is the ratio V_{PFR}/V_{BFR}. So, the advantage of using a PFR is more pronounced for larger positive values of ε_{LR}. It is less pronounced for negative values of ε_{LR}. In this context, it is imperative to note that the limiting lower value of ε_{LR}, from Equation 2.44 (Chapter 2) is –1.

The above result can also be interpreted graphically as shown in Figure 6.7. From this figure,

$$\frac{V_{PFR}}{(F_{ALR})_0} = \int_0^{(X_{ALR})_f} \frac{1}{-r_{ALR}} dX_{ALR} = \text{Area bounded by ABCEF} \tag{6.171}$$

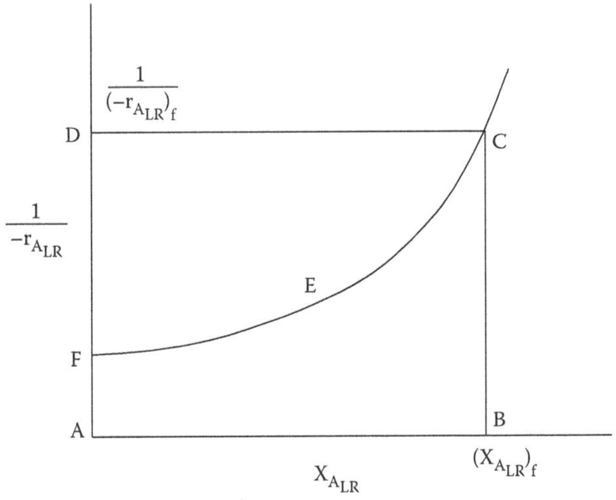

FIGURE 6.7 Graphical interpretation of performance of flow reactors.

and

$$\frac{V_{BFR}}{(F_{ALR})_0} = \frac{(X_{ALR})_f}{(-r_{ALR})_f} = \frac{1}{(-r_{ALR})_f} \times (X_{ALR})_f = \text{Area of the rectangle ABCD} \quad (6.172)$$

It is immediately evident that $V_{PFR} < V_{BFR}$, which is the result we got in Equation 6.170.

Consider a series combination of infinite number of BFRs with total volume V_{BFR} where each BFR is infinitesimally small of volume dV_{BFR} such that $dV_{BFR} \to 0$, $n \to \infty$, and $V_{BFR} = ndV_{BFR}$. After each BFR, the conversion increases by an infinitesimally small extent dX_{ALR} such that $dX_{ALR} \to 0$. The final conversion is $(X_{ALR})_f$. In Section 6.2.5.1 we have seen that for n equal-sized BFRs joined in series, from Equation 6.144

$$\frac{ndV_{BFR}}{(F_{ALR})_0} = \left[\frac{dX_{ALR}}{(-r_{ALR})_{X_{ALR}=0}} + \frac{dX_{ALR}}{(-r_{ALR})_{X_{ALR}=dX_{ALR}}} + \ldots \infty \text{ terms} \cdots + \frac{dX_{ALR}}{(-r_{ALR})_{X_{ALR}=(X_{ALR})_f}} \right]$$

$$\Rightarrow \frac{ndV_{BFR}}{(F_{ALR})_0} = \left[\left(\frac{1}{-r_{ALR}} \right)_{X_{ALR}=0} + \left(\frac{1}{-r_{ALR}} \right)_{X_{ALR}=dX_{ALR}} + \ldots \infty \text{ terms} + \left(\frac{1}{-r_{ALR}} \right)_{X_{ALR}=(X_{ALR})_f} \right] dX_{ALR} \quad (6.173)$$

Comparing this equation with Equation 6.167 we can see that

$$V_{PFR} = ndV_{BFR} \quad (6.174)$$

Thus, an infinite number of infinitesimally small BFRs joined in series approach in performance to that of a PFR of equivalent total volume.

6.3.2 Multiple Reactions

6.3.2.1 Series Reactions

Consider the series reaction

$$A \to I \to T$$

with the specific reaction rate constants for the two consecutive reaction steps being k_1 and k_2, respectively. If the species I and T are not present initially, we have from Equations 6.51 and 6.120, for maximum outgoing concentration of I

$$\frac{\tau_{PFR}}{\tau_{BFR}} = \frac{\sqrt{k_1 k_2}\ln(k_2/k_1)}{(k_2 - k_1)} \quad k_1 \neq k_2 \tag{6.175}$$

and

$$\frac{\tau_{PFR}}{\tau_{BFR}} = 1 \quad k_1 = k_2 \tag{6.176}$$

The variation in the ratio τ_{PFR}/τ_{BFR} is shown in Figure 6.8a for different values of k_1 and k_2. It is evident that the space time required for the maximum concentration of I is the same for PFR and BFR when the two specific reaction rate constants are the same. However, for all other values of k_1 and k_2, the space time required in PFR for the maximum concentration of I is less than the space time required in BFR (van de Vusse, 1964). The advantage of using a PFR becomes more and more pronounced as the difference between k_1 and k_2 increases. As shown in Figure 6.8b, the maximum attainable concentration of I is more in a PFR than that in a BFR. Interestingly, the ratio $(C_I)_{max,\ PFR}/(C_I)_{max,\ BFR}$ is highest when the two specific reaction rate constants are the same.

EXAMPLE 6.16

For the series reaction,

$$A \rightarrow I \rightarrow T$$

$k_1 = 0.3\ \text{min}^{-1}$ and $k_2 = 1\ \text{min}^{-1}$, respectively. With all other conditions remaining the same, the reaction is first carried out in an isothermal PFR and then in an isothermal BFR. The species I and T are not present initially. Both the reactors give the maximum attainable outlet concentrations of I. By how much percentage is the BFR bigger than the PFR?

Solution

For maximum outgoing concentration of I, from Equation 6.175

$$\frac{\tau_{PFR}}{\tau_{BFR}} = \frac{\sqrt{k_1 k_2}\ln(k_2/k_1)}{(k_2 - k_1)} = \frac{\sqrt{0.3 \times 1}\ln(1/0.3)}{1 - 0.3} = 0.94$$

From all other parameters being the same,

$$\frac{\tau_{PFR}}{\tau_{BFR}} = \frac{V_{PFR}}{V_{BFR}}$$

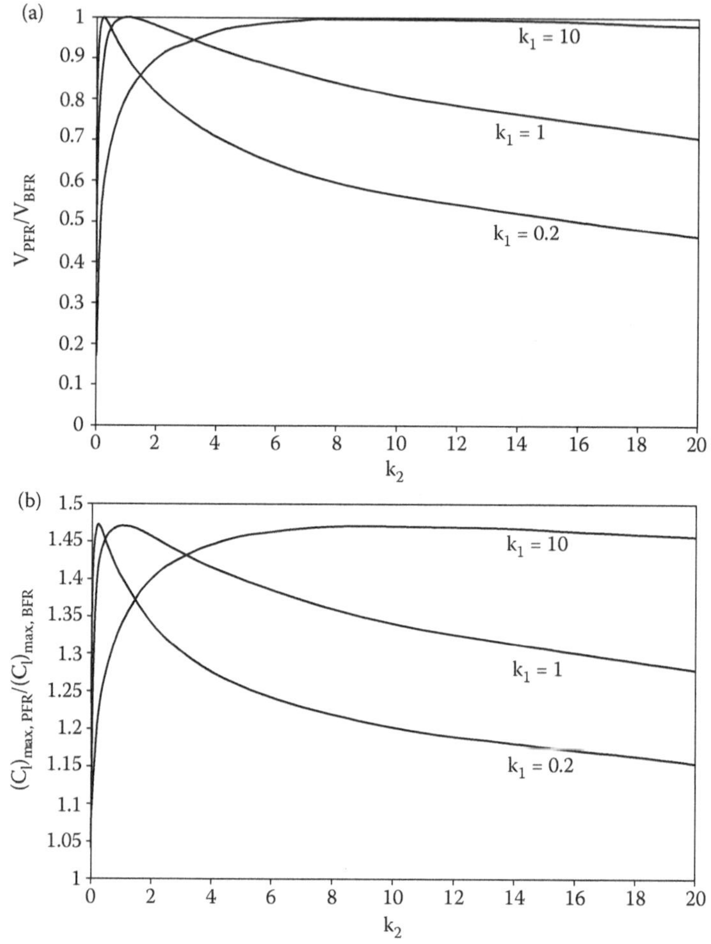

FIGURE 6.8 Comparative performance of PFR and BFR in series reaction for different values of specific reaction rate constants for maximum outlet concentration of the intermediate.

$$\Rightarrow \quad V_{BFR} = \frac{V_{PFR}}{0.94}$$

$$\Rightarrow \quad \frac{V_{BFR} - V_{PFR}}{V_{PFR}} = \frac{(V_{PFR}/0.94) - V_{PFR}}{V_{PFR}} = 0.064$$

So the BFR is 6.4% bigger in volume than the PFR.

6.3.2.2 Parallel Reactions

Consider the parallel reactions

$$\sum_{i=1}^{n} a_i A_i \rightarrow \sum_{j=1}^{l} p_j P_j$$

$$\sum_{i=1}^{n} a_i A_i \rightarrow \sum_{r=1}^{r} q_r Q_r$$

For this reacting system, Equations 6.69 and 6.133 suggest that the overall fractional yield, that is,

$$\frac{\left[\left(C_{P_j}\right)_f - \left(C_{P_j}\right)_0\right]}{\left[\left(C_{A_{LR}}\right)_0 - \left(C_{A_{LR}}\right)_f\right]}$$

is the same for both PFR and BFR. However, in general, the overall fractional yields are different for the PFR and the BFR. Let us examine a few specific cases for the following reaction:

$$A \rightarrow P \quad (-r_A)_1 = r_P = k_1 C_A^{n_1}$$

$$A \rightarrow Q \quad (-r_A)_2 = r_Q = k_2 C_A^{n_2}$$

If this reaction is carried out in a PFR, from Equation 6.68, for product P, the desired product

$$(\Psi_{ovl})_{PFR} = \frac{-1}{(C_A)_0 - (C_A)_f} \int_{(C_A)_0}^{(C_A)_f} \Psi_{inst} d(C_A)$$

$$\Rightarrow \quad (\Psi_{ovl})_{PFR} = \frac{-1}{(C_A)_0 - (C_A)_f} \int_{(C_A)_0}^{(C_A)_f} \frac{r_P}{(-r_A)_1 + (-r_A)_2} d(C_A)$$

$$\Rightarrow \quad (\Psi_{ovl})_{PFR} = \frac{-1}{(C_A)_0 - (C_A)_f} \int_{(C_A)_0}^{(C_A)_f} \frac{k_1 C_A^{n_1}}{k_1 C_A^{n_1} + k_2 C_A^{n_2}} d(C_A) \qquad (6.177)$$

If the reaction is carried out in a BFR, from Equation 6.132,

$$(\Psi_{ovl})_{BFR} = \frac{(r_P)_f}{[(-r_A)_{1,f} + (-r_A)_{2,f}]}$$

$$\Rightarrow \quad (\Psi_{ovl})_{BFR} = \frac{k_1 (C_A)_f^{n_1}}{[k_1 (C_A)_f^{n_1} + k_2 (C_A)_f^{n_2}]} \qquad (6.178)$$

Case 1; $n_1 = n_2 - 1$
From Equation 6.177,

$$(\Psi_{ovl})_{PFR} = \frac{-1}{(C_A)_0 - (C_A)_f} \int_{(C_A)_0}^{(C_A)_f} \frac{k_1 C_A^{n_1}}{k_1 C_A^{n_1} + k_2 C_A^{n_1+1}} dC_A$$

$$\Rightarrow \quad (\Psi_{ovl})_{PFR} = \frac{k_1/k_2}{(C_A)_0 - (C_A)_f} \ln\left[\frac{k_1 + k_2(C_A)_0}{k_1 + k_2(C_A)_f}\right] \tag{6.179}$$

$$\Rightarrow \quad (\Psi_{ovl})_{PFR} = \frac{k_1}{k_2(C_A)_0(X_A)_f} \ln\left[\frac{k_1 + k_2(C_A)_0}{k_1 + k_2(C_A)_0\{1-(X_A)_f\}}\right] \tag{6.180}$$

From Equation 6.178,

$$(\Psi_{ovl})_{BFR} = \frac{k_1 (C_A)_f^{n_1}}{[k_1 (C_A)_f^{n_1} + k_2 (C_A)_f^{n_1+1}]}$$

$$\Rightarrow \quad (\Psi_{ovl})_{BFR} = \frac{k_1}{[k_1 + k_2(C_A)_f]} \tag{6.181}$$

$$\Rightarrow \quad (\Psi_{ovl})_{BFR} = \frac{k_1}{[k_1 + k_2(C_A)_0\{1-(X_A)_f\}]} \tag{6.182}$$

Dividing Equation 6.180 by Equation 6.182,

$$\frac{(\Psi_{ovl})_{PFR}}{(\Psi_{ovl})_{BFR}} = \frac{[k_1 + k_2(C_A)_0\{1-(X_A)_f\}]}{k_2(C_A)_0(X_A)_f} \ln\left[\frac{k_1 + k_2(C_A)_0}{k_1 + k_2(C_A)_0\{1-(X_A)_f\}}\right]$$

$$\Rightarrow \quad \frac{(\Psi_{ovl})_{PFR}}{(\Psi_{ovl})_{BFR}} = \frac{[1 + (k_2/k_1)(C_A)_0\{1-(X_A)_f\}]}{(k_2/k_1)(C_A)_0(X_A)_f} \ln\left[\frac{1 + (k_2/k_1)(C_A)_0}{1 + (k_2/k_1)(C_A)_0\{1-(X_A)_f\}}\right] \tag{6.183}$$

Besides the fractional conversion of the reactant A, the ratio $(\Psi_{ovl})_{PFR}/(\Psi_{ovl})_{BFR}$ depends on the dimensionless parameter $k_2(C_A)_0/k_1$. The variation of the ratio $(\Psi_{ovl})_{PFR}/(\Psi_{ovl})_{BFR}$ with varying conversion and different values of the parameter $k_2(C_A)_0/k_1$ is graphically depicted in Figure 6.9.

It is evident that for any conversion the ratio $(\Psi_{ovl})_{PFR}/(\Psi_{ovl})_{BFR}$ steeply declines with increase in the parameter $k_2(C_A)_0/k_1$. If the value of this parameter is less, the PFR and the BFR give almost the same fractional yield. However, for higher values of this parameter the fractional yield obtained in a BFR far exceeds that obtained in a PFR.

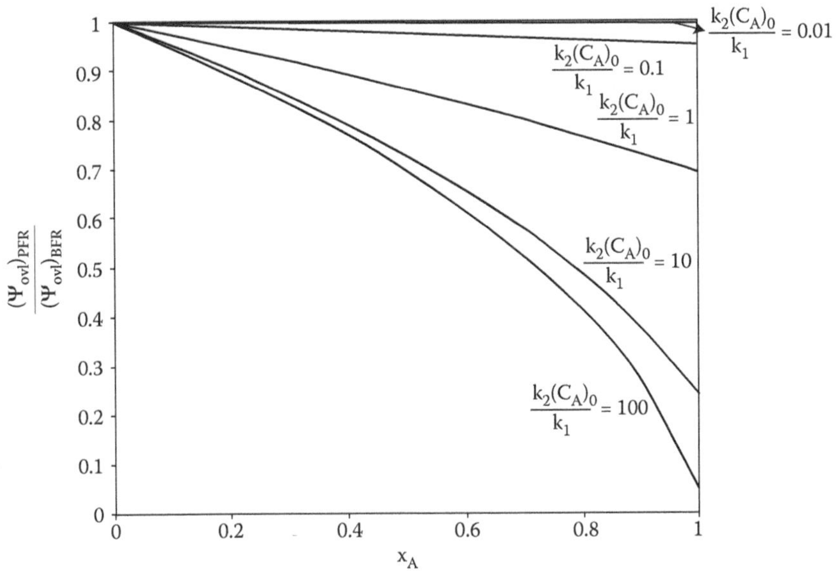

FIGURE 6.9 Overall fractional yield in PFR and BFR when the desirable reaction is of order n and the undesirable reaction is of order n + 1.

Case 2; $n_1 = n_2 - 2$

From Equation 6.177,

$$(\Psi_{ovl})_{PFR} = \frac{-1}{(C_A)_0 - (C_A)_f} \int_{(C_A)_0}^{(C_A)_f} \frac{k_1 C_A^{n_1}}{k_1 C_A^{n_1} + k_2 C_A^{n_1+2}} dC_A$$

$$\Rightarrow (\Psi_{ovl})_{PFR} = \frac{-1}{(C_A)_0 - (C_A)_f} \int_{(C_A)_0}^{(C_A)_f} \frac{1}{1 + (k_2/k_1) C_A^2} dC_A \tag{6.184}$$

$$\Rightarrow (\Psi_{ovl})_{PFR} = \frac{\sqrt{k_1/k_2}}{(C_A)_0 (X_A)_f} \tan^{-1}\left[\frac{\sqrt{k_2/k_1}(C_A)_0 (X_A)_f}{1 + (k_2/k_1)(C_A)_0^2\{1-(X_A)_f\}}\right]$$

From Equation 6.178,

$$(\Psi_{ovl})_{BFR} = \frac{k_1 (C_A)_f^{n_1}}{[k_1 (C_A)_f^{n_1} + k_2 (C_A)_f^{n_1+2}]}$$

$$\Rightarrow (\Psi_{ovl})_{BFR} = \frac{1}{[1 + (k_2/k_1)(C_A)_f^2]} \tag{6.185}$$

$$\Rightarrow (\Psi_{ovl})_{BFR} = \frac{1}{[1 + (k_2/k_1)(C_A)_0^2 [1-(X_A)_f]^2]}$$

Dividing Equation 6.184 by Equation 6.185,

$$\frac{(\Psi_{ovl})_{PFR}}{(\Psi_{ovl})_{BFR}} = \frac{[1+(k_2/k_1)(C_A)_0^2[1-(X_A)_f]^2]}{\sqrt{k_2/k_1}(C_A)_0(X_A)_f}\tan^{-1}\left[\frac{\sqrt{k_2/k_1}(C_A)_0(X_A)_f}{1+(k_2/k_1)(C_A)_0^2\{1-(X_A)_f\}}\right] \tag{6.186}$$

Besides the fractional conversion of the reactant A, the ratio $(\Psi_{ovl})_{PFR}/(\Psi_{ovl})_{BFR}$ depends on the dimensionless parameter $\sqrt{(k_2/k_1)}(C_A)_0$. The variation of the ratio $(\Psi_{ovl})_{PFR}/(\Psi_{ovl})_{BFR}$ with varying conversion and different values of the parameter $\sqrt{k_2/k_1}(C_A)_0$ is graphically depicted in Figure 6.10. If the value of this parameter is 0.1 or less, the PFR and the BFR give almost the same fractional yield. The ratio $(\Psi_{ovl})_{PFR}/(\Psi_{ovl})_{BFR}$ steeply declines with increase in the parameter $\sqrt{k_2/k_1}(C_A)_0$. So, for higher values of this parameter, the fractional yield obtained in a BFR far exceeds that obtained in a PFR. Interestingly, for intermediate values of the parameter $\sqrt{k_2/k_1}(C_A)_0$, the ratio $(\Psi_{ovl})_{PFR}/(\Psi_{ovl})_{BFR}$ passes through a minimum at some intermediate conversion.

Case 3; $n_1 = n_2 + 1$

From Equation 6.177,

$$(\Psi_{ovl})_{PFR} = \frac{-1}{(C_A)_0-(C_A)_f}\int_{(C_A)_0}^{(C_A)_f}\frac{k_1C_A^{n_1}}{k_1C_A^{n_1}+k_2C_A^{n_1-1}}dC_A$$

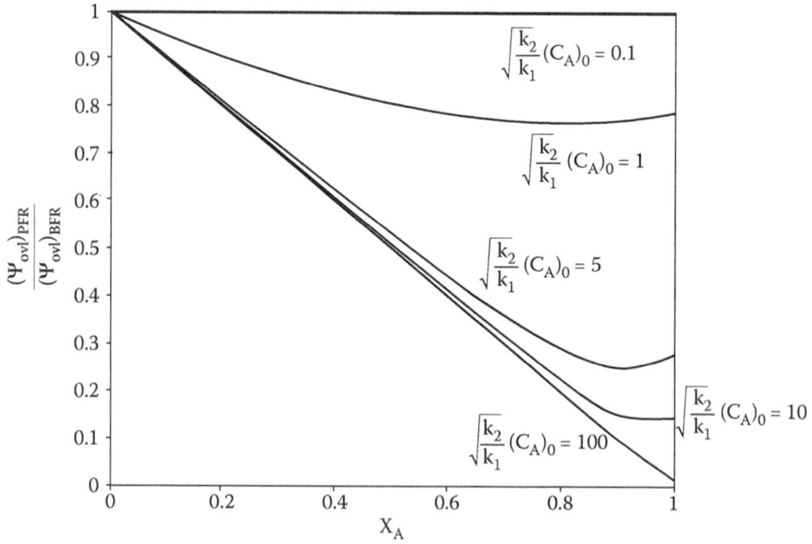

FIGURE 6.10 Overall fractional yield in PFR and BFR when the desirable reaction is of order n and the undesirable reaction is of order n + 2.

$$\Rightarrow \quad (\Psi_{ovl})_{PFR} = \frac{k_2/k_1}{(C_A)_0 - (C_A)_f}\left[\frac{k_1}{k_2}\{(C_A)_0 - (C_A)_f\} + \ln\left\{\frac{1+(k_1/k_2)(C_A)_0\{1-(X_A)_f\}}{1+(k_1/k_2)(C_A)_0}\right\}\right]$$

$$\Rightarrow \quad (\Psi_{ovl})_{PFR} = 1 + \frac{k_2/k_1}{(C_A)_0 - (C_A)_f}\ln\left\{\frac{1+(k_1/k_2)(C_A)_0\{1-(X_A)_f\}}{1+(k_1/k_2)(C_A)_0}\right\}$$

$$\Rightarrow \quad (\Psi_{ovl})_{PFR} = 1 + \frac{1}{(k_1/k_2)(C_A)_0(X_A)_f}\ln\left\{\frac{1+(k_1/k_2)(C_A)_0\{1-(X_A)_f\}}{1+(k_1/k_2)(C_A)_0}\right\} \tag{6.187}$$

From Equation 6.178,

$$(\Psi_{ovl})_{BFR} = \frac{k_1(C_A)_f^{n_1}}{[k_1(C_A)_f^{n_1} + k_2(C_A)_f^{n_1-1}]}$$

$$\Rightarrow \quad (\Psi_{ovl})_{BFR} = \frac{(k_1/k_2)(C_A)_f}{[1+(k_1/k_2)(C_A)_f]} = \frac{(k_1/k_2)(C_A)_0\{1-(X_A)_f\}}{[1+(k_1/k_2)(C_A)_0\{1-(X_A)_f\}]} \tag{6.188}$$

Dividing Equation 6.187 by Equation 6.188,

$$\frac{(\Psi_{ovl})_{BFR}}{(\Psi_{ovl})_{PFR}} = \frac{1}{\dfrac{1+\dfrac{k_1}{k_2}(C_A)_0\{1-(X_A)_f\}}{\dfrac{k_1}{k_2}(C_A)_0\{1-(X_A)_f\}} + \dfrac{1+\dfrac{k_1}{k_2}(C_A)_0\{1-(X_A)_f\}}{\left[\dfrac{k_1}{k_2}(C_A)_0\{1-(X_A)_f\}\right]\left[\dfrac{k_1}{k_2}(C_A)_0(X_A)_f\right]}\ln\left\{\dfrac{1+\dfrac{k_1}{k_2}(C_A)_0\{1-(X_A)_f\}}{1+\dfrac{k_1}{k_2}(C_A)_0}\right\}} \tag{6.189}$$

Besides the fractional conversion of the reactant A, the ratio $(\Psi_{ovl})_{BFR}/(\Psi_{ovl})_{PFR}$ depends on the dimensionless parameter $(k_1/k_2)(C_A)_0$. The variation of the ratio $(\Psi_{ovl})_{BFR}/(\Psi_{ovl})_{PFR}$ with varying conversion and different values of the parameter $(k_1/k_2)(C_A)_0$ is graphically depicted in Figure 6.11.

It is evident that in this case the PFR gives higher fractional yield than the BFR. The higher the value of the parameter $(k_1/k_2)(C_A)_0$, the higher is the ratio $(\Psi_{ovl})_{BFR}/(\Psi_{ovl})_{PFR}$ but it is always less than 1. There is a sharp decrease in this ratio at higher conversion with decrease in the parameter $(k_1/k_2)(C_A)_0$. So, for lower values of this parameter and higher conversion of the reactant, the fractional yield obtained in a PFR far exceeds that obtained in a BFR. As $(X_A)_f \rightarrow 1$ the ratio $((\Psi_{ovl})_{BFR}/(\Psi_{ovl})_{PFR}) \rightarrow 0$.

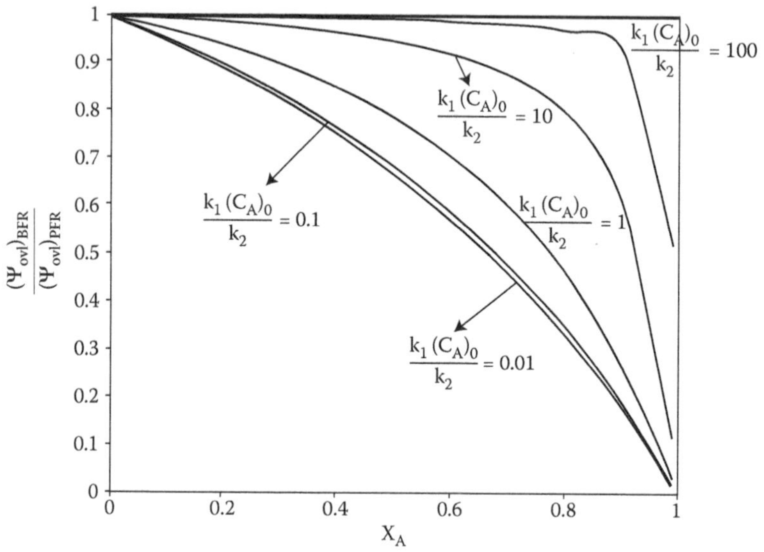

FIGURE 6.11 Overall fractional yield in PFR and BFR when the desirable reaction is of order n and the undesirable reaction is of order n − 1.

Case 4; $n_1 = n_2 + 2$
From Equation 6.177,

$$(\Psi_{ovl})_{PFR} = \frac{-1}{(C_A)_0 - (C_A)_f} \int_{(C_A)_0}^{(C_A)_f} \frac{k_1 C_A^{n_1}}{k_1 C_A^{n_1} + k_2 C_A^{n_1-2}} dC_A$$

$$\Rightarrow \quad (\Psi_{ovl})_{PFR} = \frac{1}{(C_A)_0 - (C_A)_f}[(C_A)_0 - (C_A)_f$$

$$-\sqrt{\frac{k_2}{k_1}}\tan^{-1}\left[\frac{\sqrt{k_1/k_2}\,(C_A)_0(X_A)_f}{1+(k_1/k_2)(C_A)_0^2\{1-(X_A)_f\}}\right]$$

$$\Rightarrow \quad (\Psi_{ovl})_{PFR} = 1 + \frac{1}{\sqrt{k_1/k_2}\,(C_A)_0(X_A)_f}\tan^{-1}\left[\frac{\sqrt{k_1/k_2}\,(C_A)_0(X_A)_f}{1+(k_1/k_2)(C_A)_0^2\{1-(X_A)_f\}}\right] \tag{6.190}$$

From Equation 6.178,

$$(\Psi_{ovl})_{BFR} = \frac{k_1(C_A)_f^{n_1}}{\left[k_1(C_A)_f^{n_1} + k_2(C_A)_f^{n_1-2}\right]}$$

$$\Rightarrow \quad (\Psi_{ovl})_{BFR} = \frac{(k_1/k_2)(C_A)_f^2}{[1+(k_1/k_2)(C_A)_f^2]} = \frac{(k_1/k_2)(C_A)_0^2[1-(X_A)_f]^2}{[1+(k_1/k_2)(C_A)_0^2[1-(X_A)_f]^2]} \tag{6.191}$$

Dividing Equation 6.191 by Equation 6.190,

$$\frac{(\Psi_{ovl})_{BFR}}{(\Psi_{ovl})_{PFR}} = \frac{1}{\dfrac{\left[1+\dfrac{k_1}{k_2}(C_A)_0^2[1-(X_A)_f]^2\right]}{\dfrac{k_1}{k_2}(C_A)_0^2[1-(X_A)_f]^2} + \dfrac{\left[1+\dfrac{k_1}{k_2}(C_A)_0^2[1-(X_A)_f]^2\right]}{\left\{\dfrac{k_1}{k_2}(C_A)_0^2[1-(X_A)_f]^2\right\}\left\{\sqrt{\dfrac{k_1}{k_2}}(C_A)_0(X_A)_f\right\}}\tan^{-1}\left[\dfrac{\sqrt{\dfrac{k_1}{k_2}}(C_A)_0(X_A)_f}{1+\dfrac{k_1}{k_2}(C_A)_0^2\{1-(X_A)_f\}}\right]} \quad (6.192)$$

Besides the fractional conversion of the reactant A, the ratio $(\Psi_{ovl})_{BFR}/(\Psi_{ovl})_{PFR}$ depends on the dimensionless parameter $\sqrt{k_1/k_2}\,(C_A)_0$. The variation of the ratio $(\Psi_{ovl})_{BFR}/(\Psi_{ovl})_{PFR}$ with varying conversion and different values of the parameter $\sqrt{k_1/k_2}\,(C_A)_0$ is graphically depicted in Figure 6.12.

For lower values of the parameter $\sqrt{k_1/k_2}\,(C_A)_0$ the ratio $(\Psi_{ovl})_{BFR}/(\Psi_{ovl})_{PFR}$ steeply declines even at very low conversion and thereafter the fractional yield in

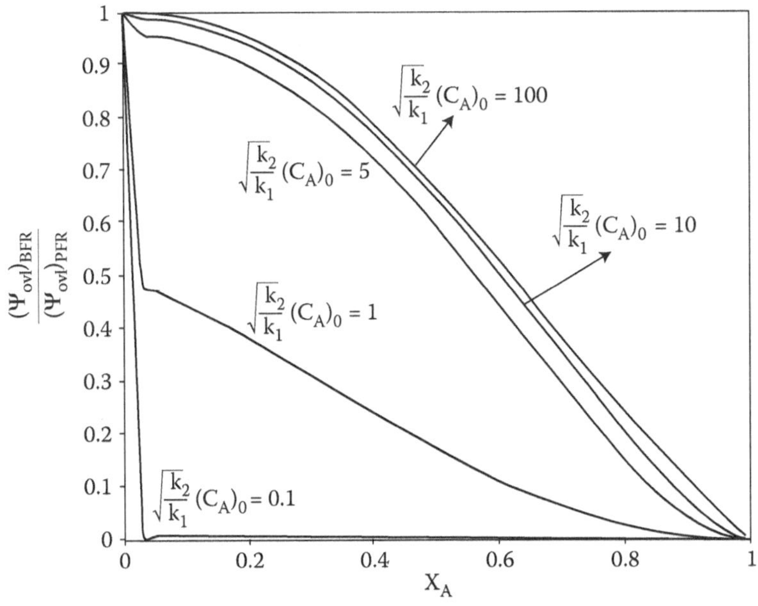

FIGURE 6.12 Overall fractional yield in PFR and BFR when the desirable reaction is of order n and the undesirable reaction is of order n − 2.

BFR is negligible compared to PFR. With increase in the parameter $\sqrt{k_1/k_2}(C_A)_0$ the decline in the ratio $(\Psi_{ovl})_{BFR}/(\Psi_{ovl})_{PFR}$ is more gradual. Therefore, for lower values of the parameter $\sqrt{k_1/k_2}(C_A)_0$ and higher conversion of the reactant, the fractional yield obtained in a PFR far exceeds that obtained in a BFR. As $(X_A)_f \to 1$ the ratio $((\Psi_{ovl})_{BFR}/(\Psi_{ovl})_{PFR}) \to 0$.

On the basis of the foregone discussion, we can generally conclude that:

1. If the order of the desirable reaction is less than the order of the undesirable reaction, BFR is preferred over PFR for higher fractional yield. The higher the conversion of reactant, the greater the benefit of using BFR.
2. If the order of the desirable reaction is more than the order of the undesirable reaction, PFR is preferred over BFR for higher fractional yield. The higher the conversion of reactant, the greater the benefit of using PFR.

In general, for parallel reactions of more complexity, the number, type, and arrangement of reactors are to be optimized (Chitra and Govind, 1981).

EXAMPLE 6.17

The parallel reaction

$$A \to P \quad (-r_A)_1 = r_P = 0.3C_A \ \text{mol/(L}\cdot\text{s)}$$

$$A \to Q \quad (-r_A)_2 = r_Q = C_A^2 \ \text{mol/(L}\cdot\text{s)}$$

The initial concentration of A is 5 mol/L. With all other conditions remaining the same, the reaction is first carried out in an isothermal PFR and then in an isothermal BFR. The species P and Q are not present initially. The reaction is carried out up to 80% conversion of A. By how much percentage is the overall fractional yield of P obtained in BFR more than that obtained in PFR?

SOLUTION

This is the case where $n_1 = n_2 - 1$.

First we evaluate the parameter $k_2(C_A)_0/k_1 = (1 \times 5)/0.3 = 16.67$

From Equation 6.183,

$$\frac{(\Psi_{ovl})_{PFR}}{(\Psi_{ovl})_{BFR}} = \frac{[1+(k_2/k_1)(C_A)_0\{1-(X_A)_f\}]}{(k_2/k_1)(C_A)_0(X_A)_f}\ln\left[\frac{1+(k_2/k_1)(C_A)_0}{1+(k_2/k_1)(C_A)_0\{1-(X_A)_f\}}\right]$$

$$\Rightarrow \quad \frac{(\Psi_{ovl})_{PFR}}{(\Psi_{ovl})_{BFR}} = \frac{[1+16.67\times\{1-0.8\}]}{16.67\times 0.8}\ln\left[\frac{1+16.67}{1+16.67\times\{1-0.8\}}\right] = 0.457$$

$$\Rightarrow \quad (\Psi_{ovl})_{BFR} = \frac{(\Psi_{ovl})_{PFR}}{0.457}$$

$$\Rightarrow \quad \frac{((\Psi_{ovl})_{PFR}/0.457) - (\Psi_{ovl})_{PFR}}{(\Psi_{ovl})_{PFR}} = 1.188$$

So, the overall fractional yield of P obtained in BFR is 118.8% more than that obtained in PFR.

6.4 COMBINATION OF PFR AND BFR

Chemical process industries often use a combination of different types of reactors. While optimization of reactor networks is a fascinating advanced topic (Balakrishna and Biegler, 1992; Lakshmanan and Biegler, 1996), the basics of combination of PFR and BFR are presented here. Consider the schematic shown in Figure 6.13a where a BFR of volume V_{BFR} is followed by a PFR of volume V_{PFR}. For the BFR, from Equation 6.86

$$V_{BFR} = \frac{\left(F_{A_{LR}}\right)_0 \left(X_{A_{LR}}\right)_1}{\left(-r_{A_{LR}}\right)_1} \tag{6.193}$$

For the PFR from Equation 6.71,

$$V_{PFR} = \left(F_{A_{LR}}\right)_0 \int_{(X_{A_{LR}})_1}^{(X_{A_{LR}})_f} \frac{dX_{A_{LR}}}{-r_{A_{LR}}} \tag{6.194}$$

Now, consider the schematic shown in Figure 6.13b where a PFR of volume V_{PFR} is followed by a BFR of volume V_{BFR}. For the PFR, from Equation 6.1,

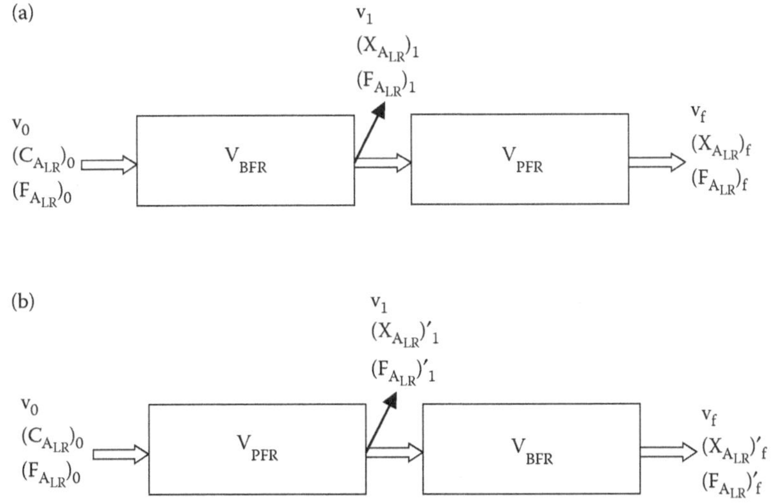

FIGURE 6.13 Combination of ideal PFR and BFR in series.

$$V_{PFR} = \left(F_{A_{LR}}\right)_0 \int_0^{(X_{A_{LR}})'_1} \frac{dX_{A_{LR}}}{-r_{A_{LR}}} \tag{6.195}$$

For the BFR, from Equation 6.135

$$V_{BFR} = \frac{\left(F_{A_{LR}}\right)_0 \left[\left(X_{A_{LR}}\right)'_f - \left(X_{A_{LR}}\right)'_1\right]}{\left(-r_{A_{LR}}\right)'_f} \tag{6.196}$$

Let us now examine in which of the two combinations we get more conversion.

Case 1, reaction order n = 1, $\varepsilon_A = 0$
From Equation 6.193,

$$V_{BFR} = \frac{(F_A)_0 (X_A)_1}{k(C_A)_0\{1-(X_A)_1\}}$$

$$\Rightarrow \quad (X_A)_1 = \frac{k(C_A)_0 V_{BFR}}{k(C_A)_0 V_{BFR} + (F_A)_0} = \frac{k\tau_{BFR}}{k\tau_{BFR}+1} \tag{6.197}$$

From Equation 6.194,

$$V_{PFR} = (F_A)_0 \int_{(X_A)_1}^{(X_A)_f} \frac{dX_A}{k(C_A)_0\{1-X_A\}}$$

$$\Rightarrow \quad V_{PFR} = \frac{(F_A)_0}{k(C_A)_0} \ln\left[\frac{1-(X_A)_1}{1-(X_A)_f}\right]$$

$$\Rightarrow \quad \ln\left[\frac{1-(X_A)_1}{1-(X_A)_f}\right] = k\tau_{PFR}$$

$$\Rightarrow \quad (X_A)_f = 1 - e^{-k\tau_{PFR}} + (X_A)_1 e^{-k\tau_{PFR}}$$

$$\Rightarrow \quad (X_A)_f = 1 - e^{-k\tau_{PFR}} + \frac{k\tau_{BFR}}{k\tau_{BFR}+1} e^{-k\tau_{PFR}}$$

$$\Rightarrow \quad (X_A)_f = 1 - \frac{e^{-k\tau_{PFR}}}{k\tau_{BFR}+1} \tag{6.198}$$

From Equation 6.195,

$$V_{PFR} = (F_A)_0 \int_0^{(X_A)_1'} \frac{dX_A}{k(C_A)_0\{1 - X_A\}}$$

$$\Rightarrow \quad V_{PFR} = -\frac{(F_A)_0}{k(C_A)_0} \ln[1 - (X_A)_1']$$

$$\Rightarrow \quad \ln[1 - (X_A)_1'] = -k\tau_{PFR}$$

$$\Rightarrow \quad (X_A)_1' = 1 - e^{-k\tau_{PFR}} \tag{6.199}$$

From Equation 6.196,

$$V_{BFR} = \frac{(F_A)_0[(X_A)_f' - (X_A)_1']}{k(C_A)_0\{1 - (X_A)_f'\}}$$

$$\Rightarrow \quad k\tau_{BFR} = \frac{[(X_A)_f' - (X_A)_1']}{\{1 - (X_A)_f'\}}$$

$$\Rightarrow \quad (X_A)_f' = \frac{k\tau_{BFR} + (X_A)_1'}{1 + k\tau_{BFR}} = \frac{k\tau_{BFR} + 1 - e^{-k\tau_{PFR}}}{1 + k\tau_{BFR}}$$

$$\Rightarrow \quad (X_A)_f' = 1 - \frac{e^{-k\tau_{PFR}}}{k\tau_{BFR} + 1} \tag{6.200}$$

From Equations 6.198 and 6.200,

$$(X_A)_f = (X_A)_f' \tag{6.201}$$

It follows that for a first-order reaction carried out in a series combination of a PFR and a BFR, the final conversion remains the same irrespective of the sequence of the reactors—PFR followed by BFR or BFR followed by PFR.

Case 2, reaction order $n = 2$, $\varepsilon_A = 0$
From Equation 6.193,

$$V_{BFR} = \frac{(F_A)_0(X_A)_1}{k(C_A)_0^2\{1 - (X_A)_1\}^2}$$

$$\Rightarrow \quad \tau_{BFR} = \frac{(C_A)_0 - (C_A)_1}{k(C_A)_1^2}$$

$$\Rightarrow \quad (C_A)_1 = \frac{\sqrt{1 + 4k\tau_{BFR}(C_A)_0} - 1}{2k\tau_{BFR}} \tag{6.202}$$

From Equation 6.194,

$$V_{PFR} = (F_A)_0 \int_{(X_A)_1}^{(X_A)_f} \frac{dX_A}{-r_A}$$

$$\Rightarrow \quad \tau_{PFR} = -\int_{(C_A)_1}^{(C_A)_f} \frac{dC_A}{kC_A^2}$$

$$\Rightarrow \quad k\tau_{PFR} = \frac{1}{(C_A)_f} - \frac{1}{(C_A)_1} \tag{6.203}$$

$$\Rightarrow \quad (C_A)_f = \frac{(C_A)_1}{1+k\tau_{PFR}(C_A)_1} = \frac{\sqrt{1+4k\tau_{BFR}(C_A)_0}-1}{k\tau_{PFR}[\sqrt{1+4k\tau_{BFR}(C_A)_0}-1]+2k\tau_{BFR}}$$

$$\Rightarrow \quad \frac{(C_A)_f}{(C_A)_0} = \frac{\sqrt{1+4k\tau_{BFR}(C_A)_0}-1}{k\tau_{PFR}(C_A)_0\left[\sqrt{1+4k\tau_{BFR}(C_A)_0}-1\right]+2k\tau_{BFR}(C_A)_0}$$

$$\Rightarrow \quad (X_A)_f = 1 - \frac{\sqrt{1+4k\tau_{BFR}(C_A)_0}-1}{k\tau_{PFR}(C_A)_0[\sqrt{1+4k\tau_{BFR}(C_A)_0}-1]+2k\tau_{BFR}(C_A)_0}$$

From Equation 6.195,

$$V_{PFR} = (F_A)_0 \int_0^{(X_A)'_1} \frac{dX_A}{kC_A^2}$$

$$\Rightarrow \quad \tau_{PFR} = -\int_{(C_A)_0}^{(C_A)'_1} \frac{dC_A}{kC_A^2} \tag{6.204}$$

$$\Rightarrow \quad k\tau_{PFR} = \frac{1}{(C_A)'_1} - \frac{1}{(C_A)_0}$$

$$\Rightarrow \quad (C_A)'_1 = \frac{(C_A)_0}{1+k\tau_{PFR}(C_A)_0}$$

From Equation 6.196,

$$V_{BFR} = \frac{(F_A)_0[(X_A)'_f - (X_A)'_1]}{k(C_A)'^2_f}$$

$$\Rightarrow \quad \tau_{BFR} = \frac{(C_A)'_1 - (C_A)'_f}{k(C_A)'^2_f}$$

$$\Rightarrow \quad (C_A)'_f = \frac{\sqrt{1+4k\tau_{BFR}(C_A)'_1}-1}{2k\tau_{BFR}} = \frac{\sqrt{1+((4k\tau_{BFR}(C_A)_0)/(1+k\tau_{PFR}(C_A)_0))}-1}{2k\tau_{BFR}}$$

$$\Rightarrow \quad \frac{(C_A)'_f}{(C_A)_0} = \frac{\sqrt{1+((4k\tau_{BFR}(C_A)_0)/(1+k\tau_{PFR}(C_A)_0))}-1}{2k\tau_{BFR}(C_A)_0}$$

$$\Rightarrow \quad (X_A)'_f = 1 - \frac{\sqrt{1+((4k\tau_{BFR}(C_A)_0)/(1+k\tau_{PFR}(C_A)_0))}-1}{2k\tau_{BFR}(C_A)_0} \qquad (6.205)$$

Comparing Equations 6.203 and 6.205, we can see that the final conversion is dependent on the sequence of the reactors—PFR followed by BFR or BFR followed by PFR. The equations themselves are mathematically complex and do not tell unequivocally which sequence would result in higher conversion. The choice should depend on the space time of the two reactors and the initial concentration of the reactant. However, Levenspiel (1999) has suggested that for reaction orders more than one, the PFR should precede the BFR and for reaction orders less than one, the PFR should follow the BFR.

EXAMPLE 6.18

The liquid-phase reaction

$$A \rightarrow \text{Products}$$

follows the rate law $-r_A = 0.5C_A^2$ mol/(L·min)

The reaction is carried out in a PFR with 1 mol/L initial concentration of reactant to achieve 70% conversion. In order to increase the conversion to 90%, a BFR is to be joined in series after the PFR. What should be the space time of the BFR?

Solution

When only the PFR is operated, from Equation 6.204

$$(C_A)'_1 = \frac{(C_A)_0}{1+k\tau_{PFR}(C_A)_0}$$

$$\Rightarrow \quad 1-(X_A)'_1 = \frac{1}{1+k\tau_{PFR}(C_A)_0}$$

$$\Rightarrow \quad 1-0.7 = \frac{1}{1+0.5\times 1\times \tau_{PFR}}$$

$$\Rightarrow \quad \tau_{PFR} = 4.67 \text{ min}$$

When the BFR is joined in series after the PFR, from Equation 6.205

$$(X_A)_f' = 1 - \frac{\sqrt{1+((4k\tau_{BFR}(C_A)_0)/(1+k\tau_{PFR}(C_A)_0))}-1}{2k\tau_{BFR}(C_A)_0}$$

$$0.9 = 1 - \frac{\sqrt{1+((4\times 0.5\times \tau_{BFR}\times 1)/(1+0.5\times 4.67\times 1))}-1}{2\times 0.5\times \tau_{BFR}\times 1}$$

$$\Rightarrow \quad \tau_{BFR} = 39.97 \text{ min}$$

EXERCISES

6.1 The liquid-phase reaction, for which the kinetic data are given in Exercise 5.1 part (a) and part (b) (Chapter 5), is to be carried out in an isothermal reactor to obtain 90% conversion.
   a. What is the space time of the PFR required in each case?
   b. What is the space time of the BFR required in each case?

6.2 For the reaction mentioned in Exercise 6.1, if product B is to be produced at the rate of 100 kmol/min
   a. What is the volume of PFR required?
   b. What is the volume of BFR required?

6.3 Repeat Example 6.2 if the feed to the reactor is a mixture of 80% A and 20% of an inert gas.

6.4 The homogeneous liquid-phase reaction

$$A + 2B \rightarrow C + D$$

is carried out with initial concentrations of A and B 2 and 3.5 mol/L, respectively. The reaction rate may be expressed as

$$-r_A = 0.2C_AC_B^2 \text{ mol/(L}\cdot\text{min)}$$

   a. Find the space time required for the PFR to get 2 mol/L outlet concentration of B.
   b. Find the space time required for the BFR to get 2 mol/L outlet concentration of B.

6.5 Repeat Exercise 6.4 if 1 mol/L outlet concentration of A is required.

6.6 In Exercise 6.4, what will be the volume of the PFR and BFR required, respectively, if the volumetric feed rate to the reactor is 100 L/min?

6.7 A liquid-phase reaction is carried out in a PFR and 51% conversion is achieved. When the reaction is carried out in a PFR double the size of the first PFR, 84% conversion is achieved. What is the order of the reaction?

6.8 Repeat Example 6.3 if the feed to the reactor consists of a gas mixture having 1 volume A and 0.8 volume B.

6.9 In Exercise 6.8, the conversion is to be increased to 90%. By what percentage will the production rate have to be reduced?

6.10 A first-order liquid-phase reaction has a half-life of 20 min. If the reactant at an initial concentration of 2 mol/L is to be processed at the rate of 50 mol/min

a. What is the size of PFR required to achieve 80% conversion?
b. What is the size of BFR required to achieve 80% conversion?

6.11 A liquid-phase reaction

$$A \rightarrow \text{Products}$$

with reaction rate $-r_A = 0.1C_A/(1 + 0.3C_A)$ mol/(L · min) is carried out in an isothermal PFR with a reactant feed rate of 50 kmol/min and 1 mol/L concentration of A in the feed.

a. Find the volume of PFR required for 75% conversion.
b. Find the volume of BFR required for 75% conversion.

6.12 The liquid-phase reversible reaction

$$A + B \rightleftharpoons C + D$$

with $k_1 = 0.05\ \text{min}^{-1}$ and $k_2 = 0.02\ \text{min}^{-1}$, is carried out with a total volumetric feed rate of 100 L/min with initial concentrations of A and B as 1 mol/L, and the initial concentrations of C and D as 0.1 mol/L, respectively.

a. Determine the volume of PFR required to achieve 60% of the equilibrium conversion.
b. Determine the volume of BFR required to achieve 60% of the equilibrium conversion.

6.13 The gas-phase reversible reaction

$$A \rightleftharpoons 2B$$

has an equilibrium constant of 10 at 400 K and 0.5 atm pressure. A feed containing 90% A and 10% inert by volume at 400 K and 0.5 atm pressure is to be reacted at a constant pressure of 0.5 atm and 400 K to achieve 50% of the equilibrium conversion. The specific reaction rate constant for the forward reaction is $0.5\ \text{min}^{-1}$.

a. What is the space time of the PFR required?
b. What is the space time of the BFR required?

6.14 The reaction

$$A \rightarrow I \rightarrow T$$

has $k_1 = 0.05\ \text{min}^{-1}$ and $k_2 = 0.02\ \text{min}^{-1}$, respectively. The reaction is carried out in an isothermal PFR of 2.5 m^3 volume with an initial concentration of A equal to 2.5 mol/L and volumetric feed rate of 100 L/min. What is the composition of the outlet stream from the reactor?

6.15 In Exercise 6.14, what will be the composition of the outlet stream if the reaction is carried out in a BFR?

6.16 Repeat Exercise 6.14 for a PFR and a BFR if I and T are initially present at 0.25 mol/L concentration.

6.17 In Exercise 6.14, what is the volume of the PFR required for maximum outlet concentration of I if I and T are initially present at 0.25 mol/L concentration? What is that maximum concentration?

6.18 Repeat Exercise 6.17 for a BFR.

6.19 The liquid-phase series reaction

$$A \rightarrow I \rightarrow T \rightarrow U$$

with the specific reaction rate constants for the consecutive reaction steps being k_1, k_2, and k_3, respectively, is carried out in an isothermal PFR. The initial concentration of A is $(C_A)_0$ and no I, T, or U is present initially. Derive an expression for $C_T/(C_A)_0$ as a function of space time in the reactor. Compare the result with that obtained in Exercise 5.19 (Chapter 5).

6.20 Repeat Exercise 6.19 for a BFR.

6.21 The liquid-phase parallel reaction

$$A \rightarrow P \quad (-r_A)_1 = r_P = 0.01C_A \text{ mol/(L} \cdot \text{min)}$$

$$A \rightarrow Q \quad (-r_A)_2 = r_Q = 0.02C_A \text{ mol/(L} \cdot \text{min)}$$

$$A \rightarrow R \quad (-r_A)_3 = r_R = 0.03C_A \text{ mol/(L} \cdot \text{min)}$$

is carried out with 1 mol/L initial concentration of A and no P, Q, and R present initially. The volumetric flow rate into the reactor is 10 L/min.

a. Determine the outlet composition from an isothermal PFR of 50-L volume.

b. Determine the outlet composition from an isothermal BFR of 50-L volume.

6.22 In Exercise 6.21, what is the volume of PFR required to get 0.25 mol/L outlet concentration of Q? All other conditions remain the same.

6.23 A liquid-phase reaction

$$A \rightarrow \text{Products}$$

with reaction rate $-r_A = 0.5\sqrt{C_A}$ mol/(L·min) is carried out in two PFRs, volumes 60 and 40 L, respectively, joined in parallel. A feed of 100 mol/min at a concentration of 1 mol/L is equally split between the two reactors. Find the conversion in the combined stream when the outlet from the two reactors is mixed together. Also, compare this conversion with the conversion that would be achieved if the entire feed of 100 mol/min were treated in a single PFR of volume 100 L.

6.24 Repeat Exercise 6.23 if the PFRs are replaced with BFRs.

6.25 A first-order reaction $P \rightarrow Q$ is carried out in two equal-sized BFRs joined in series to get 80% conversion. It is proposed to arrange another similar BFR in series with the first one. If all the parameters remain the same, calculate

a. The conversion achieved with the new arrangement.
b. If the same 80% conversion is maintained, how many times can the production rate be increased?

6.26 The homogeneous liquid-phase reaction

$$A + 2B \rightarrow C + D$$

is carried out with initial concentrations of A and B 2 and 3.5 mol/L, respectively. The reaction rate may be expressed as

$$-r_A = 0.2C_AC_B^2 \text{ mol/(L} \cdot \text{min)}$$

The reaction is carried out in two equal-sized BFRs joined in series so that the outlet concentration of B from the second reactor is 2 mol/L. Find out the space time of each BFR.

6.27 In Exercise 6.26, what will be the outlet concentration of B from the second reactor if the outlet concentration of B from the first reactor is 2 mol/L?

6.28 We have a BFR of volume 250 L and another BFR of volume 400 L. A second-order reaction with $-r_A = 1.2C_A^2$ mol/(L·min) and a reactant feed rate of 200 mol/min is carried out in the two BFRs joined in series. The initial concentration of A is 2 mol/L. What will be the conversion achieved when

a. The smaller BFR is followed by the bigger BFR.
b. The bigger BFR is followed by the smaller BFR.

6.29 Repeat Exercise 6.28 if the 400-L BFR is replaced with a 200-L PFR.

6.30 A first-order reaction with reaction rate $-r_A = 200C_A$ mol/(L·s) is carried out in four equal-sized BFRs joined in series. The final conversion is 0.84. If the third BFR is now replaced with a PFR of the same size, what will be the final conversion?

6.31 The first-order gas-phase reaction

$$A \rightarrow B + C$$

with reaction rate $-r_A = 1.5C_A$ mol/(L·min) is carried out in two BFRs joined in series. The volume of the first BFR is twice the volume of the second BFR. The final conversion is 90%. What are the space times of the two BFRs?

6.32 In Exercise 6.31, the second BFR is to be replaced with a PFR so that the final conversion remains the same. What is the space time required for the PFR?

6.33 In Exercise 6.31, the first BFR is to be replaced with a PFR so that the final conversion remains the same. What is the space time required for the PFR?

6.34 In the liquid-phase parallel reaction

$$A \rightarrow P \quad (-r_A)_1 = 0.25C_A$$

$$A \rightarrow 2Q + R \quad (-r_A)_2 = 0.1C_A^2$$

Q is the desired product. The initial concentration of A is 3 mol/L.

a. Which reactor type is to be preferred for higher fractional yield?
b. What will be the outlet concentration of Q for 80% conversion of A?
c. What will be the accompanying outlet concentration of R?

6.35 In Exercise 6.34, by how much percentage will the production rate of Q be reduced if the reactor type is changed?

6.36 Repeat Exercise 6.34 if P is the desired product.

REFERENCES

Alopaeus, V., Laavi, H., and Aittamaa, J., A dynamic model for plug flow reactor state profiles. *Comp. Chem. Eng.*, 32, 1494, 2008.

Balakrishna, S., and Biegler, L.T., Constructive targeting approaches for the synthesis of chemical reactor networks. *Ind. Eng. Chem. Res.*, 31, 300, 1992.

Chitra, S.P., and Govind, R., Yield optimization for complex reactor systems. *Chem. Eng. Sci.*, 36, 1219, 1981.

Lakshmanan, A., and Biegler, L.T., Synthesis of optimal chemical reactor networks. *Ind. Eng. Chem. Res.*, 35, 1344, 1996.

Levenspiel, O., *Chemical Reaction Engineering*, 3rd ed., John Wiley & Sons, New York, 135, 1999.

Levenspiel, O., and Bischoff, K.B., Patterns of flow in chemical process vessels, in Drew, T.B., Hoopes Jr., J.W., and Vermeulen, T. (eds.), *Advances in Chemical Engineering*, Academic Press, New York, 96, 1963.

van de Vusse, J.G., Plug-flow type reactor versus tank reactor. *Chem. Eng. Sci.*, 19, 994, 1964.

7 Nonisothermal Reactors

In Chapters 5 and 6, we have considered the temperature of the reacting system as constant, both with respect to position within the reactor and time. This simplified the mathematical analyses of the reactor performance as only the material balances were required to be considered. The performance equations developed in the previous two chapters essentially conform to the material balance over the reactor. In reality, however, isothermal reactors are exceptions rather than the rule. As a result of energy inflow and outflow into and from the reacting system and the energy transformation associated with chemical transformation, the temperature of the reacting system may vary. For a complete understanding of the reactor performance, the energy balance over the reactor is also to be taken into account, along with the material balance. In this chapter, we will analyze such situations for ideal reactors to develop performance equations and algorithms for engineering calculations.

7.1 BATCH REACTOR

7.1.1 The Energy Balance

Consider a general chemical reaction

$$\sum_{i=1}^{n} a_i A_i \rightarrow \sum_{j=1}^{l} p_j P_j \tag{7.1}$$

taking place in a batch reactor. Starting at a temperature T_0

$$\text{Energy added to the reactor} + \text{Heat absorbed in chemical reaction}$$
$$= \text{Increase in heat content of reacting system}$$

Let us now examine each of these terms.

Energy can be added to the reactor either as heat, "Q," or mechanical work, "W," or both. In an ideal batch reactor, the contents of the reactor are thoroughly mixed to make the conditions the same everywhere. This requires agitation of the reactor content. Mechanical work is done to accomplish this. The quantum of the work done depends on the amount of material present inside the reactor and the physical properties of the reactor content. We have also seen in Chapter 2 that the ratio of different chemical species in a reacting system is fixed by stoichiometry. The agitator works for the entire duration of the reaction. Therefore, as a first approximation we can consider the mechanical work done by the agitator as proportional to the amount of limiting reactant converted. Unless the reactor contents are highly viscous, the work done by the agitator is negligible in comparison to other energy terms and our approximation may not lead to much error. By a similar reasoning, the heat is added to (or removed from) the reactor for the entire duration of the reaction. Although not rigorously true, for the sake of mathematical simplicity we also consider the heat added to (or removed from) the reactor as proportional to the amount of limiting reactant converted. So,

$$\text{Energy added to the reactor} = Q + W = q\left(N_{A_{LR}}\right)_0 X_{A_{LR}} + w\left(N_{A_{LR}}\right)_0 X_{A_{LR}} \quad (7.2)$$

$$\text{Heat absorbed in chemical reaction} = -(\Delta H_r)_{LR|T_0}\left[\left(N_{A_{LR}}\right)_0 - N_{A_{LR}}\right]$$

$$\Rightarrow \text{Heat absorbed in chemical reaction} = -(\Delta H_r)_{LR|T_0}\left(N_{A_{LR}}\right)_0 X_{A_{LR}} \quad (7.3)$$

Increase in heat content of

$$\text{the reacting system} = \left[\sum_{i=1}^{n} N_{A_i}(C_p)_{A_i} + \sum_{j=1}^{1} N_{P_j}(C_p)_{P_j}\right](T - T_0)$$

⇒ Increase in heat content of

$$\text{system} = \left[\sum_{i=1}^{n}\left\{\left(N_{A_i}\right)_0 - \frac{a_i}{a_{LR}}\left(N_{A_{LR}}\right)_0 X_{A_{LR}}\right\}(C_p)_{A_i} + \sum_{j=1}^{1}\left\{\left(N_{P_j}\right)_0 + \frac{P_j}{a_{LR}}\left(N_{A_{LR}}\right)_0 X_{A_{LR}}\right\}(C_p)_{P_j}\right](T - T_0) \quad (7.4)$$

Substituting the expressions from Equations 7.2 through 7.4 into the general energy balance equation

$$[q + w - (\Delta H_r)_{LR|T_0}]\left(N_{A_{LR}}\right)_0 X_{A_{LR}} = \left[\sum_{i=1}^{n}\left\{\left(N_{A_i}\right)_0 - \frac{a_i}{a_{LR}}\left(N_{A_{LR}}\right)_0 X_{A_{LR}}\right\}(C_p)_{A_i} + \sum_{j=1}^{1}\left\{\left(N_{P_j}\right)_0 + \frac{P_j}{a_{LR}}\left(N_{A_{LR}}\right)_0 X_{A_{LR}}\right\}(C_p)_{P_j}\right](T - T_0)$$

$$\Rightarrow \quad T = T_0 + \frac{[q + w - (\Delta H_r)_{LR|T_0}](N_{A_{LR}})_0 X_{A_{LR}}}{\left[\begin{array}{l}\sum_{i=1}^{n}\left\{\left(N_{A_i}\right)_0 - (a_i/a_{LR})\left(N_{A_{LR}}\right)_0 X_{A_{LR}}\right\}(C_p)_{A_i} \\ + \sum_{j=1}^{1}\left\{\left(N_{P_j}\right)_0 + (P_j/a_{LR})\left(N_{A_{LR}}\right)_0 X_{A_{LR}}\right\}(C_p)_{P_j}\end{array}\right]} \quad (7.5)$$

Dividing the numerator and denominator by $(V_{BR})_0$,

$$T = T_0 + \frac{[q + w - (\Delta H_r)_{LR|T_0}](C_{A_{LR}})_0 X_{A_{LR}}}{\left[\sum_{i=1}^{n}\left\{(C_{A_i})_0 - (a_i/a_{LR})(C_{A_{LR}})_0 X_{A_{LR}}\right\}(C_p)_{A_i} + \sum_{j=1}^{l}\left\{(C_{P_j})_0 + (P_j/a_{LR})(C_{A_{LR}})_0 X_{A_{LR}}\right\}(C_p)_{P_j}\right]} \tag{7.6}$$

Equations 7.5 and 7.6 represent the energy balance of a batch reactor and relate the temperature of the reacting system corresponding to any conversion of the limiting reactant. We can further examine a few specific situations:

Case 1: No products are initially present

When no products are initially present, which is most likely, we have

$$\text{Increase in heat content of system} = \left[\sum_{i=1}^{n}\left\{(N_{A_i})_0 - \frac{a_i}{a_{LR}}(N_{A_{LR}})_0 X_{A_{LR}}\right\}(C_p)_{A_i} + \sum_{j=1}^{l}\frac{P_j}{a_{LR}}(N_{A_{LR}})_0 X_{A_{LR}}(C_p)_{P_j}\right](T - T_0) \tag{7.7}$$

Substituting the expressions from Equations 7.2, 7.3, and 7.7 into the general energy balance equation

$$qX_{A_{LR}} + wX_{A_{LR}} - (\Delta H_r)_{LR|T_0}(N_{A_{LR}})_0 X_{A_{LR}} = \left[\sum_{i=1}^{n}\left\{(N_{A_i})_0 - \frac{a_i}{a_{LR}}(N_{A_{LR}})_0 X_{A_{LR}}\right\}(C_p)_{A_i} + \sum_{j=1}^{l}\frac{P_j}{a_{LR}}(N_{A_{LR}})_0 X_{A_{LR}}(C_p)_{P_j}\right](T - T_0)$$

$$\Rightarrow\ T = T_0 + \frac{[q + w - (\Delta H_r)_{LR|T_0}](N_{A_{LR}})_0 X_{A_{LR}}}{\left[\sum_{i=1}^{n}\left\{(N_{A_i})_0 - (a_i/a_{LR})(N_{A_{LR}})_0 X_{A_{LR}}\right\}(C_p)_{A_i} + \sum_{j=1}^{l}(P_j/a_{LR})(N_{A_{LR}})_0 X_{A_{LR}}(C_p)_{P_j}\right]} \tag{7.8}$$

$$\Rightarrow\ T = T_0 + \frac{[q + w - (\Delta H_r)_{LR|T_0}](C_{A_{LR}})_0 X_{A_{LR}}}{\left[\sum_{i=1}^{n}\left\{(C_{A_i})_0 - (a_i/a_{LR})(C_{A_{LR}})_0 X_{A_{LR}}\right\}(C_p)_{A_i} + \sum_{j=1}^{l}(P_j/a_{LR})(C_{A_{LR}})_0 X_{A_{LR}}(C_p)_{P_j}\right]} \tag{7.9}$$

Case 2: No products are initially present and $\left(N_{A_i}\right)_0 = (a_i/a_{LR})\left(N_{A_{LR}}\right)_0$

$$\text{Increase in heat content of system} = \left[\sum_{i=1}^{n}\frac{a_i}{a_{LR}}\left(N_{A_{LR}}\right)_0\left(1-X_{A_{LR}}\right)(C_p)_{A_i} + \sum_{j=1}^{l}\frac{P_j}{a_{LR}}\left(N_{A_{LR}}\right)_0 X_{A_{LR}}(C_p)_{P_j}\right](T-T_0) \tag{7.10}$$

Substituting the expressions from Equations 7.2, 7.3, and 7.10 into the general energy balance equation

$$qX_{A_{LR}} + wX_{A_{LR}} - (\Delta H_r)_{LR|T_0}\left(N_{A_{LR}}\right)_0 X_{A_{LR}} = \left[\sum_{i=1}^{n}\frac{a_i}{a_{LR}}\left(N_{A_{LR}}\right)_0\left(1-X_{A_{LR}}\right)(C_p)_{A_i} + \sum_{j=1}^{l}\frac{P_j}{a_{LR}}\left(N_{A_{LR}}\right)_0 X_{A_{LR}}(C_p)_{P_j}\right](T-T_0)$$

$$\Rightarrow \quad T = T_0 + \frac{[q + w - (\Delta H_r)_{LR|T_0}]\left(N_{A_{LR}}\right)_0 X_{A_{LR}}}{\left[\sum_{i=1}^{n}(a_i/a_{LR})\left(N_{A_{LR}}\right)_0\left(1-X_{A_{LR}}\right)(C_p)_{A_i} + \sum_{j=1}^{l}(P_j/a_{LR})\left(N_{A_{LR}}\right)_0 X_{A_{LR}}(C_p)_{P_j}\right]}$$

$$\Rightarrow \quad T = T_0 + \frac{[q + w - (\Delta H_r)_{LR|T_0}]X_{A_{LR}}}{\left[\sum_{i=1}^{n}(a_i/a_{LR})\left(1-X_{A_{LR}}\right)(C_p)_{A_i} + \sum_{j=1}^{l}(P_j/a_{LR})X_{A_{LR}}(C_p)_{P_j}\right]} \tag{7.11}$$

Case 3: Adiabatic reactor

For an adiabatic reactor operation Q = 0, and Equation 7.5 reduces to

$$T = T_0 + \frac{[w - (\Delta H_r)_{LR|T_0}]\left(N_{A_{LR}}\right)_0 X_{A_{LR}}}{\left[\sum_{i=1}^{n}\left\{\left(N_{A_i}\right)_0 - (a_i/a_{LR})\left(N_{A_{LR}}\right)_0 X_{A_{LR}}\right\}(C_p)_{A_i} + \sum_{j=1}^{l}\left\{\left(N_{P_j}\right)_0 + (P_j/a_{LR})\left(N_{A_{LR}}\right)_0 X_{A_{LR}}\right\}(C_p)_{P_j}\right]} \tag{7.12}$$

Further, if W is negligible in relation to the heat absorbed in the chemical reaction, we have

$$T = T_0 + \frac{-(\Delta H_r)_{LR|T_0}\left(N_{A_{LR}}\right)_0 X_{A_{LR}}}{\left[\sum_{i=1}^{n}\left\{\left(N_{A_i}\right)_0 - (a_i/a_{LR})\left(N_{A_{LR}}\right)_0 X_{A_{LR}}\right\}(C_p)_{A_i} + \sum_{j=1}^{1}\left\{\left(N_{P_j}\right)_0 + (P_j/a_{LR})\left(N_{A_{LR}}\right)_0 X_{A_{LR}}\right\}(C_p)_{P_j}\right]} \tag{7.13}$$

Similarly, Equation 7.8 reduces to

$$T = T_0 + \frac{[w - (\Delta H_r)_{LR|T_0}]\left(N_{A_{LR}}\right)_0 X_{A_{LR}}}{\left[\sum_{i=1}^{n}\left\{\left(N_{A_i}\right)_0 - (a_i/a_{LR})\left(N_{A_{LR}}\right)_0 X_{A_{LR}}\right\}(C_p)_{A_i} + \sum_{j=1}^{1}(P_j/a_{LR})\left(N_{A_{LR}}\right)_0 X_{A_{LR}}(C_p)_{P_j}\right]} \tag{7.14}$$

And, for negligible W,

$$T = T_0 + \frac{-(\Delta H_r)_{LR|T_0}\left(N_{A_{LR}}\right)_0 X_{A_{LR}}}{\left[\sum_{i=1}^{n}\left\{\left(N_{A_i}\right)_0 - (a_i/a_{LR})\left(N_{A_{LR}}\right)_0 X_{A_{LR}}\right\}(C_p)_{A_i} + \sum_{j=1}^{1}(P_j/a_{LR})\left(N_{A_{LR}}\right)_0 X_{A_{LR}}(C_p)_{P_j}\right]} \tag{7.15}$$

Likewise, Equation 7.11 reduces to

$$T = T_0 + \frac{[w - (\Delta H_r)_{LR|T_0}]\left(N_{A_{LR}}\right)_0 X_{A_{LR}}}{\left[\sum_{i=1}^{n}(a_i/a_{LR})\left(N_{A_{LR}}\right)_0\left(1 - X_{A_{LR}}\right)(C_p)_{A_i} + \sum_{j=1}^{1}(P_j/a_{LR})\left(N_{A_{LR}}\right)_0 X_{A_{LR}}(C_p)_{P_j}\right]} \tag{7.16}$$

And, for negligible W,

$$T = T_0 + \frac{-(\Delta H_r)_{LR|T_0}\left(N_{A_{LR}}\right)_0 X_{A_{LR}}}{\left[\sum_{i=1}^{n}(a_i/a_{LR})\left(N_{A_{LR}}\right)_0\left(1 - X_{A_{LR}}\right)(C_p)_{A_i} + \sum_{j=1}^{1}(P_j/a_{LR})\left(N_{A_{LR}}\right)_0 X_{A_{LR}}(C_p)_{P_j}\right]} \tag{7.17}$$

Equations 7.12 through 7.17 can also be written in terms of the initial concentrations by dividing the numerator and the denominator by $(V_{BR})_0$.

Case 4: $\sum_{i=1}^{n} a_i(C_p)_{A_i} = \sum_{j=1}^{l} P_j(C_p)_{P_j}$

Under this condition, Equation 7.5 reduces to

$$T = T_0 + \frac{[q + w - (\Delta H_r)_{LR|T_0}](N_{ALR})_0 X_{ALR}}{\left[\sum_{i=1}^{n}(N_{A_i})_0(C_p)_{A_i} + \sum_{j=1}^{l}(N_{P_j})_0(C_p)_{P_j}\right]}$$

$$\Rightarrow \quad T = \frac{[q + w - (\Delta H_r)_{LR|T_0}](N_{ALR})_0}{\left[\sum_{i=1}^{n}(N_{A_i})_0(C_p)_{A_i} + \sum_{j=1}^{l}(N_{P_j})_0(C_p)_{P_j}\right]} X_{ALR} + T_0 \tag{7.18}$$

$$\Rightarrow \quad T = \frac{[q + w - (\Delta H_r)_{LR|T_0}](C_{ALR})_0}{\left[\sum_{i=1}^{n}(C_{A_i})_0(C_p)_{A_i} + \sum_{j=1}^{l}(C_{P_j})_0(C_p)_{P_j}\right]} X_{ALR} + T_0 \tag{7.19}$$

Thus, when $\sum_{i=1}^{n} a_i(C_p)_{A_i} = \sum_{j=1}^{l} P_j(C_p)_{P_j}$ the temperature of the reacting system varies linearly with conversion with a slope equal to

$$\frac{q + w - (\Delta H_r)_{LR|T_0}(N_{ALR})_0}{\left[\sum_{i=1}^{n}(N_{A_i})_0(C_p)_{A_i} + \sum_{j=1}^{l}(N_{P_j})_0(C_p)_{P_j}\right]}$$

Equation 7.18 can be modified for adiabatic conditions as

$$T = \frac{[w - (\Delta H_r)_{LR|T_0}](N_{ALR})_0}{\left[\sum_{i=1}^{n}(N_{A_i})_0(C_p)_{A_i} + \sum_{j=1}^{l}(N_{P_j})_0(C_p)_{P_j}\right]} X_{ALR} + T_0 \tag{7.20}$$

$$\Rightarrow \quad T = \frac{[w - (\Delta H_r)_{LR|T_0}](C_{ALR})_0}{\left[\sum_{i=1}^{n}(C_{A_i})_0(C_p)_{A_i} + \sum_{j=1}^{l}(C_{P_j})_0(C_p)_{P_j}\right]} X_{ALR} + T_0 \tag{7.21}$$

And, for negligible w,

$$T = \frac{-(\Delta H_r)_{LR|T_0}(N_{ALR})_0}{\left[\sum_{i=1}^{n}(N_{A_i})_0(C_p)_{A_i} + \sum_{j=1}^{l}(N_{P_j})_0(C_p)_{P_j}\right]} X_{ALR} + T_0 \tag{7.22}$$

$$\Rightarrow \quad T = \frac{-(\Delta H_r)_{LR|T_0}\left(C_{A_{LR}}\right)_0}{\left[\sum_{i=1}^{n}\left(C_{A_i}\right)_0 (C_p)_{A_i} + \sum_{j=1}^{l}\left(C_{P_j}\right)_0 (C_p)_{P_j}\right]} X_{A_{LR}} + T_0 \tag{7.23}$$

EXAMPLE 7.1

The liquid-phase irreversible reaction

$$A + 2B \rightarrow C + D$$

is carried out in a constant volume adiabatic batch reactor of 100 L volume with $(C_A)_0 = 1$ mol/L and $(C_B)_0 = 2.5$ mol/L. The initial temperature of the reactor is 50°C. The mechanical work of agitation can be neglected. Plot the reactor temperature as a function of the conversion of A. The data given are as follows:

The heat of the reaction = –10 kJ/mol A at 298 K.

$$(C_p)_A = 5\ \text{J/mol}, \quad (C_p)_B = 25\ \text{J/mol}, \quad (C_p)_C = 10\ \text{J/mol}, \quad \text{and} \quad (C_p)_D = 15\ \text{J/mol}$$

Solution

From the reaction stoichiometry and initial concentrations of A and B, we can see that A is the limiting reactant.

$$T_0 = 50\,°\text{C} = 323\ \text{K}$$

From Equation 4.9 (Chapter 4),

$$(\Delta H_r)_A|_{T_0} = (\Delta H_r)_A|_{323} = (\Delta H_r)_{298} + (323 - 298)[(C_p)_C + (C_p)_D - (C_p)_A - 2(C_p)_B]$$

$$\Rightarrow \quad (\Delta H_r)_A|_{323} = -10 + (323 - 298)[(10 + 15 - 5 - 2 \times 25) \times 10^{-3}]$$

$$\Rightarrow \quad (\Delta H_r)_A|_{323} = -10.75\ \text{kJ/mol A}$$

$$(N_A)_0 = (C_A)_0 V_{BR} = 1 \times 1000 = 1000\ \text{mol}$$

$$(N_B)_0 = (C_B)_0 V_{BR} = 2.5 \times 1000 = 2500\ \text{mol}$$

The products C and D are not present initially. So, from Equation 7.15, the reactor temperature at 10% conversion is

$$T = T_0 + \frac{-(\Delta H_r)_A|_{323}(N_A)_0 X_A}{\left[\{(N_A)_0 - (N_A)_0 X_A\}(C_p)_A + \{(N_B)_0 - 2(N_A)_0 X_A\}(C_p)_B + (N_A)_0 X_A\{(C_p)_C + (C_p)_D\}\right]}$$

$$\Rightarrow \quad T = 50 + \frac{10.75 \times 10^3 \times 1000 \times 0.1}{\left[\{1000 - 1000 \times 0.1\} \times 5 + \{2500 - 2 \times 1000 \times 0.1\} \times 25 + 1000 \times 0.1 \times 25\right]}$$

$$\Rightarrow \quad T = 66.67°\text{C}$$

TABLE EXE7.1
Temperature at Different Conversion

X_A	T (°C)
0	50
0.1	66.66667
0.2	84.95935
0.3	105.1282
0.4	127.4775
0.5	152.381
0.6	180.303
0.7	211.828
0.8	247.7011
0.9	288.8889
1	336.6667

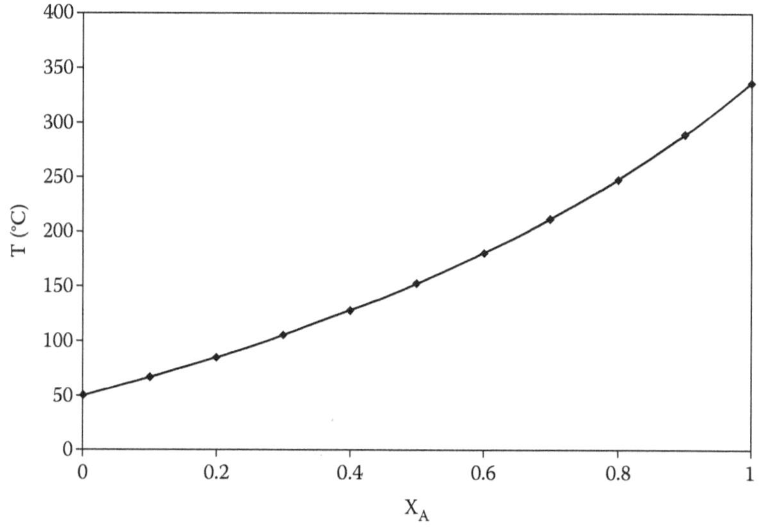

FIGURE EXE7.1 Temperature conversion plot.

Similarly, we calculate the reactor temperature at different conversions and prepare Table EXE7.1.

Next, T is plotted against X_A (Figure EXE7.1).

7.1.2 Performance Evaluation and Design for Nonisothermal Constant Volume Batch Reactor

The equations developed in the previous section, starting from the general Equation 7.5, represent the energy balance of the batch reactor and relate the temperature of the reacting system corresponding to any conversion of the limiting reactant for different operating conditions of the reactor. These equations are to be used in conjunction with the equations developed in Chapter 5, which represent the material balance for the reactor and relate the conversion in the batch reactor to the reaction time. We start with Equation 5.7 (Chapter 5), reproduced here for convenience, which represents the general mole balance equation for a batch reactor

$$\left(C_{A_{LR}}\right)_0 \int_0^{X_{A_{LR}}} \frac{dX_{A_{LR}}}{-r_{A_{LR}}} = \int_0^t dt = t \tag{5.7}$$

Referring to the discussion in Chapter 3, Section 3.2, the rate of a chemical reaction depends both on the concentration of the reactants and on the temperature of the reacting system. Therefore, in a general way, we can write

$$-r_{A_{LR}} = f\left(X_{A_{LR}}, T\right) \tag{7.24}$$

For an isothermal batch reactor, the temperature being constant, Equation 7.20 simplifies to

$$-r_{A_{LR}} = f\left(X_{A_{LR}}\right) \tag{7.25}$$

It allows for the evaluation of the integral on the left-hand side of Equation 5.7 (Chapter 5) by replacing $-r_{A_{LR}}$ with $f\left(X_{A_{LR}}\right)$. For nonisothermal reactors, the term $-r_{A_{LR}}$ is to be replaced with $f\left(X_{A_{LR}}, T\right)$ on the left-hand side of Equation 5.7 (Chapter 5). Now, for the evaluation of the integral on the left-hand side of Equation 5.7, we first need to convert the function $f\left(X_{A_{LR}}, T\right)$ into a function that only varies with $X_{A_{LR}}$. This can be done by substituting T as a function of $X_{A_{LR}}$ from the equations developed in Section 7.1.1. After all these substitutions and rearrangements, the function to be integrated on the left-hand side of Equation 5.7 (Chapter 5) becomes too complicated for analytical integration. It needs to be evaluated by numerical or graphical integration. The general algorithm is presented next:

Step 1

Using the appropriate equation from Section 7.1, estimate the reaction temperature "T" for different values of conversion $X_{A_{LR}}$. Arrange the values in a tabular form.

Step 2

Estimate $-r_{A_{LR}}$ for each $X_{A_{LR}}$ using the corresponding value of T and the appropriate rate equation. It often involves the evaluation of the specific reaction rate constant for different temperatures and estimation of $-r_{A_{LR}}$, thereafter, using the appropriate rate equation. Arrange the values of $-r_{A_{LR}}$ so estimated in the next column of the table prepared in Step 1.

Step 3

Using the values calculated in Step 2, estimate $(1/-r_{A_{LR}})$ for each $X_{A_{LR}}$. Tabulate the values in the next column of the table prepared in Step 1.

Step 4

For a graphical estimation of the integral on the left-hand side of Equation 5.7 (Chapter 5) plot $(1/-r_{A_{LR}})$ versus $X_{A_{LR}}$. The area under this curve and the $X_{A_{LR}}$ axis between origin and the desired conversion is the value of the integral. Next, evaluate the time of reaction from Equation 5.7 (Chapter 5). Alternatively, the value of the integral can also be estimated using a suitable numerical integration technique.

EXAMPLE 7.2

The reaction of Example 7.1 has a reaction rate expressed as $-r_A = 0.02C_AC_B^2$ mol/(L · s) at 50°C. The activation energy for the reaction is 10 kJ/mol. Calculate the time required for 50% conversion.

SOLUTION

Step 1: Estimate the reaction temperature "T" for different values of conversion X_A up to 50% conversion. Arrange the values in a tabular form. These values have already been calculated in Example 7.1. We start with the calculated values from Table EXE7.1 and present them in Table EXE7.2.

Step 2: Calculate the specific reaction rate constant at 0.1 conversion using the corresponding temperature and Equation 3.31 (Arrhenius equation, Chapter 3)

$$\frac{k_{T_1}}{k_{T_2}} = \exp\left\{\frac{E_a}{R}\left(\frac{1}{T_2} - \frac{1}{T_1}\right)\right\}$$

$$\Rightarrow \frac{k_{339.67}}{k_{323}} = \exp\left\{\frac{10,000}{8.314}\left(\frac{1}{323} - \frac{1}{339.67}\right)\right\}$$

$$\Rightarrow \frac{k_{339.67}}{0.02} = \exp\left\{\frac{10,000}{8.314}\left(\frac{1}{323} - \frac{1}{339.67}\right)\right\}$$

$$\Rightarrow k_{339.67} = 0.024\ L^2/mol^2 \cdot s$$

Similarly, calculate the specific reaction rate constant at other conversions. The values have been tabulated in the next column of Table EXE7.2.

TABLE EXE7.2
Calculation for Estimation of Integral

X_A	T (°C)	k ($L^2/mol^2 \cdot s$)	$-r_A$ mol/(L · s)	$(1/-r_A)$
0	50	0.02	0.125	8
0.1	66.66667	0.024009536	0.114309	8.748187
0.2	84.95935	0.02877222	0.101508	9.851402
0.3	105.1282	0.034419913	0.086979	11.49701
0.4	127.4775	0.041105862	0.071278	14.02966
0.5	152.381	0.049007878	0.055134	18.13767

Step 3: Calculate the initial reaction rate $-r_A$ at zero conversion using the corresponding specific reaction rate constant "k" as

$$-r_A = kC_AC_B^2 = k(C_A)_0(1-X_A)[(C_B)_0 - 2(C_A)_0X_A]^2$$

$$\Rightarrow \quad -r_A = 0.02\times1\times(1-0)[2.5-2\times1\times0]^2 = 0.125\ \text{mol/(L·s)}$$

Similarly, calculate the reaction rate at other conversions. The values have been tabulated in the next column of Table EXE7.2.

Step 4: Calculate $(1/-r_A)$ for different conversions and tabulate the values in the next column of Table EXE7.2.

Step 5: From Equation 5.7 (Chapter 5),

$$t = (C_A)_0\int_0^{X_A}\frac{dX_A}{-r_A} = \int_0^{0.5}\frac{dX_A}{-r_A}$$

To estimate the value of the integral, we plot $(1/-r_A)$ against X_A as given below (Figure EXE7.2).

Hence,

$$t = \int_0^{0.5}\frac{dX_A}{-r_A} = \text{Area ABCD} = 5.65\,\text{s}$$

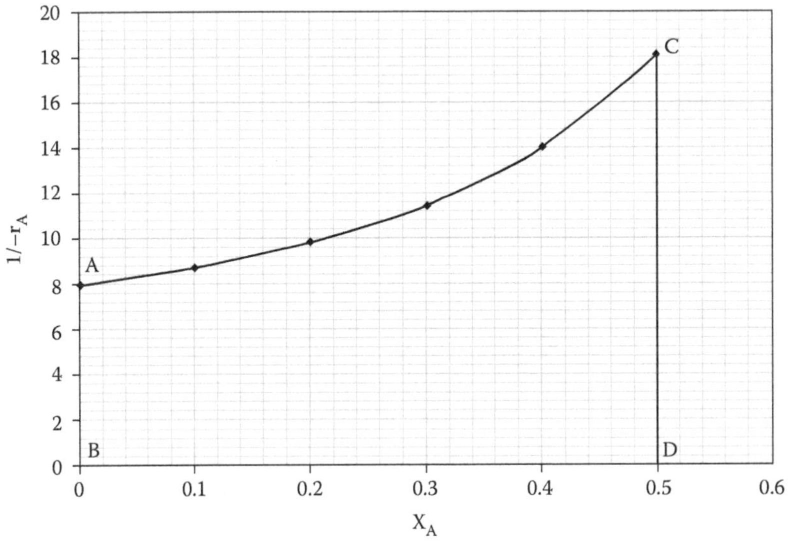

FIGURE EXE7.2 Plot for graphical integration.

7.1.3 Nonisothermal Variable Volume Batch Reactor

For a constant volume batch reactor, the quantity ε_{LR} is zero; so, Equation 5.5 (Chapter 5) reduces to Equation 5.7 (Chapter 5). This is elaborated in Example 7.2. For a variable volume batch reactor, estimating the value of ε_{LR} is essential. For an isothermal reacting system, Equation 2.44 (Chapter 2) provides the value of ε_{LR}. Here, we develop an equation for the estimation of ε_{LR} for a system with variable volume and temperature. The same can then be used for determining the reactor volume and concentration of a chemical species at a given conversion.

When no products are initially present, for a constant pressure reacting system with variable volume and temperature, from Equation 2.30 (Chapter 2), we have

$$V = \left[\sum_{i=1}^{n} N_{A_i} + \sum_{j=1}^{1} N_{P_j} + N_{inerts}\right] RT/p$$

$$\Rightarrow \quad V_f = \left[\sum_{i=1}^{n}\left\{\left(N_{A_i}\right)_0 - \frac{a_i}{a_{LR}}\left(N_{A_{LR}}\right)_0 X_{A_{LR}}\right\} + \sum_{j=1}^{1}\left\{\frac{P_j}{a_{LR}}\left(N_{A_{LR}}\right)_0 X_{A_{LR}}\right\} + N_{inerts}\right]\frac{RT_f}{p} \tag{7.26}$$

Initial volume of the reacting system

$$V_0 = \left[\sum_{i=1}^{n}\left(N_{A_i}\right)_0 + N_{inerts}\right]\frac{RT_0}{p} \tag{7.27}$$

And, at complete conversion,

$$V_{X_{A_{LR}}=1} = \left[\sum_{i=1}^{n}\left\{\left(N_{A_i}\right)_0 - \frac{a_i}{a_{LR}}\left(N_{A_{LR}}\right)_0\right\} + \sum_{j=1}^{1}\left\{\frac{P_j}{a_{LR}}\left(N_{A_{LR}}\right)_0\right\} + N_{inerts}\right]\frac{RT_{X_{A_{LR}}=1}}{p} \tag{7.28}$$

From Equation 2.44 (Chapter 2),

$$\varepsilon_{LR} = \frac{V_f - V_0}{V_0} = \frac{\left(N_{A_{LR}}\right)_0\left[\sum_{j=1}^{1}(p_j/a_{LR}) - \sum_{i=1}^{n}(a_i/a_{LR})\right]}{\sum_{i=1}^{n}\left(N_{A_i}\right)_0 + N_{inerts}}\left(\frac{T_{X_{A_{LR}}=1}}{T_0}\right) + \left(\frac{T_{X_{A_{LR}}=1} - T_0}{T_0}\right) \tag{7.29}$$

From Equations 7.26 and 7.29,

$$\frac{V_f}{V_0} = \frac{T_f}{T_0} + \left(1 + \varepsilon_{LR} - \frac{T_{X_{A_{LR}}=1}}{T_0}\right)\left(\frac{T_f}{T_{X_{A_{LR}}=1}}\right) X_{A_{LR}} \tag{7.30}$$

Also,

$$\left(C_{A_{LR}}\right)_f = \frac{\left(N_{A_{LR}}\right)_f}{V_f} = \frac{\left(N_{A_{LR}}\right)_0\left(1 - X_{A_{LR}}\right)}{V_0\left[(T_f/T_0) + \left(1 + \varepsilon_{LR} - \left(T_{X_{A_{LR}}=1}/T_0\right)\right)\left(T_f/T_{X_{A_{LR}}=1}\right)X_{A_{LR}}\right]}$$

$$\left(C_{A_{LR}}\right)_f = \frac{\left(C_{A_{LR}}\right)_0\left(1 - X_{A_{LR}}\right)}{\left[(T_f/T_0) + \left(1 + \varepsilon_{LR} - \left(T_{X_{A_{LR}}=1}/T_0\right)\right)\left(T_f/T_{X_{A_{LR}}=1}\right)X_{A_{LR}}\right]} \tag{7.31}$$

From Equations 5.1 (Chapter 5) and 7.30,

$$\frac{dN_{A_{LR}}}{dt} = r_{A_{LR}}(V_{BR})_0\left[\frac{T_f}{T_0} + \left(1 + \varepsilon_{LR} - \frac{T_{X_{A_{LR}}=1}}{T_0}\right)\left(\frac{T_f}{T_{X_{A_{LR}}=1}}\right)X_{A_{LR}}\right]$$

$$\Rightarrow -\left(N_{A_{LR}}\right)_0\frac{dX_{A_{LR}}}{dt} = r_{A_{LR}}(V_{BR})_0\left[\frac{T_f}{T_0} + \left(1 + \varepsilon_{LR} - \frac{T_{X_{A_{LR}}=1}}{T_0}\right)\left(\frac{T_f}{T_{X_{A_{LR}}=1}}\right)X_{A_{LR}}\right]$$

$$\Rightarrow \left(C_{A_{LR}}\right)_0\frac{dX_{A_{LR}}}{dt} = -r_{A_{LR}}\left[\frac{T_f}{T_0} + \left(1 + \varepsilon_{LR} - \frac{T_{X_{A_{LR}}=1}}{T_0}\right)\left(\frac{T_f}{T_{X_{A_{LR}}=1}}\right)X_{A_{LR}}\right]$$

$$\Rightarrow \left(C_{A_{LR}}\right)_0\int_0^{X_{A_{LR}}}\frac{dX_{A_{LR}}}{-r_{A_{LR}}\left[(T_f/T_0) + \left(1 + \varepsilon_{LR} - \left(T_{X_{A_{LR}}=1}/T_0\right)\right)\left(T_f/T_{X_{A_{LR}}=1}\right)X_{A_{LR}}\right]} = \int_0^t dt = t \tag{7.32}$$

This is the performance equation of a variable volume batch reactor under nonisothermal operation.

EXAMPLE 7.3

The second-order gas-phase reaction

$$A \rightarrow B + C$$

is carried out in a variable volume adiabatic batch reactor starting with pure A at 700°C and 5 atm pressure. The specific reaction rate constant at 700°C is 10 L/(mol · min). The following data are given:

The heat of reaction = 5 kJ/mol A at 300 K.

$$(C_p)_A = 10\ \text{J/mol}, \quad (C_p)_B = 5\ \text{J/mol}, \quad (C_p)_C = 5\ \text{J/mol}$$

Activation energy = 25 kJ/mol

Calculate the time required for 60% conversion. The mechanical work of mixing can be neglected.

Solution

Since pure A is used initially, $(x_A)_0 = 1$, so from Equation 2.27 (Chapter 2), $(p_A)_0 = 5$ atm

From Equation 2.26 (Chapter 2),

$$(C_A)_0 = \frac{(p_A)_0}{RT_0} = \frac{5}{0.082 \times (273+700)} = 0.063\ \text{mol/L}$$

The products B and C are not present initially. Also,

$$(C_p)_A = (C_p)_B + (C_p)_C$$

Therefore, from Equation 7.19, the reactor temperature at 10% conversion is

$$T = T_0 + \frac{-(\Delta H_r)_A|_{300}(N_A)_0 X_A}{(N_A)_0 (C_p)_A} = T_0 + \frac{-(\Delta H_r)_A|_{300} X_A}{(C_p)_A}$$

$$\Rightarrow \quad T = 973 - \frac{5 \times 10^3 \times 0.1}{10}$$

$$\Rightarrow \quad T = 923\ \text{K}$$

Similarly, we calculate the reactor temperature at different conversions and prepare Table EXE7.3.
Also,

$$T_{X_A=1} = 473\ \text{K}$$

Next, we calculate the specific reaction rate constant at each of these temperatures using Equation 3.31 (Chapter 3)

TABLE EXE7.3
Calculation Table

X_A	T (°C)	k ($L^2/mol^2 \cdot min$)	$\frac{\left[(T/T_0)+\left(1+\varepsilon_A-\left(T_{X_A=1}/T_0\right)\right)\left(T/T_{X_A=1}\right)X_A\right]}{k(C_A)_0^2(1-X_A)^2}$
0	973	10	25.19526
0.1	923	8.458514184	38.40669
0.2	873	7.018751719	60.48787
0.3	823	5.693502044	99.52947
0.4	773	4.495129818	173.6518
0.5	723	3.434859472	328.0943
0.6	673	2.52182903	693.5754

$$\frac{k_{T_1}}{k_{T_2}} = \exp\left\{\frac{E_a}{R}\left(\frac{1}{T_2} - \frac{1}{T_1}\right)\right\}$$

and tabulate the values in the next column of Table EXE7.3.

From Equation 7.29,

$$\varepsilon_A = \frac{(N_A)_0[2-1]}{(N_A)_0}\left(\frac{T_{X_A=1}}{T_0}\right) + \left(\frac{T_{X_A=1} - T_0}{T_0}\right) = 1\times\left(\frac{473}{973}\right) + \left(\frac{473-973}{973}\right)$$

$$\Rightarrow \quad \varepsilon_A = -0.023$$

From Equation 7.32,

$$t = (C_A)_0 \int_0^{X_A} \frac{dX_A}{-r_A\left[(T/T_0) + \left(1 + \varepsilon_A - (T_{X_A=1}/T_0)\right)(T/T_{X_A=1})X_A\right]} \qquad (7.3.1)$$

The rate of reaction

$$-r_A = kC_A^2 = k\frac{(C_A)_0^2(1-X_A)^2}{\left[(T/T_0) + \left(1 + \varepsilon_A - (T_{X_A=1}/T_0)\right)(T/T_{X_A=1})X_A\right]^2} \qquad (7.3.2)$$

From Equations 7.3.1 and 7.3.2,

$$t = (C_A)_0 \int_0^{X_A} \frac{\left[(T/T_0) + \left(1 + \varepsilon_A - (T_{X_A=1}/T_0)\right)(T/T_{X_A=1})X_A\right]}{k(C_A)_0^2(1-X_A)^2} dX_A$$

$$\Rightarrow \quad t = 0.063 \times \int_0^{0.6} f(X_A)dX_A \qquad (7.3.3)$$

where

$$f(X_A) = \frac{\left[(T/T_0) + \left(1 + \varepsilon_A - (T_{X_A=1}/T_0)\right)(T/T_{X_A=1})X_A\right]}{k(C_A)_0^2(1-X_A)^2}$$

Next, we calculate $f(X_A)$ and tabulate the values in the next column.

To estimate the value of the integral, we plot $f(X_A)$ against X_A as given below (Figure EXE7.3).

Hence,

$$t = 0.063 \times \int_0^{0.6} f(X_A)dX_A = 0.063 \times \text{Area ABCD} = 0.063 \times 102.8 = 6.48 \text{ min}$$

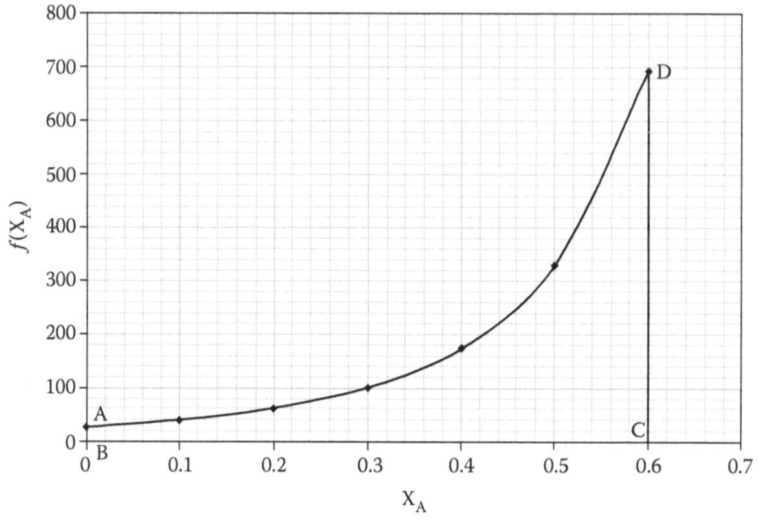

FIGURE EXE7.3 Graphical integration for estimation of reaction time.

7.2 PLUG FLOW REACTOR

7.2.1 Energy Balance for the Differential Volume Element

Referring to Chapter 6, Section 6.1, a PFR is one where every differential fluid element entering the reactor spends the same amount of time inside the reactor. To achieve this, it is imperative that every differential fluid element, as it enters the PFR and travels through the length of the PFR, does not exchange any material with any fluid element that precedes it or succeeds it. Consider one such differential fluid element of infinitesimal volume dV that enters the PFR. We call this differential fluid element a "plug," to borrow the term from PFR. Since this plug does not exchange any material with its surrounding (the preceding or succeeding plugs), it can be considered a differential batch reactor in which the reaction takes place for a time equal to the residence time of the plug in the PFR. We can apply the energy balance equations developed in Section 7.1.1 to this plug to relate the temperature with conversion. Since every plug that enters the PFR and traverses through it to exit from the other end is identical to every other plug, the equations can be applied to the PFR by extension. Here, we exclude the possibility of products being present in the feed to the PFR. So, in the "plug batch reactor" of our consideration, no products are initially present. From Equation 7.8,

$$T = T_0 + \frac{[q + w - (\Delta H_r)_{LR|T_0}]\left(\dot{N}_{ALR}\right)_0 X_{ALR}}{\left[\sum_{i=1}^{n}\left\{\left(\dot{N}_{A_i}\right)_0 - (a_i/a_{LR})\left(\dot{N}_{ALR}\right)_0 X_{ALR}\right\}(C_p)_{A_i} + \sum_{j=1}^{l}(P_j/a_{LR})\left(\dot{N}_{ALR}\right)_0 X_{ALR}(C_p)_{P_j}\right]}$$

$$\Rightarrow T = T_0 + \frac{[q + w - (\Delta H_r)_{LR|T_0}]X_{A_{LR}}}{\left[\sum_{i=1}^{n}\left\{(\dot{N}_{A_i})_0/(\dot{N}_{A_{LR}})_0 - (a_i/a_{LR})X_{A_{LR}}\right\}(C_p)_{A_i} + \sum_{j=1}^{l}(P_j/a_{LR})X_{A_{LR}}(C_p)_{P_j}\right]}$$

$$\Rightarrow T = T_0 + \frac{[q + w - (\Delta H_r)_{LR|T_0}]X_{A_{LR}}}{\left[\sum_{i=1}^{n}\left\{(C_{A_i})_0/(C_{A_{LR}})_0 - (a_i/a_{LR})X_{A_{LR}}\right\}(C_p)_{A_i} + \sum_{j=1}^{l}(P_j/a_{LR})X_{A_{LR}}(C_p)_{P_j}\right]} \tag{7.33}$$

We can now examine the specific cases.

Case 1: $(C_{A_i})_0 = (a_i/a_{LR})(C_{A_{LR}})_0$

In this case, Equation 7.11 is applicable for the plug under consideration and, by extension, to the PFR.

Case 2: Adiabatic PFR

For q = 0, from Equation 7.33,

$$T = T_0 + \frac{[w - (\Delta H_r)_{LR|T_0}]X_{A_{LR}}}{\left[\sum_{i=1}^{n}\left\{(C_{A_i})_0/(C_{A_{LR}})_0 - (a_i/a_{LR})X_{A_{LR}}\right\}(C_p)_{A_i} + \sum_{j=1}^{l}(P_j/a_{LR})X_{A_{LR}}(C_p)_{P_j}\right]} \tag{7.34}$$

Further, if W is negligible in relation to the heat absorbed in the chemical reaction, we have

$$T = T_0 + \frac{-(\Delta H_r)_{LR|T_0}X_{A_{LR}}}{\left[\sum_{i=1}^{n}\left\{(C_{A_i})_0/(C_{A_{LR}})_0 - (a_i/a_{LR})X_{A_{LR}}\right\}(C_p)_{A_i} + \sum_{j=1}^{l}(P_j/a_{LR})X_{A_{LR}}(C_p)_{P_j}\right]} \tag{7.35}$$

For adiabatic PFR with $(C_{A_i})_0 = (a_i/a_{LR})(C_{A_{LR}})_0$ Equation 7.11 can be suitably modified with q = 0.

Case 3: $\sum_{i=1}^{n} a_i(C_p)_{A_i} = \sum_{j=1}^{l} P_j(C_p)_{P_j}$

From Equation 7.33,

$$T = \frac{[q + w - (\Delta H_r)_{LR|T_0}](C_{A_{LR}})_0}{\left[\sum_{i=1}^{n}(C_{A_i})_0(C_p)_{A_i}\right]}X_{A_{LR}} + T_0 \tag{7.36}$$

Equation 7.36 can also be modified for adiabatic condition.

7.2.2 Performance Evaluation and Design for Nonisothermal PFR

Just as in the case of a batch reactor, the equations developed in the previous section—Equations 7.33 through 7.36—are to be used in conjunction with the equations developed in Chapter 6, which represent the material balance for the reactor and relate the conversion in the PFR to the space time or reactor volume. We start with Equation 6.1 (Chapter 6), reproduced below for convenience, which represents the general mole balance equation for a PFR:

$$V_{PFR} = \left(F_{A_{LR}}\right)_0 \int_0^{(X_{A_{LR}})_f} \frac{dX_{A_{LR}}}{-r_{A_{LR}}} \tag{6.1}$$

As explained in Section 7.1.2, the rate of reaction $-r_{A_{LR}}$ is a function of both conversion and temperature. In order to evaluate the integral on the RHS of Equation 6.1 (Chapter 6), we need to first convert the function $-r_{A_{LR}}$ into a function that only varies with $X_{A_{LR}}$. This can be done by substituting T as a function of $X_{A_{LR}}$ from the equations developed in Section 7.2.1. The integral is then evaluated by graphical or numerical integration. The general algorithm is presented below:

Step 1
Using the appropriate equation from Section 7.2.1, estimate the reaction temperature "T" for different values of conversion $X_{A_{LR}}$. Arrange the values in a tabular form.

Step 2
For gas-phase reactions, calculate ε_{LR} from Equation 7.29. Write the rate equation in terms of conversion, ε_{LR}, and temperature. Use Equation 7.31 to relate concentration with conversion. For liquid-phase reactions, $\varepsilon_{LR} = 0$.

Step 3
Estimate $-r_{A_{LR}}$ for each $X_{A_{LR}}$ using the corresponding value of T and the appropriate rate equation. It often involves the evaluation of a specific reaction rate constant for different temperature and estimation of $-r_{A_{LR}}$, thereafter using the appropriate rate equation. Arrange the values of $-r_{A_{LR}}$ so estimated in the next column of the table prepared in Step 1.

Step 4
Using the values calculated in Step 3, estimate $(1/-r_{A_{LR}})$ for each $X_{A_{LR}}$. Tabulate the values in the next column of the table prepared in Step 1.

Step 5
For a graphical estimation of the integral on the RHS of Equation 6.1 (Chapter 6) plot $(1/-r_{A_{LR}})$ versus $X_{A_{LR}}$. The area under this curve and the $X_{A_{LR}}$ axis between origin and the desired conversion is the value of the integral. Next, evaluate the volume of PFR required from Equation 6.1 (Chapter 6). Alternatively, the value of the integral can also be estimated using a suitable numerical integration technique.

Alternatively, we can use Equation 6.6 (Chapter 6) if the space time of the PFR is desired. Equation 6.6 (Chapter 6) is reproduced below for convenience:

$$\tau_{PFR} = \left(C_{A_{LR}}\right)_0 \int_0^{(X_{ALR})_f} \frac{dX_{A_{LR}}}{-r_{A_{LR}}} \tag{6.6}$$

EXAMPLE 7.4

Calculate the volume of the PFR required for the reaction of Examples 7.1 and 7.2 carried out with 100 mol/min of A fed to the reactor to achieve 50% conversion.

Solution

Step 1

From the reaction stoichiometry and initial concentrations of A and B, we can see that A is the limiting reactant.

From the solution to Example 7.1,

$$T_0 = 50°C = 323\text{ K}$$

$$(\Delta H_r)_A|_{323} = -10.75\text{ kJ/mol A}$$

$$(N_A)_0 = (C_A)_0 V_{BR} = 1 \times 1000 = 1000\text{ mol}$$

$$(N_B)_0 = (C_B)_0 V_{BR} = 2.5 \times 1000 = 2500\text{ mol}$$

From Equation 7.35, the reactor temperature at 10% conversion is

$$T = T_0 + \frac{-(\Delta H_r)_{LR}|_{T_0} X_A}{\left[\sum_{i=1}^{n}\left\{\left(C_{A_i}\right)_0/(C_A)_0 - (a_i/a_A)X_A\right\}(C_p)_{A_i} + \sum_{j=1}^{l}(P_j/a_A)X_A(C_p)_{P_j}\right]}$$

$$\Rightarrow \quad T = 50 + \frac{10.75 \times 10^3 \times 0.1}{[\{1-0.1\} \times 5 + \{2.5 - 2 \times 0.1\} \times 25 + 0.1 \times 25]}$$

$$\Rightarrow \quad T = 66.67°C$$

Similarly, we calculate the reactor temperature at different conversions and prepare Table EXE7.4.

Step 2

Calculate the specific reaction rate constant at 0.1 conversion using the corresponding temperature and Equation 3.31 (Arrhenius equation, Chapter 3)

TABLE EXE7.4
Calculation Table

X_A	T (°C)	k ($L^2/mol^2 \cdot s$)	$-r_A$ mol/(L · s)	($1/-r_A$)
0	50	0.02	0.125	8
0.1	66.66667	0.024009536	0.114309	8.748187
0.2	84.95935	0.02877222	0.101508	9.851402
0.3	105.1282	0.034419913	0.086979	11.49701
0.4	127.4775	0.041105862	0.071278	14.02966
0.5	152.381	0.049007878	0.055134	18.13767

$$\frac{k_{T_1}}{k_{T_2}} = \exp\left\{\frac{E_a}{R}\left(\frac{1}{T_2} - \frac{1}{T_1}\right)\right\}$$

$$\Rightarrow \quad \frac{k_{339.67}}{k_{323}} = \exp\left\{\frac{10{,}000}{8.314}\left(\frac{1}{323} - \frac{1}{339.67}\right)\right\}$$

$$\Rightarrow \quad \frac{k_{339.67}}{0.02} = \exp\left\{\frac{10{,}000}{8.314}\left(\frac{1}{323} - \frac{1}{339.67}\right)\right\}$$

$$\Rightarrow \quad k_{339.67} = 0.024\, L^2/(mol^2 \cdot s)$$

Similarly, calculate the specific reaction rate constant at other conversions. The values have been tabulated in the next column of Table EXE7.4.

Step 3

Calculate the initial reaction rate $-r_A$ at zero conversion using the corresponding specific reaction rate constant "k" as

$$-r_A = kC_AC_B^2 = k(C_A)_0(1-X_A)[(C_B)_0 - 2(C_A)_0X_A]^2$$

$$\Rightarrow \quad -r_A = 0.02 \times 1 \times (1-0)[2.5 - 2 \times 1 \times 0]^2 = 0.125\, mol/(L \cdot s)$$

Similarly, calculate the reaction rate at other conversions. The values have been tabulated in the next column of Table EXE7.4.

Step 4

Calculate ($1/-r_A$) for different conversions and tabulate the values in the next column of Table EXE7.4.

Step 5

From Equation 6.1 (Chapter 6),

$$V_{PFR} = (F_{ALR})_0 \int_0^{(X_{ALR})_f} \frac{dX_{ALR}}{-r_{ALR}} = \frac{100}{60}\int_0^{0.5} \frac{dX_{ALR}}{-r_{ALR}}$$

To estimate the value of the integral, we plot ($1/-r_A$) against X_A as given below (Figure EXE7.4).

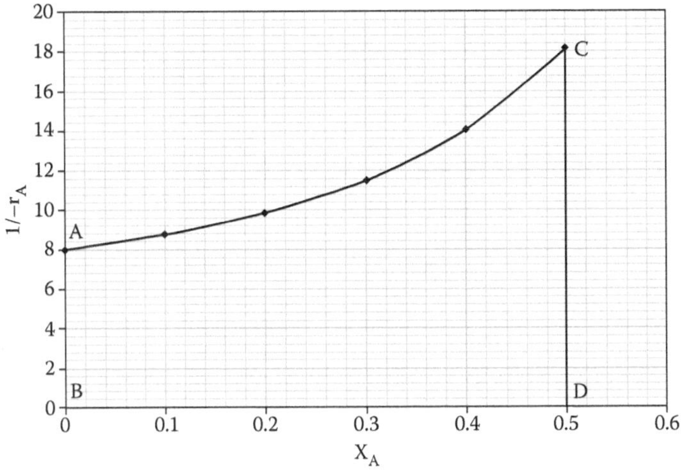

FIGURE EXE7.4 Graphical integration for estimation of volume of PFR.

Hence,

$$V_{PFR} = \frac{100}{60} \times \text{Area ABCD} = \frac{100}{60} \times 5.65 = 9.42\text{L}$$

EXAMPLE 7.5

Repeat Example 7.3 if the reaction is carried out in a PFR to calculate the space time required for 60% conversion for 100 mol/min feed of A.

Solution

Step 1

From the solution to Example 7.3,

$$(C_A)_0 = 0.063\,\text{mol/L}$$

The products B and C are not present initially. Also,

$$(C_p)_A = (C_p)_B + (C_p)_C$$

Therefore, from Equation 7.36, modified for adiabatic reactor with no mechanical work, the reactor temperature at 10% conversion is

$$T = T_0 + \frac{-(\Delta H_r)_A|_{300} X_A}{(C_p)_A}$$

$$\Rightarrow \quad T = 973 - \frac{5 \times 10^3 \times 0.1}{10}$$

$$\Rightarrow \quad T = 923\,\text{K}$$

TABLE EXE7.5

X_A	T (°C)	k (L²/mol² · min)	$-r_A$ mol/(L · min)	(1/ $-r_A$)
0	973	10	0.03969	25.19526
0.1	923	8.458514184	0.02493	41.85833
0.2	873	7.018751719	0.01533	70.44667
0.3	823	5.693502044	0.009117	122.2045
0.4	773	4.495129818	0.005163	222.6069
0.5	723	3.434859472	0.002726	435.9547
0.6	673	2.52182903	0.001298	949.9845

Similarly, we calculate the reactor temperature at different conversions and prepare Table EXE7.5.

Also, $T_{X_A=1} = 473\,K$

Step 2

Calculate the specific reaction rate constant at each of these temperatures using Equation 3.31 (Chapter 3)

$$\frac{k_{T_1}}{k_{T_2}} = \exp\left\{\frac{E_a}{R}\left(\frac{1}{T_2}-\frac{1}{T_1}\right)\right\}$$

and tabulate the values in the next column of Table EXE7.3.

From Equation 7.29,

$$\varepsilon_A = \frac{(N_A)_0[2-1]}{(N_A)_0}\left(\frac{T_{X_A=1}}{T_0}\right)+\left(\frac{T_{X_A=1}-T_0}{T_0}\right) = 1\times\left(\frac{473}{973}\right)+\left(\frac{473-973}{973}\right)$$

$$\Rightarrow \quad \varepsilon_A = -0.023$$

Step 3

From the solution to Example 7.3,

$$-r_A = kC_A^2 = k\frac{(C_A)_0^2(1-X_A)^2}{\left[(T/T_0)+\left(1+\varepsilon_A-\left(T_{X_A=1}/T_0\right)\right)\left(T/T_{X_A=1}\right)X_A\right]^2}$$

Hence,

Initial rate of reaction

$$-r_A\big|_{initial} = 10\times 0.063\times 0.063 = 0.03969\text{ mol/(L}\cdot\text{min)}$$

Similarly we calculate the reaction rate for each conversion and tabulate the values in the next column of Table EXE7.5.

Step 4

From Equation 6.6 (Chapter 6),

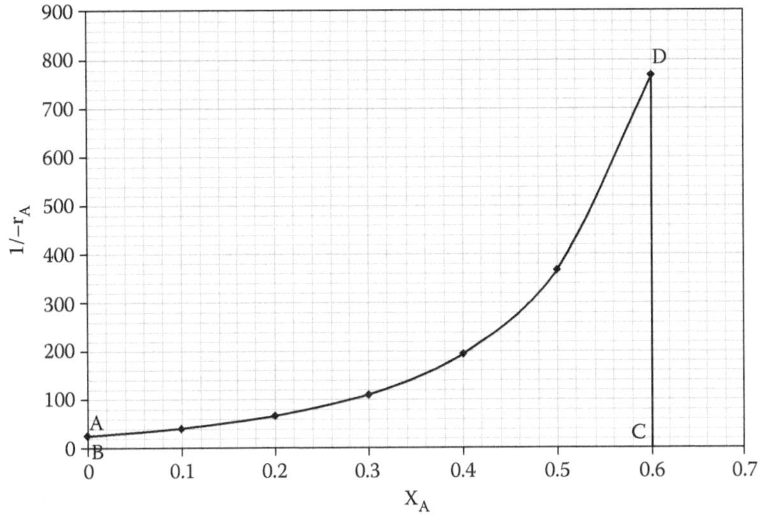

FIGURE EXE7.5 Graphical integration for estimation of space time of PFR.

$$\tau_{PFR} = (C_A)_0 \int_0^{(X_A)_f} \frac{dX_A}{-r_A} = 0.063 \times \int_0^{0.6} \frac{dX_A}{-r_A}$$

To evaluate the integral graphically, we calculate the values of $(1/-r_A)$ for different conversions. The calculated values are tabulated in the next column of Table EXE7.5. Next, we plot $(1/-r_A)$ against X_A as shown in Figure EXE7.5.

Hence,

$$\tau_{PFR} = 0.063 \times \int_0^{0.6} \frac{dX_A}{-r_A} = 0.063 \times \text{Area ABCD} = 0.063 \times 114.4 = 7.21\,\text{min}$$

7.3 BACKMIX FLOW REACTOR

7.3.1 Energy Balance

At every point inside an ideal BFR, the concentration of each chemical species and the temperature is the same. The outgoing stream from the reactor has the same concentration of the chemical species and the temperature as everywhere inside the reactor. By this argument, the conversion of the limiting reactant in the outgoing stream from the BFR is the same as everywhere inside the reactor. A fluid element that enters the reactor instantly attains the conversion as well as the temperature prevailing inside the reactor. Consider one such fluid element in the entering feed, which does not contain any product chemical species. At the instant it enters the BFR, we can write an energy balance over the fluid element as

Energy added + Heat absorbed in chemical reaction = Increase in heat content

$$\text{Energy added} = \dot{Q} + \dot{W} = q\left(\dot{N}_{A_{LR}}\right)_0 X_{A_{LR}} + w\left(\dot{N}_{A_{LR}}\right)_0 X_{A_{LR}} \tag{7.37}$$

$$\text{Heat absorbed in chemical reaction} = -(\Delta H_r)_{LR|T_0}\left[\left(\dot{N}_{A_{LR}}\right)_0 - \dot{N}_{A_{LR}}\right]$$

$$\Rightarrow \text{Heat absorbed in chemical reaction} = -(\Delta H_r)_{LR|T_0}\left(\dot{N}_{A_{LR}}\right)_0 X_{A_{LR}} \tag{7.38}$$

$$\text{Increase in heat content} = \left[\sum_{i=1}^{n} \dot{N}_{A_i}(C_p)_{A_i} + \sum_{j=1}^{1} \dot{N}_{P_j}(C_p)_{P_j}\right](T - T_0)$$

$$\Rightarrow \text{Increase in heat content} = \left[\sum_{i=1}^{n}\left\{\left(\dot{N}_{A_i}\right)_0 - \frac{a_i}{a_{LR}}\left(\dot{N}_{A_{LR}}\right)_0 X_{A_{LR}}\right\}(C_p)_{A_i} + \sum_{j=1}^{1}\frac{P_j}{a_{LR}}\left(\dot{N}_{A_{LR}}\right)_0 X_{A_{LR}}(C_p)_{P_j}\right](T - T_0) \tag{7.39}$$

Substituting the respective expressions from Equations 7.37 through 7.39 into the energy balance statement

$$[q + w - (\Delta H_r)_{LR|T_0}]\left(\dot{N}_{A_{LR}}\right)_0 X_{A_{LR}} = \left[\sum_{i=1}^{n}\left\{\left(\dot{N}_{A_i}\right)_0 - \frac{a_i}{a_{LR}}\left(\dot{N}_{A_{LR}}\right)_0 X_{A_{LR}}\right\}(C_p)_{A_i} + \sum_{j=1}^{1}\frac{P_j}{a_{LR}}\left(\dot{N}_{A_{LR}}\right)_0 X_{A_{LR}}(C_p)_{P_j}\right](T - T_0)$$

$$\Rightarrow T = T_0 + \frac{[q + w - (\Delta H_r)_{LR|T_0}]\left(\dot{N}_{A_{LR}}\right)_0 X_{A_{LR}}}{\left[\sum_{i=1}^{n}\left\{\left(\dot{N}_{A_i}\right)_0 - (a_i/a_{LR})\left(\dot{N}_{A_{LR}}\right)_0 X_{A_{LR}}\right\}(C_p)_{A_i} + \sum_{j=1}^{1}(P_j/a_{LR})\left(\dot{N}_{A_{LR}}\right)_0 X_{A_{LR}}(C_p)_{P_j}\right]}$$

$$\Rightarrow T = T_0 + \frac{[q + w - (\Delta H_r)_{LR|T_0}]\left(C_{A_{LR}}\right)_0 X_{A_{LR}}}{\left[\sum_{i=1}^{n}\left\{\left(C_{A_i}\right)_0 - (a_i/a_{LR})\left(C_{A_{LR}}\right)_0 X_{A_{LR}}\right\}(C_p)_{A_i} + \sum_{j=1}^{1}(P_j/a_{LR})\left(C_{A_{LR}}\right)_0 X_{A_{LR}}(C_p)_{P_j}\right]} \tag{7.40}$$

Equation 7.40 is identical to Equation 7.9. This relates the conversion in BFR with the temperature in the reactor.

Case 1: $(C_{A_i})_0 = (a_i/a_{LR})(C_{A_{LR}})_0$
In this case, Equation 7.11 is applicable for the BFR.

Case 2: Adiabatic BFR
For q = 0, from Equation 7.40

$$T = T_0 + \frac{[w - (\Delta H_r)_{LR|T_0}]X_{A_{LR}}}{\left[\sum_{i=1}^{n}\left\{\left(C_{A_i}\right)_0/\left(C_{A_{LR}}\right)_0 - (a_i/a_{LR})X_{A_{LR}}\right\}(C_p)_{A_i} + \sum_{j=1}^{1}(P_j/a_{LR})X_{A_{LR}}(C_p)_{P_j}\right]} \tag{7.41}$$

Further, if w is negligible in relation to the heat absorbed in the chemical reaction, we have

$$T = T_0 + \frac{-(\Delta H_r)_{LR|T_0}X_{A_{LR}}}{\left[\sum_{i=1}^{n}\left\{\left(C_{A_i}\right)_0/\left(C_{A_{LR}}\right)_0 - (a_i/a_{LR})X_{A_{LR}}\right\}(C_p)_{A_i} + \sum_{j=1}^{1}(P_j/a_{LR})X_{A_{LR}}(C_p)_{P_j}\right]} \tag{7.42}$$

For adiabatic BFR with $(C_{A_i})_0 = (a_i/a_{LR})(C_{A_{LR}})_0$, Equation 7.11 can be suitably modified with q = 0. Here, it is again evident that Equations 7.41 and 7.42 are identical to the corresponding equations for the batch reactor and PFR.

Case 3: $\sum_{i=1}^{n} a_i(C_p)_{A_i} = \sum_{j=1}^{1} P_j(C_p)_{P_j}$
From Equation 7.40,

$$T = \frac{[q + w - (\Delta H_r)_{LR|T_0}]\left(C_{A_{LR}}\right)_0}{\left[\sum_{i=1}^{n}\left(C_{A_i}\right)_0 (C_p)_{A_i}\right]}X_{A_{LR}} + T_0 \tag{7.43}$$

Again, Equation 7.43 is identical to the corresponding equation for batch reactor and PFR.

7.3.2 Performance Evaluation and Design for Nonisothermal BFR

Here also, the conversion temperature relation—Equations 7.40 through 7.43—are to be used in conjunction with the material balance relation given by Equation 6.86 or 6.88 (Chapter 6), reproduced below for convenience.

$$V_{BFR} = \frac{\left(F_{A_{LR}}\right)_0\left(X_{A_{LR}}\right)_f}{\left(-r_{A_{LR}}\right)_f} \tag{6.86}$$

$$\tau_{BFR} = \frac{\left(C_{A_{LR}}\right)_0\left(X_{A_{LR}}\right)_f}{\left(-r_{A_{LR}}\right)_f} \tag{6.88}$$

For determining the required volume of BFR for a given conversion, the appropriate energy balance equation and Equation 6.86 (Chapter 6) is to be solved simultaneously. The general algorithm is presented next:

Step 1
Using the appropriate equation from Section 7.3.1, estimate the reaction temperature "T" for the given conversion "$(X_{A_{LR}})_f$."

Step 2
Substituting the values of "T," calculated in Step 1, and "$(X_{A_{LR}})_f$" in the appropriate rate expression, calculate $(-r_{A_{LR}})_f$.

Step 3
Using Equation 6.86 or 6.88, calculate the volume of the BFR or the space time of the BFR.

EXAMPLE 7.6

Repeat Example 7.4 if the reaction is carried out in a BFR.

Solution

From the solution to Example 7.4,

$$T_0 = 50°C = 323\,K$$

$$(\Delta H_r)_A|_{323} = -10.75\ \text{kJ/mol A}$$

Step 1
From Equation 7.42, the reactor temperature at 50% conversion is

$$T = T_0 + \frac{-(\Delta H_r)_{LR}\,|_{T_0}\,(X_A)_f}{\left[\sum_{i=1}^{n}\left\{(C_{A_i})_0/(C_A)_0 - (a_i/a_A)(X_A)_f\right\}(C_p)_{A_i} + \sum_{j=1}^{l}(P_j/a_A)(X_A)_f(C_p)_{P_j}\right]}$$

$$\Rightarrow\quad T = 50 + \frac{10.75\times10^3\times0.5}{[\{1-0.5\}\times5+\{2.5-2\times0.5\}\times25+0.5\times25]}$$

$$\Rightarrow\quad T = 152.381°C = 425.381K$$

Step 2
Calculate the specific reaction rate constant at 425.381 K using Equation 3.31 (Chapter 3):

$$\frac{k_{425.381}}{k_{323}} = \exp\left\{\frac{E_a}{R}\left(\frac{1}{T_2}-\frac{1}{T_1}\right)\right\}$$

$$\Rightarrow\quad \frac{k_{425.381}}{0.02} = \exp\left\{\frac{10{,}000}{8.314}\left(\frac{1}{323}-\frac{1}{425.381}\right)\right\}$$

$$\Rightarrow\quad k_{425.381} = 0.049\ (L^2/mol^2\cdot s)$$

Step 3

Calculate the reaction rate corresponding to temperature "T" and $k_{425.381}$

$$(-r_A)_f = kC_AC_B^2 = k_{425.381}(C_A)_0[1-(X_A)_f][(C_B)_0 - 2(C_A)_0(X_A)_f]^2$$

$$\Rightarrow \quad (-r_A)_f = 0.049\times 1\times(1-0.5)[2.5-2\times 1\times 0.5]^2 = 0.055 \text{ mol/(L}\cdot\text{s)}$$

Step 4

From Equation 6.86 (Chapter 6)

$$V_{BFR} = \frac{(F_A)_0(X_A)_f}{(-r_A)_f} = \frac{100}{60}\times\frac{0.5}{0.055} = 15.15 \text{ L}$$

7.3.3 Stability of Nonisothermal BFR Operation

We have seen in the preceding section—Equations 7.40 through 7.43—that for any reaction the conversion and the temperature of the reacting system are uniquely related. Consider an existing BFR with volume V_{BFR} operating with a molar feed rate $(F_{A_{LR}})_0$ of the limiting reactant. For this operation, from Equation 6.86 (Chapter 6),

$$\frac{(X_{A_{LR}})_f}{(-r_{A_{LR}})_f} = \frac{V_{BFR}}{(F_{A_{LR}})_0} = \text{constant} \tag{7.44}$$

Since $(-r_{A_{LR}})_f$ is dependent on the conversion and temperature, Equation 7.44 translates into a unique relationship between conversion and reactor temperature. In addition, the energy balance of the reactor—the appropriate equation from among Equations 7.40 through 7.43—provides a unique relationship between conversion and reactor temperature. These two conversion temperature relationships must be compatible with each other. It implies that for an existing BFR operating under a given feed condition, the reactor temperature and conversion achieved cannot be independently fixed. Rather, we can only have one (or may be more than one!) combination of temperature and conversion that simultaneously satisfies both the energy balance (Equations 7.40 through 7.43) and material balance (Equation 7.44) (Uppal et al., 1974, 1976). This permissible set (or sets) of temperature and conversion allows for the BFR operation to be stable. Any attempt to operate the BFR with a combination of temperature and conversion other than the permissible one will lead to an unstable condition. The reacting system will automatically self adjust to attain the permissible combination of temperature and conversion. Alternatively, the feed conditions will have to be manipulated in order to have a new permissible combination of temperature and conversion that matches with the operating conditions.

The permissible set of values for the reactor temperature and conversion can be estimated by simultaneously solving the energy balance—the appropriate equation from among Equations 7.40 through 7.43—and the material balance—Equation 7.44—of the reactor. A graphical procedure is often suitable where we plot T_f as a function of $(X_A)_f$ (or $(X_A)_f$ as a function of T_f) using the energy balance and the material balance equations. The point (or points) where the two plots intersect provides the permissible combination of T_f and $(X_A)_f$. We examine a few specific cases for gas-phase reactions, when Equation 7.43 is applicable and q and w are zero:

Case 1: Single-reactant first-order reaction
From Equation 7.43 (energy balance relationship),

$$T_f = \frac{-(\Delta H_r)_{A|T_0}}{(C_p)_A}(X_A)_f + T_0 \tag{7.45}$$

From Equation 7.44 (material balance relationship),

$$(X_A)_f = (-r_A)_f \frac{V_{BFR}}{(F_A)_0}$$

$$\Rightarrow (X_A)_f = k_0\left\{\exp\left(-\frac{E_a}{RT_f}\right)\right\}\frac{(C_A)_0[1-(X_A)_f]}{\left[(T_f/T_0)+\left(1+\varepsilon_A-\left(T_{X_A=1}/T_0\right)\right)\left(T_f/T_{X_A=1}\right)(X_A)_f\right]}\frac{\tau_{BFR}}{(C_A)_0}$$

$$\Rightarrow \exp\left(-\frac{E_a}{RT_f}\right) = \frac{(X_A)_f\left[(T_f/T_0)+\left(1+\varepsilon_A-\left(T_{X_A=1}/T_0\right)\right)\left(T_f/T_{X_A=1}\right)(X_A)_f\right]}{k_0\tau_{BFR}[1-(X_A)_f]}$$

$$\Rightarrow T_f = -\frac{E_a}{R}\left[\frac{1}{\ln\left\{\frac{\left((X_A)_f\left[(T_f/T_0)+\left(1+\varepsilon_A-\left(T_{X_A=1}/T_0\right)\right)\left(T_f/T_{X_A=1}\right)(X_A)_f\right]\right)}{\left(k_0\tau_{BFR}[1-(X_A)_f]\right)}\right\}}\right] \tag{7.46}$$

For the desired solution, we must plot T_f versus $(X_A)_f$ from Equations 7.45 and 7.46 on the same graph. The point (or points) of intersection of the two plots gives the desired combination of temperature and conversion.

Case 2: Single-reactant second-order reaction
The energy balance relationship, that is, Equation 7.45, holds good.
From Equation 7.44 (material balance relationship),

$$(X_A)_f = (-r_A)_f \frac{V_{BFR}}{(F_A)_0}$$

$$\Rightarrow (X_A)_f = k_0\left\{\exp\left(-\frac{E_a}{RT_f}\right)\right\}\left[\frac{(C_A)_0[1-(X_A)_f]}{\left[(T_f/T_0)+\left(1+\varepsilon_A-\left(T_{X_A=1}/T_0\right)\right)\left(T_f/T_{X_A=1}\right)(X_A)_f\right]}\right]^2\frac{\tau_{BFR}}{(C_A)_0}$$

$$\Rightarrow \exp\left(-\frac{E_a}{RT_f}\right) = \frac{(X_A)_f\left[(T_f/T_0)+\left(1+\varepsilon_A-\left(T_{X_A=1}/T_0\right)\right)\left(T_f/T_{X_A=1}\right)(X_A)_f\right]^2}{k_0\tau_{BFR}(C_A)_0[1-(X_A)_f]^2}$$

$$\Rightarrow \quad T_f = -\frac{E_a}{R}\left[\frac{1}{\ln\left\{\dfrac{\left((X_A)_f\left[(T_f/T_0)+\left(1+\varepsilon_A-\left(T_{X_A=1}/T_0\right)\right)\left(T_f/T_{X_A=1}\right)(X_A)_f\right]^2\right)}{\left(k_0\tau_{BFR}(C_A)_0[1-(X_A)_f]^2\right)}\right\}}\right] \tag{7.47}$$

For the desired solution, we must plot T_f versus $(X_A)_f$ from Equations 7.45 and 7.47 on the same graph. The point (or points) of intersection of the two plots gives the desired combination of temperature and conversion.

Case 3: Single-reactant nth-order reaction

Here, Equation 7.45 also holds good. For the material balance relationship using the rate equation $(-r_A)_f = k_0\{\exp-(E_a/RT_f)\}(C_A)_0^n[1-(X_A)_f]^n$ yields

$$T_f = -\frac{E_a}{R}\left[\frac{1}{\ln\left\{\dfrac{\left((X_A)_f\left[(T_f/T_0)+\left(1+\varepsilon_A-\left(T_{X_A=1}/T_0\right)\right)\left(T_f/T_{X_A=1}\right)(X_A)_f\right]^n\right)}{\left(k_0\tau_{BFR}(C_A)_0^{n-1}[1-(X_A)_f]^n\right)}\right\}}\right] \tag{7.48}$$

For the desired solution, we must plot T_f versus $(X_A)_f$ from Equations 7.45 and 7.48 on the same graph. The point (or points) of intersection of the two plots gives the desired combination of temperature and conversion.

For a liquid-phase reaction, a volume change of the reacting system is negligible and the material balance relationships are

$$T_f = -\frac{E_a}{R}\left[\frac{1}{\ln\{(X_A)_f/(k_0\tau_{BFR}[1-(X_A)_f])\}}\right] \quad \text{for first-order reaction} \tag{7.49}$$

$$T_f = -\frac{E_a}{R}\left[\frac{1}{\ln\{(X_A)_f/(k_0\tau_{BFR}(C_A)_0[1-(X_A)_f]^2)\}}\right] \quad \text{for second-order reaction} \tag{7.50}$$

$$T_f = -\frac{E_a}{R}\left[\frac{1}{\ln\{(X_A)_f/(k_0\tau_{BFR}(C_A)_0^{n-1}[1-(X_A)_f]^n)\}}\right] \quad \text{for nth-order reaction} \tag{7.51}$$

EXAMPLE 7.7

The first-order liquid-phase reaction

$$A \rightarrow B + C$$

is carried out in an adiabatic BFR with a space time of 100 min and negligible work done by the agitator. Reactor feed has 1 mol/L concentration of A. The feed temperature is 17°C. The following data is given:

Heat of reaction = –3 kJ/mol

$$(C_p)_A = 10 \text{ J/mol}, \quad (C_p)_B = 5 \text{ J/mol}, \quad (C_p)_C = 5 \text{ J/mol}$$

Activation energy = 25 kJ/mol
Arrhenius frequency factor = 10 min^{-1}

Calculate the conversion achieved in the reactor and the reactor temperature.

Solution

The reaction is liquid phase and $(C_p)_A = (C_p)_B + (C_p)_C$, so from Equation 7.49

$$T_f = -\frac{E_a}{R}\left[\frac{1}{\ln\{(X_A)_f/(k_0\tau_{BFR}[1-(X_A)_f])\}}\right]$$

$$\Rightarrow \quad T_f = -\frac{25{,}000}{8.314}\left[\frac{1}{\ln\{(X_A)_f/(10\times 100\times[1-(X_A)_f])\}}\right] \qquad (7.7.1)$$

From Equation 7.45,

$$T_f = \frac{-(\Delta H_r)_A|_{T_0}}{(C_p)_A}(X_A)_f + T_0$$

$$\Rightarrow \quad T_f = \frac{3000}{10}(X_A)_f + (17+273) \qquad (7.7.2)$$

To estimate the conversion achieved in the reactor and the reactor temperature, we must simultaneously solve Equations 7.7.1 and 7.7.2. For this purpose, we plot the two equations as given below (Figure EXE7.7). From this Figure we can see that there are three solution sets for the system of equations 7.7.1 and 7.7.2.

Corresponding to point A:

$$(X_A)_f = 0.73 \quad \text{and} \quad T_f = 510\text{ K}$$

Corresponding to point B:

$$(X_A)_f = 0.41 \quad \text{and} \quad T_f = 413\text{ K}$$

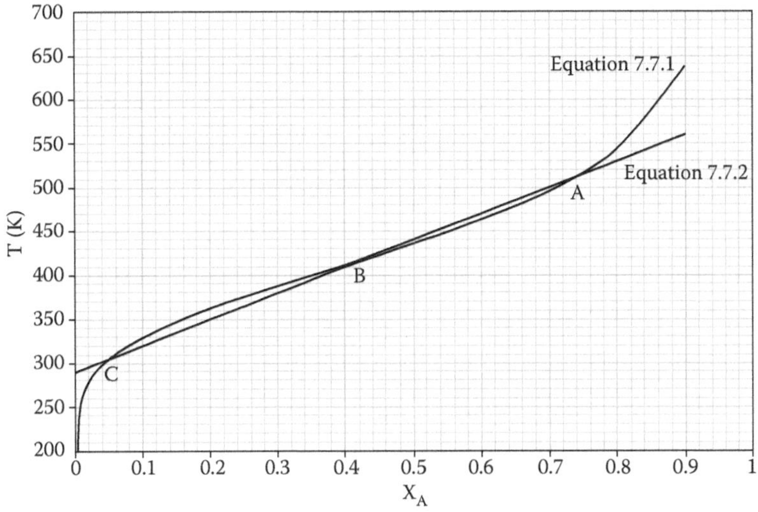

FIGURE EXE7.7 Graphical solution of Equations 7.7.1 and 7.7.2.

Corresponding to point C:

$$(X_A)_f = 0.05 \quad \text{and} \quad T_f = 305\ \text{K}$$

From the solution to Example 7.7 and Figure EXE7.7 we can see that in the BFR operation under consideration we can have three different combinations of conversion and temperature that simultaneously satisfies the energy balance as well as the material balance. Any of these three combinations should give a stable reactor operation. If the BFR operating at stable condition corresponding to point C undergoes some perturbations—slight increase in temperature, for example—the reacting system will self adjust to bring the reactor conditions back to the stable condition corresponding to point C. The same holds for the stable reactor operation at a condition corresponding to point A. The conversion temperature combinations represented by point A and point C, therefore, provide a *truly stable reactor operation*. However, the same is not true for the condition represented by point B. If the BFR is operating with the conversion temperature combination precisely maintained at that corresponding to point B, its operation will be stable. However, even the slightest of perturbations in the operating conditions will drift the reactor operation to a state where the conditions are either matching with point A or matching with point C. For the BFR operating at point B, the slightest increase in temperature will bring the reactor condition to that represented by point A. The slightest decrease in temperature will bring the reactor condition to that represented by point C. Point B, therefore, represents a *metastable condition of reactor operation*. Looking at Figure EXE7.7, the BFR should preferably be operated at conditions of point A as this would give higher conversion. But the energy costs would be more as the reactor has to be operated at a higher temperature.

7.4 REVERSIBLE REACTIONS UNDER NONISOTHERMAL CONDITIONS

For a reversible reaction, the composition of the reacting system at equilibrium, and hence, the conversion at equilibrium, $(X_{A_{LR}})_{eq}$ is uniquely related to the equilibrium constant. This relationship is represented by Equation 4.20 (Chapter 4), reproduced below for convenience.

$$K_C = \left[a_{LR}\left(1+\varepsilon_{LR}\left(X_{A_{LR}}\right)_{eq}\right)\right]^{\left(\Sigma_{i=1}^{n} a_i - \Sigma_{j=1}^{l} p_j\right)} \times \frac{\prod_{j=1}^{l}\left[a_{LR}\left(C_{P_j}\right)_0 + p_j\left(C_{A_{LR}}\right)_0 X_{A_{LR}}\right]_{eq}^{p_j}}{\prod_{i=1}^{n}\left[a_{LR}\left(C_{A_i}\right)_0 - a_i\left(C_{A_{LR}}\right)_0 X_{A_{LR}}\right]_{eq}^{a_i}} \tag{4.20}$$

In Chapter 4, we have seen that the equilibrium constant for a reversible reaction is dependent on temperature. This dependence is given by Equation 4.28 (Chapter 4), reproduced below for convenience.

$$\ln\frac{(K_C)_2}{(K_C)_1} + \left(\sum_{j=1}^{l} p_j - \sum_{i=1}^{n} a_i\right)\ln\frac{T_2}{T_1} = \frac{\Delta H_r^0}{R}\left[\frac{1}{T_1} - \frac{1}{T_2}\right] \tag{4.28}$$

So, a change in temperature of the reacting system will be accompanied by a change in $(X_{A_{LR}})_{eq}$. Therefore, if a reversible reaction is to be carried out, the design and operation of the reactor requires the following to be considered:

1. Temperature conversion relationship as predicted by the reactor energy balance.
2. Temperature equilibrium conversion relationship as mandated for the reacting system by thermodynamic consideration.
3. The material balance for the reactor.

The general algorithm to be followed is presented next.

7.4.1 Batch Reactor

Step 1

Using the appropriate equation from Section 7.1, estimate the reaction temperature "T" for different values of conversion $X_{A_{LR}}$. Arrange the values in a tabular form.

Step 2

Using Equation 4.28 (Chapter 4), determine the equilibrium constant K_C for each temperature obtained in Step 1. Tabulate the values in the next column of the table prepared in Step 1.

Step 3

For each value of K_C determined in Step 2, calculate the value of $(X_{A_{LR}})_{eq}$ from Equation 4.20 (Chapter 4). Check if $(X_{A_{LR}})_{eq} > X_{A_{LR}}$. *The actual conversion in the reactor can never exceed the equilibrium conversion. This imposes a limit on the maximum conversion that can be achieved in the reactor.* Alternatively, on the same graph, plot $(X_{A_{LR}})_{eq}$ versus T and $X_{A_{LR}}$ versus T. The point of intersection of the two plots gives the value of the maximum achievable conversion.

Once the maximum achievable conversion from the reactor is determined, the reaction time can be estimated for any conversion below the maximum achievable conversion.

Step 4

Estimate $-r_{A_{LR}}$ for each $X_{A_{LR}}$ using the corresponding value of T and the appropriate rate equation. Reaction rates are to be calculated only up to the desired conversion, which should be less than the maximum achievable conversion estimated in Step 3. Arrange the values of $-r_{A_{LR}}$ so estimated in the next column of the table prepared in Step 1.

Step 5

Using the values calculated in Step 4, estimate $(1/-r_{A_{LR}})$ for each $X_{A_{LR}}$. Tabulate the values in the next column of the table prepared in Step 1.

Step 6

For graphical estimation of the integral on the left-hand side of Equation 5.7 (Chapter 5) plot $(1/-r_{A_{LR}})$ versus $X_{A_{LR}}$. The area under this curve and the $X_{A_{LR}}$ axis between the origin and the desired conversion is the value of the integral. Next, evaluate the time of the reaction from Equation 5.7 (Chapter 5). Alternatively, the value of the integral can also be estimated using a suitable numerical integration technique.

7.4.2 Plug Flow Reactor

Step 1

Using the appropriate equation from Section 7.2.1, estimate the reaction temperature "T" for different values of conversion $X_{A_{LR}}$. Arrange the values in a tabular form.

Step 2

Using Equation 4.28 (Chapter 4) determine the equilibrium constant K_C for each temperature obtained in Step 1. Tabulate the values in the next column of the table prepared in Step 1.

Step 3

For each value of K_C determined in Step 2, calculate the value of $(X_{A_{LR}})_{eq}$ from Equation 4.20 (Chapter 4). Check if $(X_{A_{LR}})_{eq} > X_{A_{LR}}$. *The actual conversion in the reactor can never exceed the equilibrium conversion. This imposes a limit on the maximum conversion that can be achieved in the reactor.* Alternatively, on the same graph, plot $(X_{A_{LR}})_{eq}$ versus T and $X_{A_{LR}}$ versus T. The point of intersection of the two plots gives the value of the maximum achievable conversion.

Once the maximum achievable conversion from the reactor is determined, the volume or space time for the PFR can be estimated for any conversion below the maximum achievable conversion.

Step 4

Estimate $-r_{A_{LR}}$ for each $X_{A_{LR}}$ using the corresponding value of T and the appropriate rate equation. Reaction rates are to be calculated only up to the desired conversion, which should be less than the maximum achievable conversion estimated in Step 3. Arrange the values of $-r_{A_{LR}}$ so estimated in the next column of the table prepared in Step 1.

Step 5

Using the values calculated in Step 4, estimate $(1/-r_{A_{LR}})$ for each $X_{A_{LR}}$. Tabulate the values in the next column of the table prepared in Step 1.

Step 6

For graphical estimation of the integral on the RHS of Equation 6.1 (Chapter 6), plot $(1/-r_{A_{LR}})$ versus $X_{A_{LR}}$. The area under this curve and the $X_{A_{LR}}$ axis between origin and the desired conversion is the value of the integral. Next, evaluate the volume of PFR required from Equation 6.1 (Chapter 6). The value of the integral can also be estimated using a suitable numerical integration technique. Alternatively, we can use Equation 6.6 (Chapter 6) if the space time of the PFR is desired.

7.4.3 Backmix Flow Reactor

Step 1

Using the appropriate equation from Section 7.3.1, estimate the reaction temperature "T" for different values of conversion $X_{A_{LR}}$. Arrange the values in a tabular form.

Step 2

Using Equation 4.28 (Chapter 4), determine the equilibrium constant K_C for each temperature obtained in Step 1. Tabulate the values in the next column of the table prepared in Step 1.

Step 3

For each value of K_C determined in Step 2, calculate the value of $(X_{A_{LR}})_{eq}$ from Equation 4.20 (Chapter 4). Check if $(X_{A_{LR}})_{eq} > X_{A_{LR}}$. *The actual conversion in the reactor can never exceed the equilibrium conversion. This imposes a limit on the maximum conversion that can be achieved in the reactor.* Alternatively, on the same graph plot $(X_{A_{LR}})_{eq}$ versus T and $X_{A_{LR}}$ versus T. The point of intersection of the two plots gives the value of the maximum achievable conversion.

Once the maximum achievable conversion from the reactor is determined, the volume or space time for the BFR can be estimated for any conversion below the maximum achievable conversion.

Step 4

Substituting the values of "T_f," calculated in Step 1, and "$(X_{A_{LR}})_f$" in the appropriate rate expression, calculate $(-r_{A_{LR}})_f$.

Step 5

Using Equation 6.86 or 6.88, calculate the volume of the BFR or the space time of the BFR.

EXAMPLE 7.8

The gas-phase reversible reaction

$$A \rightleftharpoons B$$

is to be carried out in a PFR with pure A entering as the feed to the reactor at 50°C. Determine the maximum conversion that can be achieved. The following data are given:

Heat of reaction = –5 kJ/mol

$$(C_p)_A = 10 \text{ J/mol}, \quad (C_p)_B = 10 \text{ J/mol}, \quad K_C(\text{at } 50°\text{C}) = 5$$

Solution

From Equation 7.36, reactor temperature at 0.1 conversion

$$T = T_0 + \frac{-(\Delta H_r)_A |_{323} X_A}{(C_p)_A} = 323 + \frac{5000 \times 0.1}{10} = 373\,\text{K}$$

Similarly, we calculate the reactor temperature at different conversions and tabulate the values in Table EXE7.8.

From Equation 4.28 (Chapter 4), we determine the equilibrium constant at 373 K as

$$\ln\frac{(K_C)_{373}}{(K_C)_{323}} = \frac{\Delta H_r^0}{R}\left[\frac{1}{T_1} - \frac{1}{T_2}\right] = \frac{-5000}{8.314}\left[\frac{1}{323} - \frac{1}{373}\right]$$

$$\Rightarrow \quad (K_C)_{373} = (K_C)_{323} \exp\left\{\frac{-5000}{8.314}\left[\frac{1}{323} - \frac{1}{373}\right]\right\}$$

$$\Rightarrow \quad (K_C)_{373} = 5 \times \exp\left\{\frac{-5000}{8.314}\left[\frac{1}{323} - \frac{1}{373}\right]\right\} = 3.89562$$

TABLE EXE7.8
Calculation Table for Equilibrium Constant and Equilibrium Conversion

X_A	T (K)	K_C	$(X_A)_{eq}$
0	323	5	0.833333
0.1	373	3.89562	0.795736
0.2	423	3.219646	0.763013
0.3	473	2.770373	0.734774
0.4	523	2.453286	0.710421
0.5	573	2.219071	0.689351
0.6	623	2.039806	0.671032
0.7	673	1.898638	0.65501
0.8	723	1.784857	0.640915
0.9	773	1.691363	0.628441
1	823	1.613279	0.617339

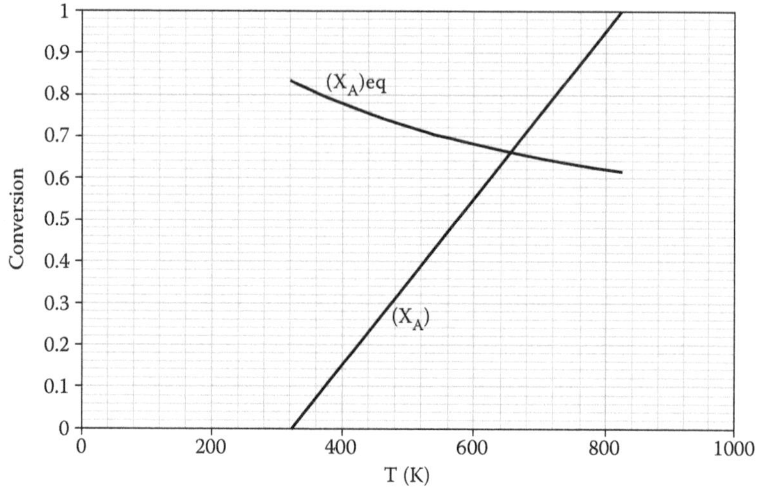

FIGURE EXE7.8 Graphical estimation of maximum achievable conversion.

Similarly, we calculate the equilibrium constant at each temperature and tabulate the values in the next column of Table EXE7.8

From Equation 2.44 (Chapter 2),

$$\varepsilon_A = \frac{1-1}{1} = 0$$

From Equation 4.20 (Chapter 4),

$$K_C = \frac{(X_A)_{eq}}{1-(X_A)_{eq}}$$

$$\Rightarrow \quad (X_A)_{eq} = \frac{K_C}{1+K_C} \tag{7.8.1}$$

From Equation 7.8.1, we calculate $(X_A)_{eq}$ for different values of K_C and tabulate them in the next column of Table EXE7.8.

Next we plot both X_A and $(X_A)_{eq}$ against temperature as shown in (Figure EXE7.8).

The two curves intersect at 0.66 conversion. This is the maximum conversion that can be achieved.

7.5 OPTIMUM TEMPERATURE

7.5.1 Exothermic Irreversible Reaction

If an exothermic irreversible reaction is carried out under adiabatic conditions, the temperature of the reacting system increases with conversion. However, the temperature conversion profile is dependent on the initial temperature. This makes for interesting optimization situations (Denbigh, 1958; Omoleye, 1989; Smets et al., 2002). A general discussion is presented here.

As shown in Figure 7.1a, the temperature conversion profile shifts to higher temperatures with increase in the feed temperature. This results in the $(1/-r_A)$ versus X_A plot (Figure 7.1b) to shift closer and closer to the conversion axis. This progressively reduces the area under the $(1/-r_A)$ versus X_A plot. Therefore, the time required in the batch reactor and the volume or space time required for the PFR or BFR decreases for a given conversion. *For the shortest reaction time in a batch reactor and the smallest size of PFR or BFR, the feed temperature should be as high as*

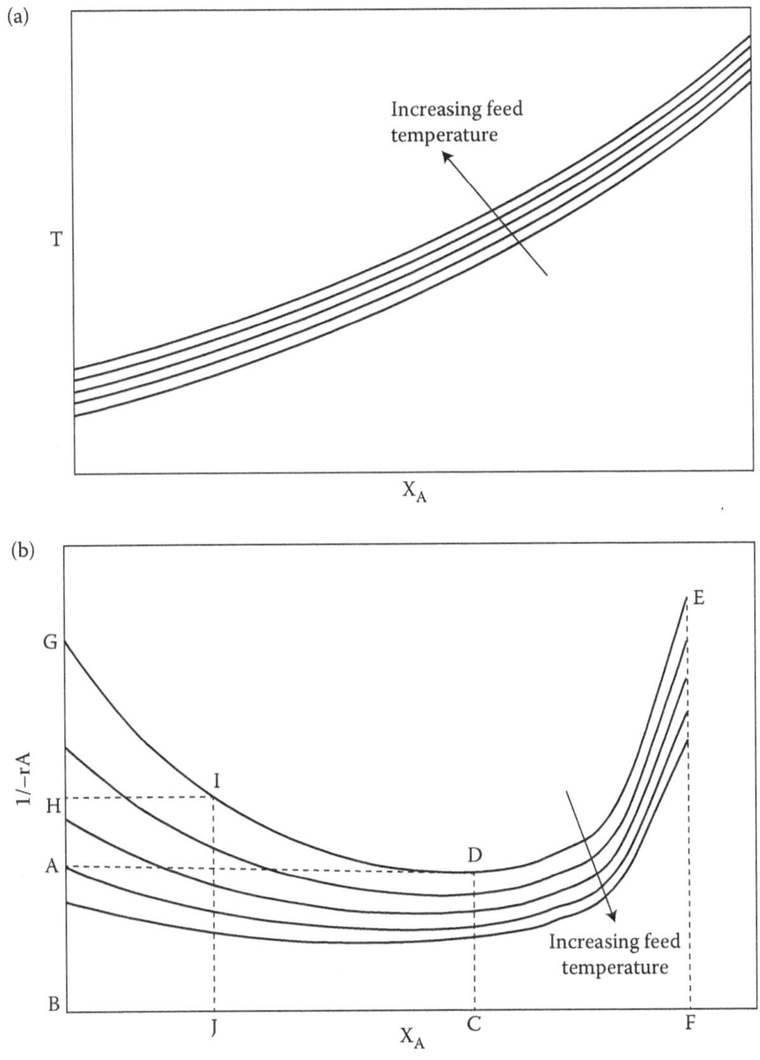

FIGURE 7.1 Energy and material balance considerations for exothermic irreversible reaction.

possible to give the highest possible temperature inside the reactor compatible with the material of construction.

For any feed temperature, with the progress of reaction the specific reaction rate constant increases whereas the reactant concentration decreases. In the initial stages of the reaction, the effect of increasing the rate constant predominates. At higher conversions, the diminishing reactant concentration masks the influence of a higher rate constant. Consequently, *the reaction rate first increases and then decreases. So, the plot, $(1/-r_A)$ versus X_A passes through a minimum,* as shown by point D in Figure 7.1b. Let the conversion corresponding to this minimum be X_{min}.

Case 1: Desired conversion is less than X_{min}

Consider the point I on the $(1/-r_A)$ versus X_A plot where the conversion is less than X_{min}. If this conversion is to be achieved in a PFR, the volume of the PFR required will be

$$V_{PFR} = (F_A)_0 \times \text{Area GIJB}$$

And, if the conversion is to be achieved in a BFR, the volume of the BFR required will be

$$V_{BFR} = (F_A)_0 \times \text{Area HIJB}$$

So, the BFR required will be smaller than the PFR.

Case 2: Desired conversion is more than X_{min}

Beyond the conversion X_{min} there exists a conversion for which the PFR or the BFR will require the same volume. If we aim for still higher conversion, a PFR will be preferred over a BFR. Consider the point E on the $(1/-r_A)$ versus X_A plot where the conversion is more than X_{min}. Clearly, the volume of PFR required to achieve this conversion will be less than that required for a BFR. However, the total reactor volume can still be minimized if we use a BFR followed by a PFR. For this, we need a BFR that operates at an outlet conversion of X_{min}. The outlet from this BFR is then to be fed to the PFR, which will give the desired conversion corresponding to point E. The total volume of this reactor assembly will be

$$V_{total} = V_{BFR} + V_{PFR} = (F_A)_0 \times \text{Area ABCD} + (F_A)_0 \times \text{Area CDEF}$$

This will be less than the volume of any single PFR or BFR for the same duty.

7.5.2 Endothermic Irreversible Reaction

If an endothermic irreversible reaction is carried out under adiabatic conditions, the temperature of the reacting system decreases with conversion. However, the temperature conversion profile is dependent on the feed temperature.

As shown in Figure 7.2a, the temperature conversion profile shifts to higher temperatures with increase in the feed temperature. This results in the $(1/-r_A)$ versus X_A

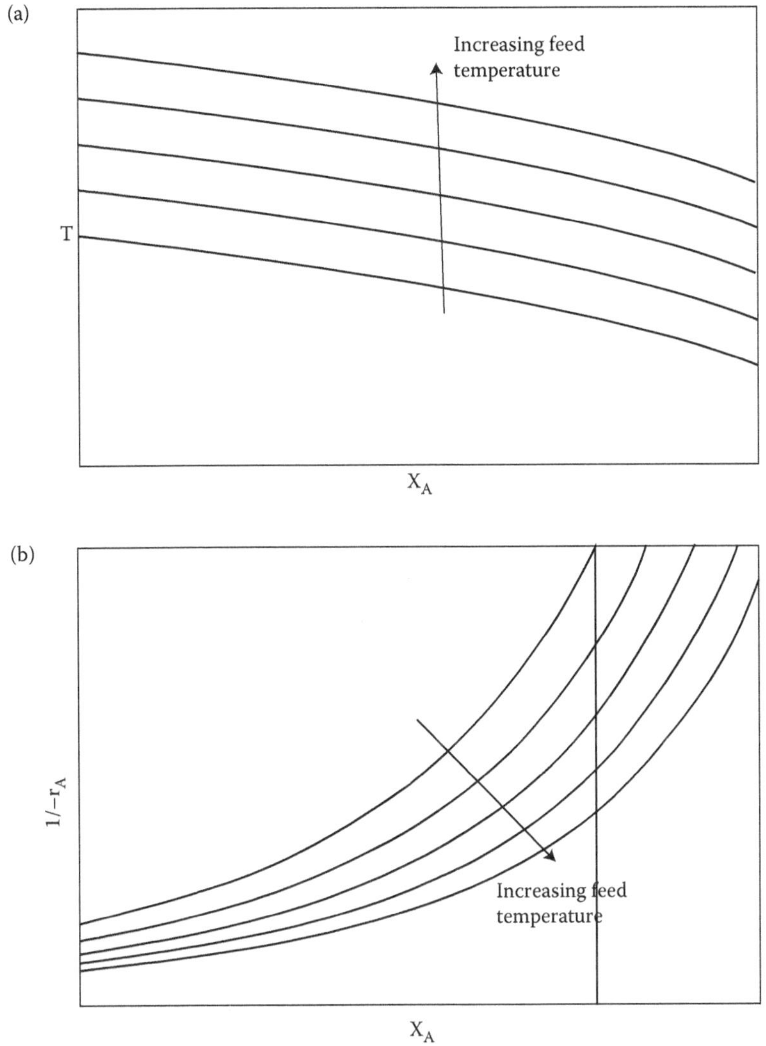

FIGURE 7.2 Energy and material balance considerations for endothermic irreversible reactions.

plot (Figure 7.2b) to shift closer and closer to the conversion axis. This progressively reduces the area under the $(1/-r_A)$ versus X_A plot. Therefore, the time required in the batch reactor and the volume or space time required for the PFR decreases for a given conversion. *For the shortest reaction time in a batch reactor and the smallest size of PFR, the feed temperature should be as high as possible* to give the highest possible temperature inside the reactor.

Figure 7.2b also shows that for any conversion the quantity $(1/-r_A)$ is the least for the highest feed temperature. Therefore, from Equation 6.86 (Chapter 6), the *volume of the BFR required is also the least for the highest feed temperature.*

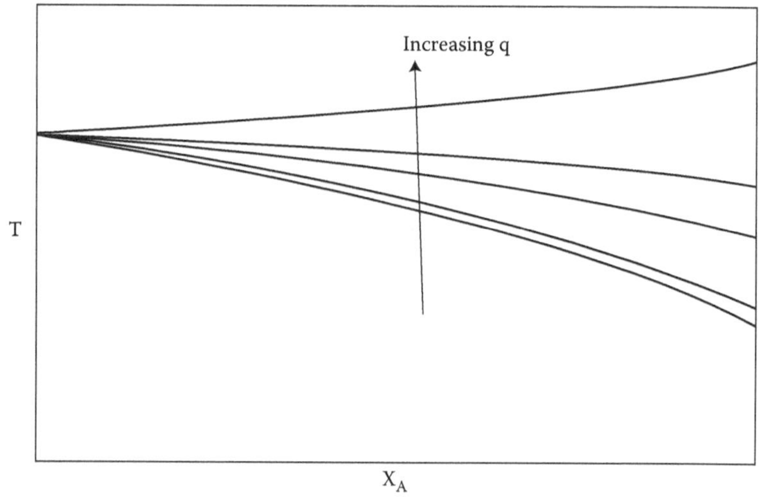

FIGURE 7.3 Temperature conversion profiles for endothermic reactions for changing heating rates.

For a strongly endothermic reaction, just increasing the feed temperature may not be sufficient to maintain a reasonably high temperature in the reactor. In addition, the reactor must be heated to prevent the reactor temperature from dropping considerably. Figure 7.3 shows the changing temperature conversion profiles for different heating rates for the reactor.

We may conclude that a high feed temperature and heating the reactor are both desirable to lower the required size of the reactor for a given conversion.

7.5.3 Exothermic Reversible Reaction

For an exothermic reversible reaction, the equilibrium constant, and hence, the equilibrium conversion decreases with temperature. Figure 7.4a shows that the maximum achievable conversion decreases with increasing feed temperature.

In practice, we rarely intend to achieve the equilibrium conversion. For any feed temperature, with the progress of reaction the net reaction rate, $r_{fwd} - r_{bwd}$, first increases and then decreases. So the plot $(1/-r_A)$ versus X_A passes through a minimum, as shown by point D in Figure 7.4b. Let the conversion corresponding to this minimum be X_{min}.

Case 1: Desired conversion is less than X_{min}

Consider the point I on the $(1/-r_A)$ versus X_A plot where the conversion is less than X_{min}. If this conversion is to be achieved in a PFR, the volume of the PFR required will be

$$V_{PFR} = (F_A)_0 \times \text{Area GIJB}$$

And, if the conversion is to be achieved in a BFR, the volume of the BFR required will be

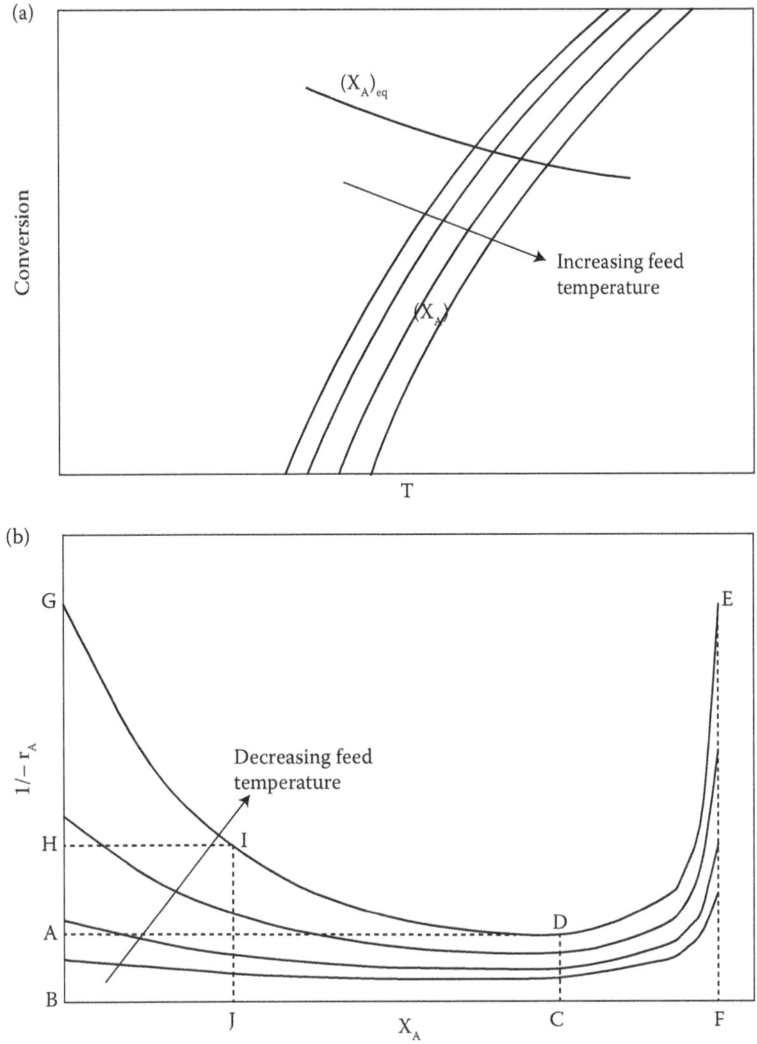

FIGURE 7.4 Effect of feed temperature on reversible exothermic reaction.

$$V_{BFR} = (F_A)_0 \times \text{Area HIJB}$$

So, the BFR required will be smaller than the PFR.

Case 2: Desired conversion is more than X_{min}

Beyond the conversion X_{min} there exists a conversion for which the PFR or the BFR will require the same volume. If we aim for still higher conversion, a PFR will be preferred over a BFR. Consider the point E on the $(1/-r_A)$ versus X_A plot where

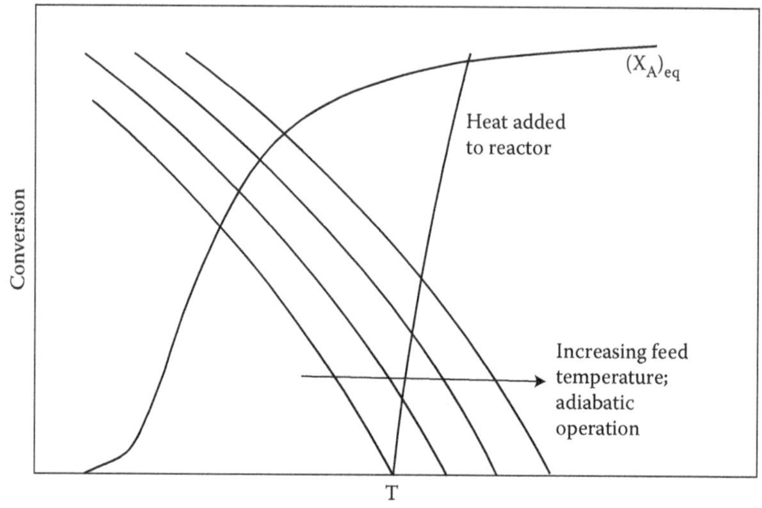

FIGURE 7.5 Maximum attainable conversion for endothermic reversible reaction.

the conversion is more than X_{min}. Clearly, the volume of PFR required to achieve this conversion will be less than that required for a BFR. But the total reactor volume can still be minimized if we use a BFR followed by a PFR. For this we need a BFR that operates at an outlet conversion of X_{min}. The outlet from this BFR is then to be fed to the PFR, which will give the desired conversion corresponding to point E. The total volume of this reactor assembly will be

$$V_{total} = V_{BFR} + V_{PFR} = (F_A)_0 \times \text{Area ABCD} + (F_A)_0 \times \text{Area CDEF}$$

This will be less than the volume of any single PFR or BFR for the same duty.

7.5.4 Endothermic Reversible Reaction

For an endothermic reversible reaction, the equilibrium constant, and hence, the equilibrium conversion increases with temperature. Figure 7.5 shows that the maximum achievable conversion in a batch reactor or a PFR increases with increasing feed temperature. It can be further increased by supplying heat to the reactor to maintain an increasing temperature conversion profile.

EXERCISES

7.1 Repeat Example 7.1 if heat is withdrawn from the reactor at the rate of 5 kJ/mol A reacted.

7.2 In Exercise 7.1, what is the rate at which the heat should be withdrawn for a reactor operation to be isothermal?

7.3 Repeat Example 7.1 for $(C_B)_0 = 2$ mol/L.

7.4 Repeat Example 7.1 for $(C_B)_0 = 1.8$ mol/L.

7.5 The liquid-phase irreversible reaction

$$A + B \rightarrow 2C$$

is carried out in a constant volume adiabatic batch reactor with $(C_A)_0 = 1$ mol/L and $(C_B)_0 = 1$ mol/L. The initial temperature of the reactor is 250°C. The mechanical work of agitation can be neglected. The reaction rate is expressed as $-r_A = 0.2C_AC_B$ mol/(L · min) at 250°C. The activation energy for the reaction is 25 kJ/mol. Calculate the time required for 50% conversion of A. The data given are

The heat of reaction = 2 kJ/mol A at 298 K.

$$(C_p)_A = 5 \text{ J/mol}, \quad (C_p)_B = 7 \text{ J/mol}, \quad (C_p)_C = 8 \text{ J/mol}$$

7.6 Repeat Exercise 7.5 if $(C_B)_0 = 1.25$ mol/L

7.7 Repeat Exercise 7.5 if $(C_B)_0 = 0.75$ mol/L

7.8 Repeat Exercise 7.5 if the initial temperature of the reactor is 300°C.

7.9 The reaction of Exercise 7.5 is to be carried out in a PFR. The feed temperature is 250°C. The volumetric flow rate to the reactor is 100 L/min. What is the volume of the PFR required for 70% conversion of A?

7.10 Repeat Exercise 7.9 if the reaction is to be carried out in a BFR.

7.11 The gas-phase reaction

$$A + B \rightarrow C$$

is carried out in a variable volume adiabatic batch reactor operating at a constant pressure of 5 atm. The reaction is started with an equimolar mixture of A and B at 700°C. The specific reaction rate constant at 700°C is 5 L/(mol · min). The following data are given:

The heat of reaction = 5 kJ/mol A at 298 K.

$$(C_p)_A = 10 \text{ J/mol}, \quad (C_p)_B = 5 \text{ J/mol}, \quad (C_p)_C = 5 \text{ J/mol}$$

Activation energy = 20 kJ/mol

Calculate the time required for 60% conversion. The mechanical work of mixing can be neglected.

7.12 The reaction of Exercise 7.11 is to be carried out in a PFR operating at a constant pressure of 5 atm. The feed to the reactor is a mixture of 60% A and 40% B at 700°C. The volumetric flow rate to the reactor is 100 L/min. What is the space time of the PFR required for 50% conversion of A?

7.13 Repeat Exercise 7.12 if the reaction is to be carried out in a BFR.

7.14 The second-order liquid-phase reaction

$$A \rightarrow B$$

is carried out in an adiabatic BFR with a space time of 100 min and negligible work done by the agitator. Reactor feed has 1 mol/L concentration of A. The feed temperature is 400 K. The following data are given:

Heat of reaction = –5 kJ/mol at 400 K

$$(C_p)_A = 15 \text{ J/mol}, \quad (C_p)_B = 15 \text{ J/mol}$$

Activation energy = 30 kJ/mol
Arrhenius frequency factor = 10 min^{-1}
Calculate the conversion achieved in the reactor and the reactor temperature.

7.15 Repeat Exercise 7.14 if the initial temperature is 500 K.
7.16 Repeat Exercise 7.14 if the reactor space time is 130 min.
7.17 The gas-phase reversible reaction

$$A + B \rightleftharpoons 2C$$

is to be carried out in a constant pressure PFR, operating at 5 atm pressure, with an equimolar mixture of A and B entering as the feed to the reactor at 50°C. Determine the maximum conversion that can be achieved. The following data are given:
Heat of reaction = –4 kJ/mol (at 50°C)

$$(C_p)_A = 5 \text{ J/mol}, \quad (C_p)_B = 7 \text{ J/mol}, \quad (C_p)_C = 6 \text{ J/mol}$$

$$K_C \text{ (at 50°C)} = 8$$

7.18 In Exercise 7.17 what will be the space time of PFR required to get 70% of the maximum attainable conversion if the specific reaction rate constant for the forward reaction at 50°C is 5 L/(mol · min) and the activation energy is 15 kJ/mol?
7.19 The liquid-phase reversible reaction

$$A \rightleftharpoons B + C$$

is to be carried out in a batch reactor, starting with 10 M concentration of A at 150°C. Determine the maximum conversion that can be achieved. The following data are given:
Heat of reaction = 2.5 kJ/mol (at 150°C)

$$(C_p)_A = 15 \text{ J/mol}, \quad (C_p)_B = 7 \text{ J/mol}, \quad (C_p)_C = 8 \text{ J/mol}$$

$$K_C \text{ (at 150°C)} = 10$$

7.20 In Exercise 7.19, what will be the reaction time required to get 70% of the maximum attainable conversion if the specific reaction rate constant for the forward reaction at 150°C is 0.1 min^{-1} and the activation energy is 10 kJ/mol?

7.21 In Exercise 7.19, what will be the space time required to get 70% of the maximum attainable conversion if the reaction is carried out in a constant pressure BFR?

7.22 The liquid-phase reaction

$$A + 2B \rightarrow C$$

has the following thermochemical and kinetic parameters:

Heat of reaction = –10.75 kJ/mol at 323 K

$$(C_p)_A = 5\ \text{J/mol}, \quad (C_p)_B = 25\ \text{J/mol}, \quad (C_p)_C = 25\ \text{J/mol}$$

Activation energy = 30 kJ/mol

Reaction rate constant at 323 K = 0.2 $L^2/(mol^2 \cdot min)$

The reaction is to be carried out at a molar feed rate of 50 mol/min to get 80% conversion. The initial concentrations of A and B are 1 and 2 mol/L, respectively. Should we use a PFR, a BFR, or a combination of PFR and BFR so that the total reactor volume is minimum? If your answer is a combination of PFR and BFR, what should be the volumes of the individual reactors and their sequence? Reactors are to be used under adiabatic conditions.

REFERENCES

Denbigh, K.G., Optimum temperature sequences in reactors. *Chem. Eng. Sci.*, 8, 125, 1958.

Omoleye, J.A., Adesina, A.A., and Udegbunam, E.O., Optimal design of nonisothermal reactors: Derivation of equations for the rate-temperature-conversion profile and the optimum temperature progression for a general class of reversible reactions. *Chem. Eng. Commun.*, 79, 95, 1989.

Smets, I.Y., Dochain, D., and van Impe, J.F., Optimal temperature control of a steady state exothermic plug flow reactor. *AIChE J.*, 48, 279, 2002.

Uppal, A., Ray, W.H., and Poore, A.B., On the dynamic behaviour of continuous stirred tank reactors. *Chem. Eng. Sci.*, 29, 967, 1974.

Uppal, A., Ray, W.H., and Poore, A.B., The classification of the dynamic behaviour of continuous stirred tank reactors – influence of reactor residence time. *Chem. Eng. Sci.*, 31, 205, 1976.

8 Nonideal Flow in Reactors

In Chapter 6, we considered the flow of fluid through the reactor as conforming to one of the two ideal types—plug flow and complete backmix flow. In the former, every fluid element entering the reactor spends the same amount of time inside the reactor. The movement of material in the axial direction of the reactor is so orderly that even diffusion is excluded. In the latter, on the other hand, a fluid element that enters the reactor instantly mixes with the entire fluid present within the reactor. Thus, for some fluid elements there is no time lag between entry and exit, whereas some fluid elements endlessly remain inside the reactor and never come out. In between there are different fluid elements that spend different amounts of time in the reactor. Both of these flow conditions—the plug flow and the complete backmix flow—are conceptually ideal extreme propositions. In a real reactor, the flow will certainly lack the complete order hypothesized in a plug flow. As a result, different fluid elements will spend different times in the reactor. However, neither will there be any fluid element that spends no time in the reactor nor will there be any fluid element that never comes out of the reactor. Briefly, different fluid elements spend different finite amounts of time inside a real flow reactor. Analysis of the distribution of residence times, therefore, is central to the understanding of the performance of real reactors (Danckwerts, 1953). We define the *residence time distribution (RTD) function "E" as fraction of material coming out of the reactor per unit time between times t and t + dt, as dt → 0.*

8.1 RTD FOR IDEAL REACTORS

8.1.1 Ideal PFR

By definition of a plug flow, any quantum of material that enters the reactor at a single instant of time t = 0 will entirely come out of the reactor in a single instant after time τ_{PFR}. From this quantum of material, neither anything will come out before time τ_{PFR} nor will anything come out after time τ_{PFR}. So, the fraction of material leaving the reactor between $t = \tau_{PFR}$ and $t = \tau_{PFR} + dt$, where $dt \to 0$, is 1. By the definition of E, at time τ_{PFR},

$$(E)_{\tau_{PFR}} = \lim_{dt \to 0} \frac{1}{dt} = \infty \tag{8.1}$$

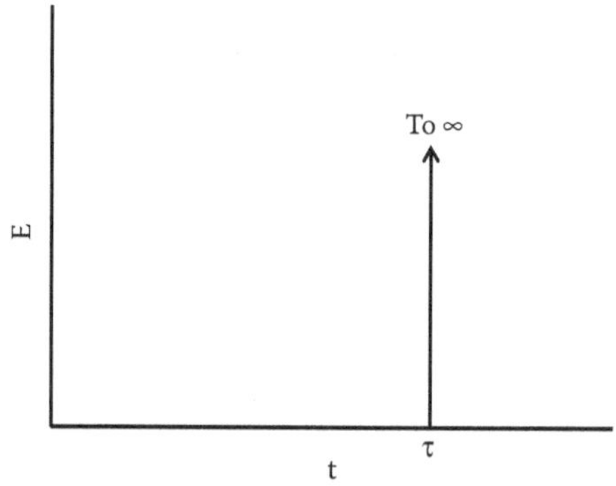

FIGURE 8.1 RTD function for ideal PFR.

At any other time, the fraction of material leaving the reactor is zero, and so is the RTD function E. Mathematically, the RTD function for the ideal PFR can be represented by the Dirac delta function as

$$E = \infty, \quad \text{at } t = \tau_{PFR} \tag{8.2}$$

$$E = 0, \quad \text{at } t \neq \tau_{PFR} \tag{8.3}$$

Graphically it can be represented as shown in Figure 8.1.

8.1.2 Ideal Backmix Flow Reactor

Consider a BFR of volume V_{BFR} through which fluid is flowing at a constant volumetric flow rate of "v". At the instant of time t = 0, a "packet" of nonreacting fluid—called tracer in technical parlance—of mass "m_0" is fed to the BFR. By definition of complete backmix flow, this "packet" will instantly mix into the entire volume of the BFR so that the concentration of the tracer in the reactor at t = 0 is $C_0 = (m_0/V_{BFR})$. At time "t," let the mass of the tracer left in the reactor be "m" so that the concentration of the tracer in the reactor at time "t" is $C = (m/V_{BFR})$. Applying a material balance over the reactor for the tracer, we have

$$\text{Input} = \text{Output} + \text{Accumulation}$$

$$\Rightarrow \quad 0 = \frac{mv}{V_{BFR}} + \frac{dm}{dt}$$

$$\Rightarrow \quad \frac{dm}{dt} = -\frac{m}{\tau_{BFR}}$$

$$\Rightarrow \quad \int_{m_0}^{m} \frac{dm}{m} = -\int_{0}^{t} \frac{dt}{\tau_{BFR}} = -\frac{t}{\tau_{BFR}}$$

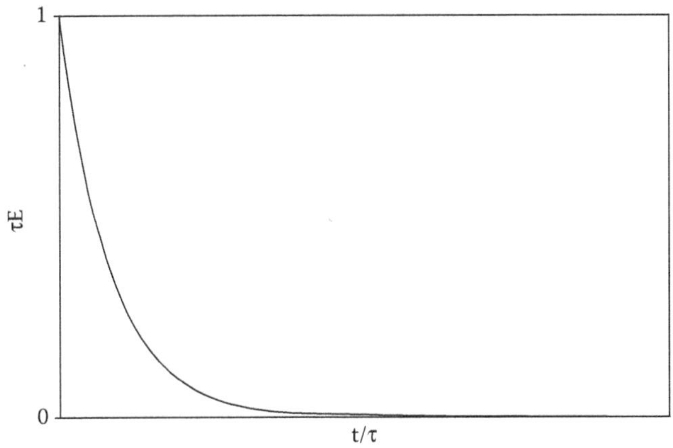

FIGURE 8.2 RTD function for ideal BFR.

$$\Rightarrow \quad \ln \frac{m}{m_0} = -\frac{t}{\tau_{BFR}}$$

$$\Rightarrow \quad \frac{m}{m_0} = e^{-(t/\tau_{BFR})}$$

$$\Rightarrow \quad \frac{(mv/V_{BFR})}{m_0} = \frac{v}{V_{BFR}} e^{-(t/\tau_{BFR})}$$

$$\Rightarrow \quad \frac{\text{Mass flow rate of tracer leaving the reactor at time t}}{m_0} = \frac{1}{\tau_{BFR}} e^{-(t/\tau_{BFR})} \tag{8.4}$$

$$\Rightarrow \quad E = \frac{1}{\tau_{BFR}} e^{-(t/\tau_{BFR})}$$

Graphically, it can be represented as shown in Figure 8.2.

8.2 RTD FOR REAL REACTORS

Flow of fluid in a real reactor deviates from that hypothesized for an ideal PFR or an ideal BFR. This can be attributed to numerous factors. Some of the possible reasons for nonideality in flow are

- Molecular and eddy diffusion
- Channeling
- Short circuiting
- Recycling
- Dead or stagnant pockets in the reactor
- Agitator malfunction

Therefore, real reactors are characterized by RTD functions different from those presented above for ideal reactors. Some of the possible examples are presented in Figure 8.3.

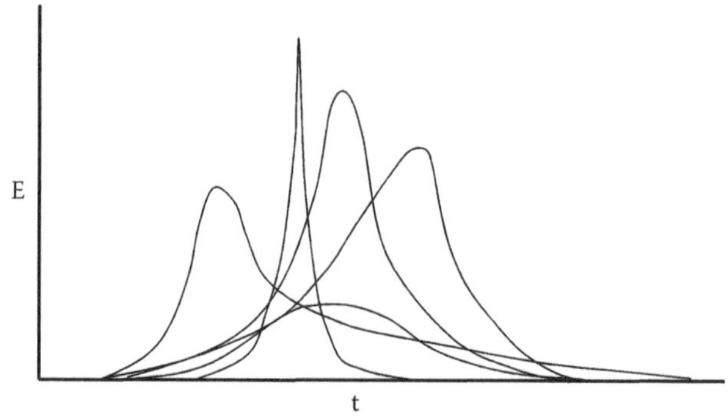

FIGURE 8.3 RTD in real reactors.

For a real reactor, the RTD function is to be determined experimentally. This is done by feeding a known amount of tracer into the reactor in a known fashion and measuring the amount of tracer that comes out of the reactor at different times. The desirable characteristics in a tracer are

- It should be nonreacting
- It should be easily detectable so that its amount leaving the reactor can be measured

8.2.1 Experimental Determination of RTD by Pulse Tracer Input

In pulse tracer input, a fixed amount of tracer is let into the reactor in a single shot at a precise instant of time. There is no input of tracer before or after this instant. In an ideal tracer input, the time interval over which the entire tracer shot is fed into the reactor is infinitesimally small. However, because the volumetric flow rate into the reactor is finite, the volume element of fluid entering the reactor within this infinitesimally small time duration, though small, is still finite. As a result, the inlet tracer concentration jumps from zero to a very high but finite concentration exactly at this instant and immediately falls to zero (certainly a hypothetical proposition). In practice, the time duration over which the tracer is fed into the reactor, though finite in value, is negligibly small compared to the space time of the reactor. The result is a sharp spike in the inlet concentration of the tracer. Mathematically, if the instant of tracer input is taken as t = 0 we have, for the ideal pulse tracer input,

$$C_{in} = C_0, \quad \text{at } t = 0 \tag{8.5}$$

$$C_{in} = 0, \quad \text{at } t \neq 0 \tag{8.6}$$

In practice, however, starting from zero the inlet concentration C_{in} rapidly reaches its maximum value C_0 and equally rapidly falls to zero again.

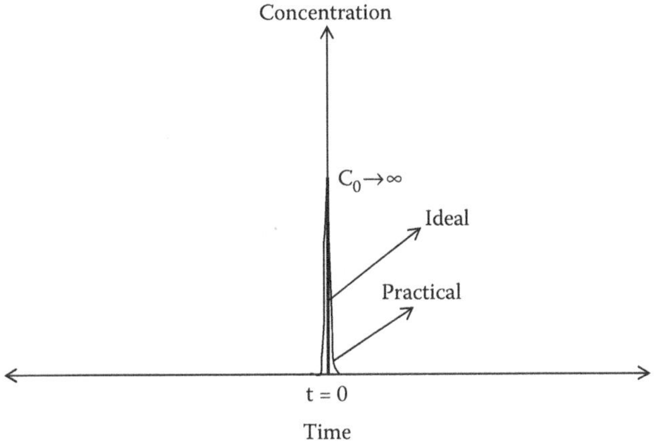

FIGURE 8.4 Concentration of tracer at reactor inlet for pulse input.

Graphically, this can be shown as in Figure 8.4.

The concentration of the tracer at the reactor outlet depends on the RTD of the reactor. For the ideal PFR, we get an identical tracer concentration pulse at the outlet of the reactor at $t = \tau$. Thus, an ideal PFR is one that faithfully transmits the tracer concentration pulse without distortion. For the ideal BFR, the moment we feed the tracer pulse we immediately notice an outlet tracer concentration. The outlet tracer concentration decreases exponentially and asymptotically drops to zero at infinite time. For real reactors, different types of outlet tracer concentration profiles will be obtained characteristic of the RTD of the reactor. Figure 8.5 shows some typical outlet tracer concentration profiles for an ideal pulse input of tracer.

Because the outlet tracer concentration profile is characteristic of the RTD of the reactor, a pulse tracer experiment can be utilized to determine the RTD of a reactor. For this, a pulse tracer input is given to the reactor and the outlet concentration of the

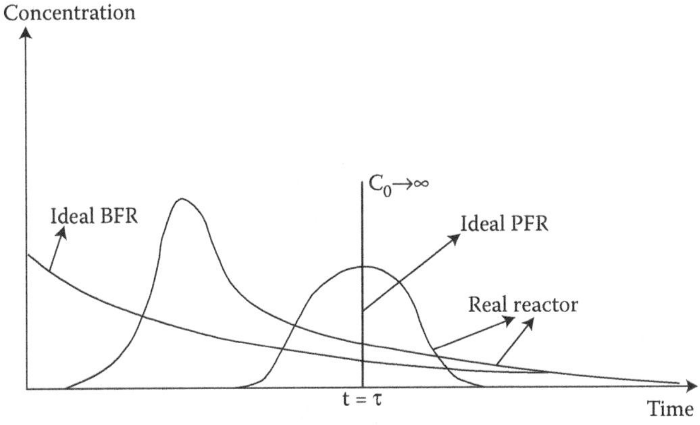

FIGURE 8.5 Concentration of tracer at reactor outlet for ideal pulse input.

tracer is measured at different times. The RTD function at any time is related to the outlet concentration at that time. When dt → 0,

Amount of tracer leaving the reactor between time t and t + dt = $C_{out}vdt$

⇒ Fraction of tracer leaving the reactor between time t and t + dt

$$= \frac{C_{out}vdt}{\text{Amount of tracer fed to the reactor}}$$

⇒ Fraction of tracer leaving the reactor per unit time between time t and t + dt

$$= \frac{C_{out}v}{\text{Amount of tracer fed to the reactor}}$$

$$\Rightarrow \quad E = \frac{C_{out}v}{\text{Amount of tracer fed to the reactor}}$$

$$\Rightarrow \quad E = \frac{C_{out}v}{\text{Amount of tracer leaving the reactor from } t = 0 \text{ to } t = \infty}$$

$$\Rightarrow \quad E = \frac{C_{out}v}{\int_0^\infty C_{out}vdt} = \frac{C_{out}}{\int_0^\infty C_{out}dt} \tag{8.7}$$

The integral in the denominator of Equation 8.7 is estimated graphically by measuring the area under the outlet concentration–time curve. Alternatively, the integral can be determined using a suitable numerical integration technique.

EXAMPLE 8.1

In a pulse tracer experiment, the following outlet concentration data are obtained:

T (min)	**0**	**2**	**4**	**6**	**8**	**10**	**12**	**14**	**16**
C_{out} (mol/L)	0	0.05	0.4	0.9	3	1.2	0.5	0.1	0

Plot the E versus t graph.

Solution

From the given data, we plot the C_{out} versus t graph as follows (Figure EXE8.1a):

$$\int_0^\infty C_{out}dt = \text{Area ABCDEFG} = 12.32\ \frac{\text{mol} \cdot \text{min}}{\text{L}}$$

Next, we calculate the RTD function at different times as
at t = 0,

$$E = \frac{C_{out} \text{ at } t = 0}{\int_0^\infty C_{out}dt} = \frac{0}{12.32} = 0$$

Similarly, we calculate E for all other values of t. The calculated values are tabulated below (Table EXE8.1).

From this, we plot the E versus t graph (Figure EXE8.1b).

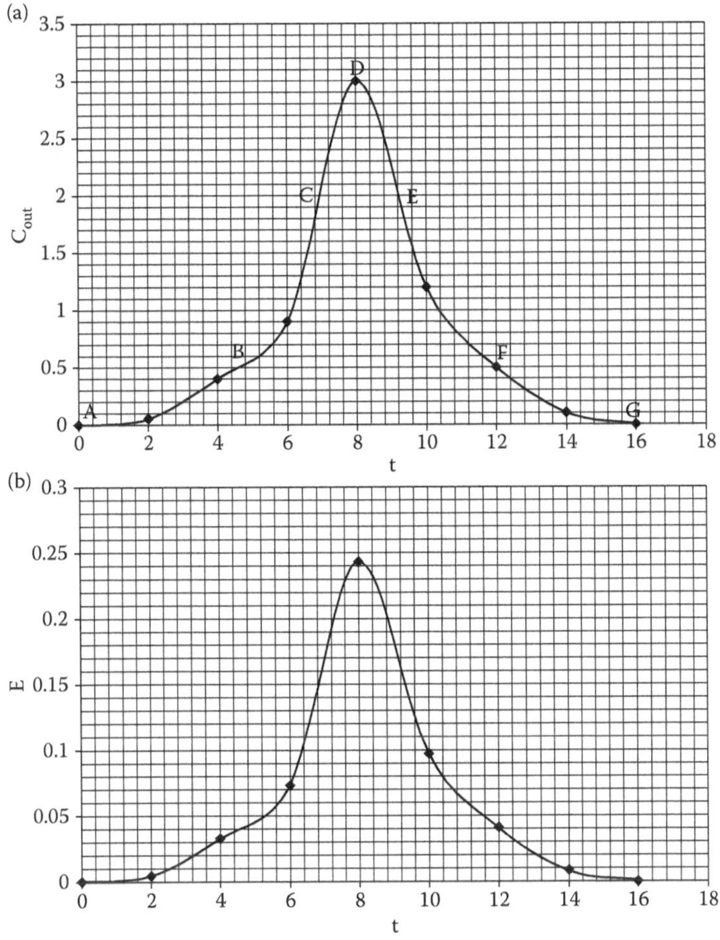

FIGURE EXE8.1 Outlet concentration and RTD function at different times.

TABLE EXE8.1
Calculated Values of RTD

t (min)	E
0	0
2	0.004058
4	0.032468
6	0.073052
8	0.243506
10	0.097403
12	0.040584
14	0.008117
16	0

8.2.2 Experimental Determination of RTD by Step Tracer Input

In step tracer input, at a precise instant of time we abruptly start feeding the tracer into the reactor at a constant inlet concentration and keep feeding the tracer at this constant concentration until the end of the experiment. There is no input of tracer before this instant. In an ideal tracer input, the time interval over which the tracer inlet concentration rises from zero to its constant finite value is infinitesimally small. As a result, the inlet tracer concentration jumps from zero to a finite concentration exactly at a precise instant and remains at this constant value thereafter until the end of the experiment. In practice, the time duration over which the tracer inlet concentration rises from zero to its constant value, though finite in value, is negligibly small. The result is a sudden and sharp rise in the inlet concentration of tracer from zero to a finite value. Mathematically, if the instant of starting the step tracer input is taken as t = 0 we have, for the ideal step tracer input,

$$C_{in} = 0, \quad \text{at } t < 0 \tag{8.8}$$

$$C_{in} = C_0, \quad \text{at } t \geq 0 \tag{8.9}$$

In practice, however, starting from zero the inlet concentration C_{in} rapidly reaches its constant value C_0 and remains so thereafter.

Graphically, this can be shown as follows in Figure 8.6.

The concentration of the tracer at the reactor outlet depends on the RTD of the reactor. For the ideal PFR, we get an identical tracer concentration step at the outlet of the reactor at $t = \tau$. Thus, an ideal PFR is one that faithfully transmits the tracer concentration step without distortion. For the ideal BFR, the moment we start feeding the tracer we immediately notice an outlet tracer concentration. The outlet tracer concentration increases exponentially and asymptotically attains the value C_0 at infinite time. For real reactors, different types of outlet tracer concentration profiles will

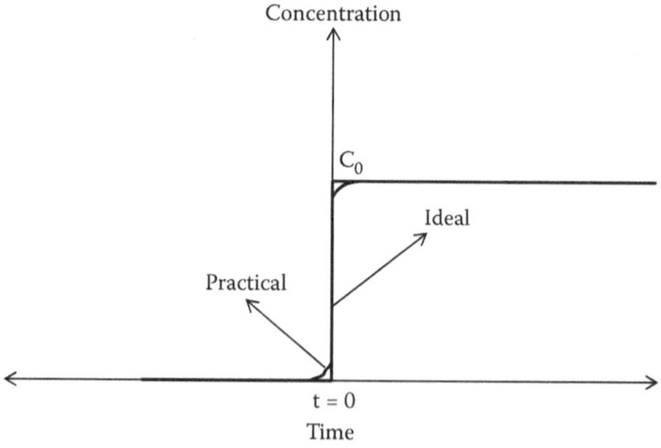

FIGURE 8.6 Concentration of tracer at reactor inlet for pulse input.

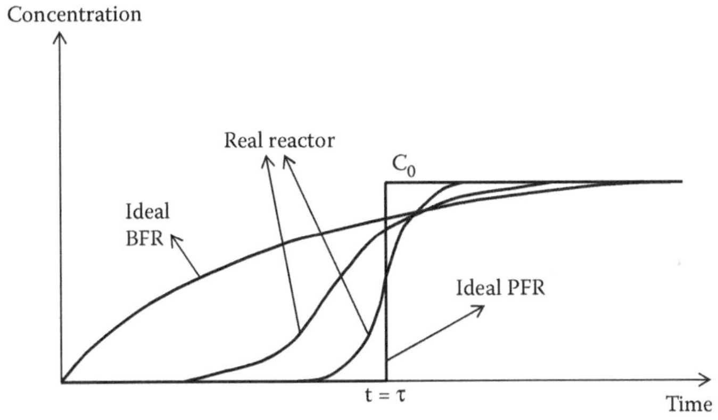

FIGURE 8.7 Concentration of tracer at reactor outlet for ideal step input.

be obtained characteristic of the RTD of the reactor. Figure 8.7 shows some typical outlet tracer concentration profiles for an ideal step input of tracer.

Since the outlet tracer concentration profile is characteristic of the RTD of the reactor, a step tracer experiment can be utilized to determine the RTD of a reactor. For this, a step tracer input is given to the reactor and the outlet concentration of the tracer is measured at different times. The experiment is carried out until the outlet concentration becomes equal to the inlet concentration. Remember, for an ideal BFR this will require infinite time. However, for a real reactor this will be achieved within a finite time. The RTD function at any time is related to the outlet concentration at that time. When $dt \to 0$,

$$\text{Amount of tracer leaving the reactor between time t and t} + dt = C_{out}vdt \quad (8.10)$$

This amount of tracer leaving the reactor consists of different fractions of tracer that had entered the reactor at different times. Because the inlet concentration of the tracer is constant, the differential amount of tracer entering the reactor within any time interval dt is C_0vdt. Consider the differential amount of tracer that enters the reactor between time t and t + dt. Within this differential time interval,

$$\text{Amount of tracer entering the reactor between time t and t} + dt = C_0vdt$$

A fraction of this differential tracer amount will immediately leave the reactor and contribute to the amount of tracer leaving the reactor between time t and t + dt. By the definition of an RTD function, this quantity would be

$$m_{out}\Big|_{t=t} = C_0vdt \times E\Big|_{t=0} dt \quad (8.11)$$

Next, consider the differential amount of tracer that enters the reactor between time t − dt and t. A fraction of this differential tracer amount will leave the reactor after time dt and contribute to the amount of tracer leaving the reactor between time t and t + dt. By the definition of an RTD function, this quantity would be

$$m_{out}\Big|_{t=t-dt} = C_0 v dt \times E\Big|_{t=dt} dt \tag{8.12}$$

Next, consider the differential amount of tracer that enters the reactor between time t − 2dt and t − dt. A fraction of this differential tracer amount will leave the reactor after time 2dt and contribute to the amount of tracer leaving the reactor between time t and t + dt. By the definition of an RTD function, this quantity would be

$$m_{out}\Big|_{t=t-2dt} = C_0 v dt \times E\Big|_{t=2dt} dt \tag{8.13}$$

Similarly, we can determine the fractional quantities $m_{out}\Big|_{t=t-3dt}$, $m_{out}\Big|_{t=t-4dt}$, and so on until $m_{out}\Big|_{t=0}$. The sum of all these fractions is equal to the amount of tracer leaving the reactor between time t and t + dt. Therefore,

$$m_{out}\Big|_{t=t} + m_{out}\Big|_{t=t-dt} + m_{out}\Big|_{t=t-2dt} + \ldots + m_{out}\Big|_{t=0} = C_{out} v dt$$

$$\Rightarrow \quad C_0 v dt \times E\Big|_{t=0} dt + C_0 v dt \times E\Big|_{t=dt} dt + C_0 v dt \times E\Big|_{t=2dt} dt + \ldots$$
$$+ C_0 v dt \times E\Big|_{t=t} dt = C_{out} v dt$$

$$\Rightarrow \quad C_0\left[E\Big|_{t=0} dt + E\Big|_{t=dt} dt + E\Big|_{t=2dt} dt + \ldots + E\Big|_{t=t} dt\right] = C_{out}$$

$$\Rightarrow \quad \left[E\Big|_{t=0} dt + E\Big|_{t=dt} dt + E\Big|_{t=2dt} dt + \ldots + E\Big|_{t=t} dt\right] = \frac{C_{out}}{C_0}$$

$$\Rightarrow \quad \int_0^t E dt = \frac{C_{out}}{C_0} \tag{8.14}$$

$$\Rightarrow \quad E = \frac{1}{C_0}\frac{dC_{out}}{dt} \tag{8.15}$$

EXAMPLE 8.2

In a step tracer experiment with 6.5 g/L inlet tracer concentration, the following outlet concentration data are obtained:

t (min)	0	2	4	6	8	10	12	14	16
C_{out} (g/L)	0	0.05	0.2	0.9	3	5.7	6.4	6.5	6.5

Plot the E versus t graph.

SOLUTION

Here, $C_0 = 6.5$ g/L

From the given data, we plot the C_{out} versus t curve. Next, we draw tangents on the C_{out} versus t curve at different times (Figure EXE8.2a). The slopes of these tangents provide us the values of (dC_{out}/dt) at different times. These values are shown in Table EXE8.2.

Next, we calculate the RTD function at t = 0 as

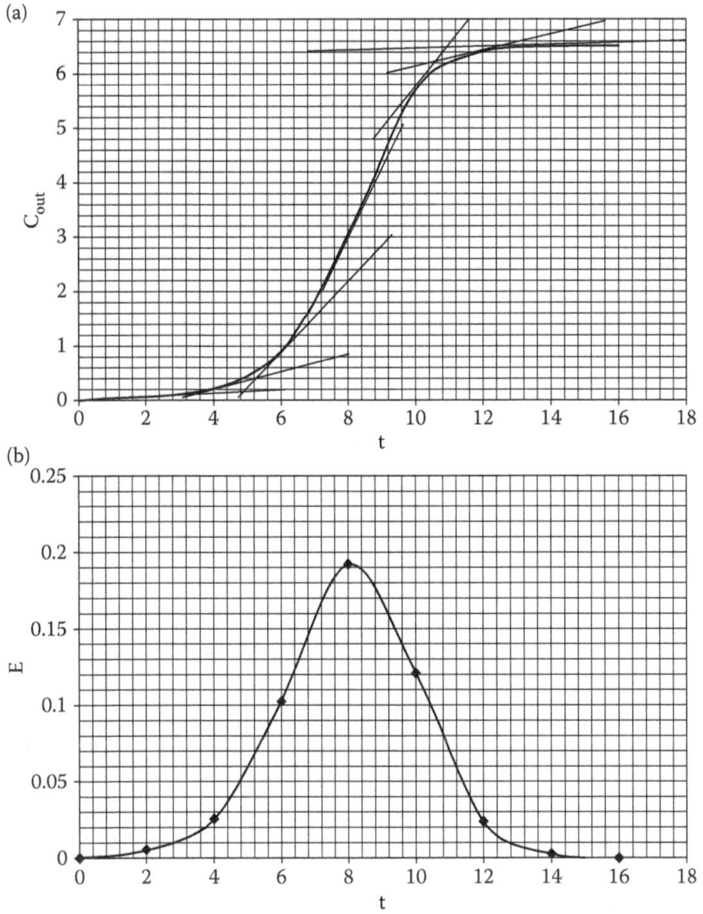

FIGURE EXE8.2 (a) Graphical procedure for determination of RTD, (b) RTD versus time.

TABLE EXE8.2
Calculation Table for RTD

t (min)	0	2	4	6	8	10	12	14	16
$\frac{dC_{out}}{dt}$	0	0.0375	0.1667	0.6667	1.25	0.7857	0.1563	0.0179	0
E	0	0.0057	0.0256	0.1026	0.1923	0.1209	0.0240	0.0027	0

$$E = \frac{1}{C_0}\frac{dC_{out}}{dt} = \frac{0}{6.5} = 0\ \text{min}^{-1}$$

Similarly we calculate E for different times. The values are shown in Table EXE8.2. From these values, we plot the E versus t (Figure EXE8.2b).

8.2.3 General Relation between Input and Outlet Tracer Concentration and RTD Function

We have discussed above two ideal tracer input types—the pulse input and the step input. In a real experiment, the tracer input will deviate from these ideal patterns. Let us deduce a relationship from first principles that relates the input and output tracer concentration and the RTD function of a reactor. Consider a general case of tracer inlet concentration profile and its corresponding outlet tracer concentration profile for a particular reactor as shown in Figure 8.8.

When $dt' \rightarrow 0$,

$$\text{Amount of tracer leaving the reactor between time t and } t + dt' = C_{out}\Big|_{t} vdt' \tag{8.16}$$

This amount of tracer leaving the reactor consists of different fractions of tracer that had entered the reactor at different times. The differential amount of tracer entering the reactor within any time interval dt' is $C_{in}vdt'$, where C_{in} is the corresponding inlet concentration at that time. Consider the differential amount of tracer that enters the reactor between time $t' = 0$ and $t' = dt'$. Within this differential time interval,

$$\text{Amount of tracer entering the reactor between time } t' = 0 \text{ and } t' = dt' = C_{in}\Big|_{t'=0} vdt'$$

The fraction of this differential tracer amount that leaves the reactor after time $t' = t$ will contribute to the amount of tracer leaving the reactor between time t and $t + dt$. By the definition of an RTD function, this quantity would be

$$m_{out}\Big|_{t'=0} = C_{in}\Big|_{t'=0} vdt' \times E\Big|_{t'=t} dt'$$

$$m_{out}\Big|_{t'=0} = C_{in}\Big|_{t'=0} E\Big|_{t'=t} vdt' \times dt' \tag{8.17}$$

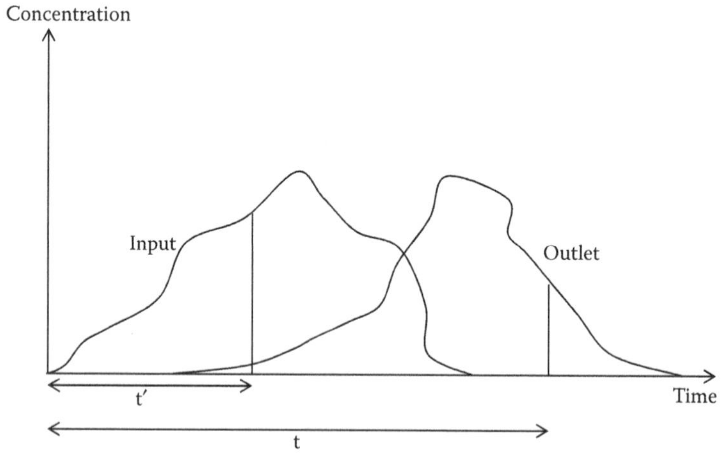

FIGURE 8.8 Tracer concentration profiles in a general case.

Next, consider the differential amount of tracer that enters the reactor between time $t' = dt'$ and $t' = 2dt'$. A fraction of this differential tracer amount will leave the reactor after time $t' = t - dt'$ and contribute to the amount of tracer leaving the reactor between time t and t + dt. By the definition of an RTD function, this quantity would be

$$\begin{aligned} m_{out}\Big|_{t'=dt'} &= C_{in}\Big|_{t'=dt'} vdt' \times E\Big|_{t'=t-dt'} dt' \\ \Rightarrow \quad m_{out}\Big|_{t'=dt'} &= C_{in}\Big|_{t'=dt'} E\Big|_{t'=t-dt'} vdt' \times dt' \end{aligned} \tag{8.18}$$

Next, consider the differential amount of tracer that enters the reactor between time $t' = 2dt'$ and $t' = 3dt'$. A fraction of this differential tracer amount will leave the reactor after time $t' = t - 2dt'$ and contribute to the amount of tracer leaving the reactor between time t and t + dt. By the definition of an RTD function, this quantity would be

$$\begin{aligned} m_{out}\Big|_{t'=2dt'} &= C_{in}\Big|_{t'=2dt'} vdt' \times E\Big|_{t'=t-2dt'} dt' \\ \Rightarrow \quad m_{out}\Big|_{t'=2dt'} &= C_{in}\Big|_{t'=2dt'} E\Big|_{t'=t-2dt'} vdt' \times dt' \end{aligned} \tag{8.19}$$

Similarly, we can determine the fractional quantities $m_{out}\Big|_{t'=3dt'}$, $m_{out}\Big|_{t'=4dt'}$, and so on until $m_{out}\Big|_{t'=t}$. The sum of all these fractions is equal to the amount of tracer leaving the reactor between time t and t + dt. Therefore,

$$m_{out}\Big|_{t'=0} + m_{out}\Big|_{t'=dt'} + m_{out}\Big|_{t'=2dt'} + \ldots + m_{out}\Big|_{t'=t} = C_{out}\Big|_{t} vdt'$$

$$\Rightarrow \quad C_{in}\Big|_{t'=0} E\Big|_{t'=t} vdt' \times dt' + C_{in}\Big|_{t'=dt'} E\Big|_{t'=t-dt'} vdt' \times dt' + C_{in}\Big|_{t'=2dt'} E\Big|_{t'=t-2dt'} vdt' \times dt'$$

$$+ \ldots + C_{in}\Big|_{t'=t} E\Big|_{t'=0} vdt' \times dt' = C_{out}\Big|_{t} vdt'$$

$$\Rightarrow \quad C_{in}\Big|_{t'=0} E\Big|_{t'=t} dt' + C_{in}\Big|_{t'=dt'} E\Big|_{t'=t-dt'} dt' + C_{in}\Big|_{t'=2dt'} E\Big|_{t'=t-2dt'} dt'$$

$$+ \ldots + C_{in}\Big|_{t'=t} E\Big|_{t'=0} dt' = C_{out}\Big|_{t}$$

$$\Rightarrow \quad \int_0^t C_{in}\Big|_{t'} E\Big|_{(t-t')} dt' = C_{out}\Big|_{t} \tag{8.20}$$

This is known as the convolution integral (Levenspiel, 1999).

EXAMPLE 8.3

Tracer is fed into the reactor of Example 8.2 with the following input concentration profile:

Time (min)	0	2	4	6	8
C_{in} (g/L)	0	0.8	7.2	0.9	0

What will be the outlet concentration of the tracer after 6 min?

TABLE EXE8.3
Calculation Table for Numerical Integration

t'	0	2	4	6
$6 - t'$	6	4	2	0
$C_{in}\big\|_{t'}$	0	0.8	7.2	0.9
$E\big\|_{(t-t')}$	0.1026	0.0256	0.0057	0
$C_{in}\big\|_{t'} E\big\|_{(t-t')}$	0	0.0205	0.041	0

Solution

Since an RTD function is a characteristic of the reactor flow, the RTD function in this case is the same as that computed in Example 8.2. We reproduce it below for convenience.

Time (min)	0	2	4	6	8	10	12	14	16
E	0	0.0057	0.0256	0.1026	0.1923	0.1209	0.0240	0.0027	0

Referring to Figure 8.8, the time given in input concentration profile data is t′ and the time given corresponding to the RTD function is t − t′. From Equation 8.20

$$C_{out}\big|_t = \int_0^t C_{in}\big|_{t'} E\big|_{(t-t')} dt'$$

$$\Rightarrow \quad C_{out}\big|_6 = \int_0^6 C_{in}\big|_{t'} E\big|_{(6-t')} dt'$$

To compute the integral we prepare Table EXE8.3.
By Simpson's one-third rule,

$$C_{out}\big|_6 = \int_0^6 C_{in}\big|_{t'} E\big|_{(6-t')} dt' = \frac{2}{3}[0 + 0 + 4 \times 0.0205 + 2 \times 0.041] = 0.109 \text{ g/L}$$

8.3 PROPERTIES OF RTD FUNCTION

8.3.1 Fraction of Material Leaving the Reactor in a Time Interval

From the definition of an RTD function,

Fraction of material leaving the reactor per unit time between time t and t + dt = E

⇒ Fraction of material leaving the reactor between time t and t + dt = Edt

⇒ Fraction of material leaving the reactor between time 0 and t = $\int_0^t Edt$

We define a new function *F as fraction of material leaving the reactor between time 0 and t.* So,

$$F = \int_0^t E\,dt \tag{8.21}$$

$$\Rightarrow \quad E = \frac{dF}{dt} \tag{8.22}$$

By the same reasoning,

$$\text{Fraction of material leaving the reactor in time interval } t_1 \text{ to } t_2 = \int_{t_1}^{t_2} E\,dt \tag{8.23}$$

Because the sum of the fraction of material left in the reactor at time t and the fraction of material that has left the reactor up to time t is 1.

$$\text{Fraction of material left in the reactor at time } t = 1 - \int_0^t E\,dt = 1 - F \tag{8.24}$$

Further, because the entire material will leave the reactor in infinite time, the fraction of material leaving the reactor in infinite time is 1. Hence, we have (Danckwerts, 1953)

$$\int_0^\infty E\,dt = 1 \tag{8.25}$$

From Equations 8.24 and 8.25,

$$1 - F = \int_0^\infty E\,dt - \int_0^t E\,dt = \int_0^t E\,dt + \int_t^\infty E\,dt - \int_0^t E\,dt$$

$$\Rightarrow \quad 1 - F = \int_t^\infty E\,dt \tag{8.26}$$

$$\Rightarrow \quad F = 1 - \int_t^\infty E\,dt \tag{8.27}$$

EXAMPLE 8.4

For the reactor reported in Example 8.1, plot F versus t.

Solution

Since an RTD function is a characteristic of the reactor flow, the RTD function in this case is the same as that plotted in Example 8.1. For convenience, we reproduce it as Figure EXE8.4a. From this plot,

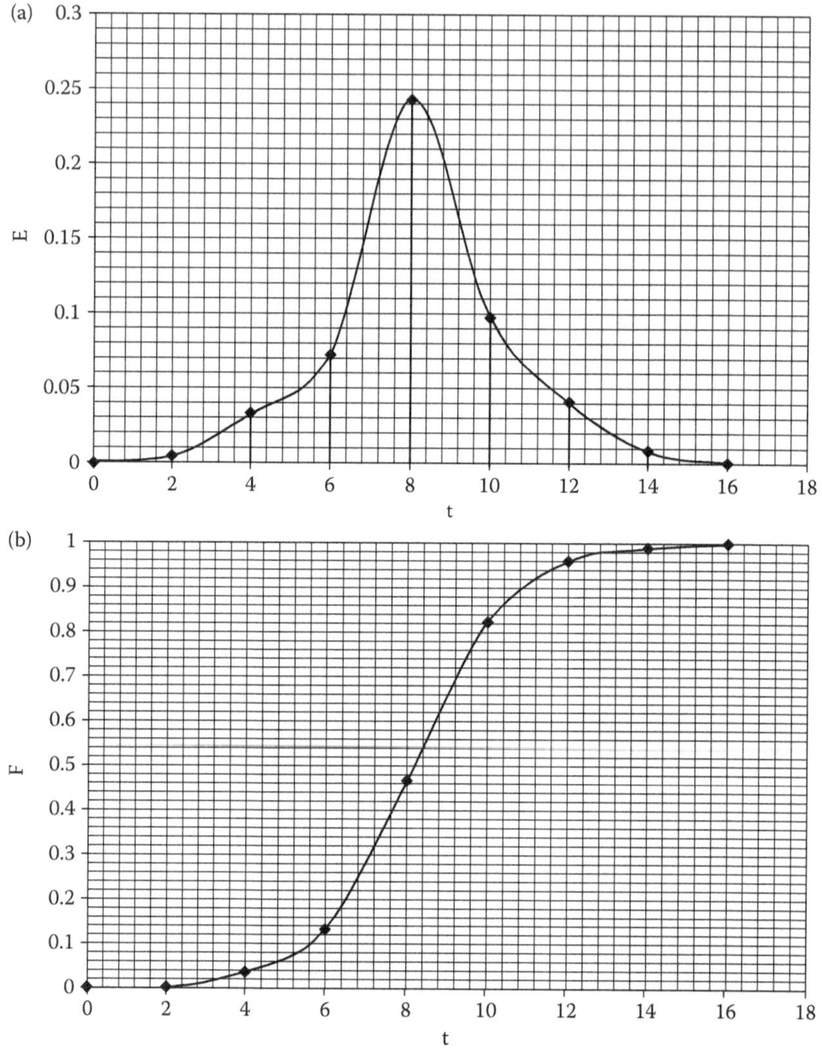

FIGURE EXE8.4 Plots for E versus t, and F versus t.

$F|_{t=2}$ = Area under the E versus t curve from 0 to 2 = 0.001
$F|_{t=4}$ = Area under the E versus t curve from 0 to 4 = 0.036
$F|_{t=6}$ = Area under the E versus t curve from 0 to 6 = 0.132
$F|_{t=8}$ = Area under the E versus t curve from 0 to 8 = 0.466
$F|_{t=10}$ = Area under the E versus t curve from 0 to 10 = 0.824
$F|_{t=12}$ = Area under the E versus t curve from 0 to 12 = 0.96
$F|_{t=14}$ = Area under the E versus t curve from 0 to 14 = 0.99
$F|_{t=16}$ = Area under the E versus t curve from 0 to 16 = 1

From these data, we plot the required F versus t graph as shown in Figure EXE8.4b.

8.3.2 Mean Residence Time

The previous sections in this chapter make it clear that, except for ideal PFR, different fractions of material spend different amounts of time in the reactor. The weighted average of all these different times is the mean residence time of material in the reactor. So,

$$t_{mean} = \frac{\sum_{i=0}^{\infty} t_i E\big|_{t_i} dt}{\sum_{i=0}^{\infty} E\big|_{t_i} dt}$$

$$\Rightarrow \quad t_{mean} = \frac{\int_0^{\infty} tEdt}{\int_0^{\infty} Edt} \quad \text{as } dt \rightarrow 0$$

$$\Rightarrow \quad t_{mean} = \int_0^{\infty} tEdt \tag{8.28}$$

For pulse input of tracer, from Equations 8.7 and 8.28,

$$t_{mean} = \int_0^{\infty} t \frac{C_{out}}{\int_0^{\infty} C_{out} dt} dt$$

$$\Rightarrow \quad t_{mean} = \frac{\int_0^{\infty} tC_{out} dt}{\int_0^{\infty} C_{out} dt} \tag{8.29}$$

For step input of tracer, from Equations 8.15 and 8.28,

$$t_{mean} = \int_0^{\infty} t \frac{1}{C_0} \frac{dC_{out}}{dt} dt$$

$$\Rightarrow \quad t_{mean} = \frac{\int_0^{C_0} tdC_{out}}{C_0} \tag{8.30}$$

8.3.3 Variance of RTD

The spread of residence times about the mean residence time is represented by the variance. Mathematically, it is represented as

$$\sigma^2 = \frac{\sum_{i=0}^{\infty} [t_i - t_{mean}]^2 E\big|_{t_i} dt}{\sum_{i=0}^{\infty} E\big|_{t_i} dt}$$

$$\Rightarrow \quad \sigma^2 = \frac{\int_0^{\infty} [t - t_{mean}]^2 E dt}{\int_0^{\infty} E dt}$$

$$\Rightarrow \quad \sigma^2 = \int_0^{\infty} [t - t_{mean}]^2 E dt \tag{8.31}$$

For pulse input of tracer, from Equations 8.7 and 8.31,

$$\sigma^2 = \int_0^{\infty} [t - t_{mean}]^2 \frac{C_{out}}{\int_0^{\infty} C_{out} dt} dt$$

$$\Rightarrow \quad \sigma^2 = \frac{\int_0^{\infty} [t - t_{mean}]^2 C_{out} dt}{\int_0^{\infty} C_{out} dt} \tag{8.32}$$

For step input of tracer, from Equations 8.15 and 8.31,

$$\sigma^2 = \int_0^{\infty} [t - t_{mean}]^2 \frac{1}{C_0} \frac{dC_{out}}{dt} dt$$

$$\Rightarrow \quad \sigma^2 = \frac{\int_0^{C_0} [t - t_{mean}]^2 dC_{out}}{C_0} \tag{8.33}$$

EXAMPLE 8.5

Calculate the mean residence time and variance for the reactor system described in Example 8.1.

SOLUTION

From the solution to Example 8.1, the RTD function is represented as Table EXE8.5.

We multiply each E value with the corresponding t value and tabulate the results in the next column of Table EXE8.5.

From Equation 8.28,

$$t_{mean} = \int_0^{\infty} tEdt$$

TABLE EXE8.5
Calculation Table for Mean Residence Time and Variance

t (min)	E	tE	$[t - t_{mean}]^2$	$[t - t_{mean}]^2$ E
0	0	0	67.8976	0
2	0.004058	0.008117	38.9376	0.158026
4	0.032468	0.12987	17.9776	0.583688
6	0.073052	0.438312	5.0176	0.366545
8	0.243506	1.948052	0.0576	0.014026
10	0.097403	0.974026	3.0976	0.301714
12	0.040584	0.487013	14.1376	0.573766
14	0.008117	0.113636	33.1776	0.269299
16	0	0	60.2176	0

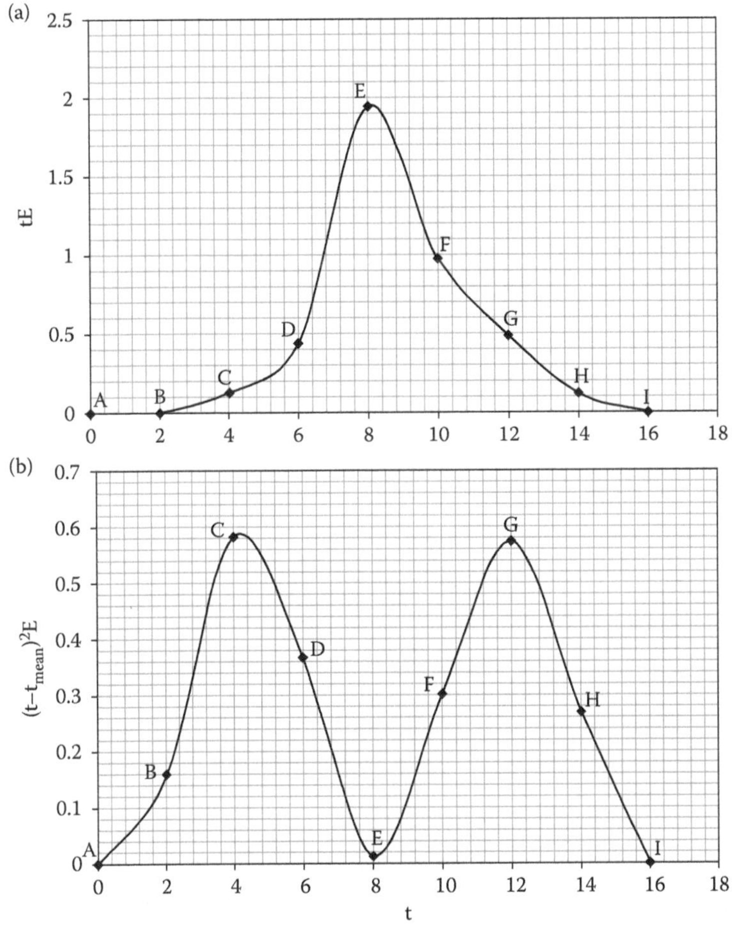

FIGURE EXE 8.5 Plots for graphical integration.

To evaluate the integral on the RHS, we plot tE versus t as shown in Figure EXE8.5a. From this plot,

$$t_{mean} = \text{Area ABCDEFGHI} = 8.24 \text{ min}$$

Next, we calculate $[t - t_{mean}]^2$ for each t and tabulate the values in the next column of Table EXE8.5. Next we multiply each E value with the corresponding $[t - t_{mean}]^2$ and tabulate the results in the next column of Table EXE8.5.

From Equation 8.31

$$\sigma^2 = \int_0^\infty [t - t_{mean}]^2 E dt$$

To evaluate the integral on the RHS, we plot $[t - t_{mean}]^2$ E versus t as shown in Figure EXE8.5b. From this plot,

$$\sigma^2 = \text{Area ABCDEFGHI} = 4.616 \text{min}^2$$

8.4 CONVERSION FROM RTD

Chemical engineers would be concerned about the conversion achieved in a real reactor. For engineering calculations, we must have means to predict the conversion to be achieved if a reaction is carried out in a real reactor for which the RTD is known. Here the readers are once again reminded that the RTD function described in the previous sections is based on information on the material entering the reactor and the material leaving the reactor. In addition, to predict the conversion from a real reactor, we also need to have some inside information about the reactor. RTD tells us about the time the different fluid elements spend inside the reactor. It does not tell us anything about the manner in which this time is spent. A fluid element with a particular residence time, during its stay inside the reactor, can interact with other fluid elements of different residence times by exchanging material with them. This interaction can be of various degrees (Danckwerts, 1958; Zwietering, 1959). On one extreme, we can have a situation where fluid elements of different residence times remain completely aloof from each other—complete segregation. The other extreme is when fluid elements with different residence times, during their stay inside the reactor, completely mix with each other—complete intermixing. In between these two extremes, there can be numerous other ways in which the fluid elements of different residence times can intermix inside the reactor. Zwietering has argued, however, that complete segregation can be realized in a real reactor but complete intermixing is not.

8.4.1 Conversion under Complete Segregation

Consider a reactor in which the fluid elements with different residence times, as they pass through the reactor, remain completely aloof from each other. The time spent by different fluid elements inside the reactor is according to the RTD function for the reactor. Let the molar feed rate of the limiting reactant to the reactor be "$(F_{ALR})_0$." In a time interval between t = 0 and t = dt, where dt → 0, the moles of limiting reactant entering the reactor is $(F_{ALR})_0\,dt$. Different fractions of this entering amount will spend different lengths of time inside the reactor and undergo different conversions. Consider a fraction of this feed that comes out from the reactor between time t_i and $t_i + dt$. From the definition of RTD function this fraction would be

$$fr\big|_{t_i} = E\big|_{t_i}\,dt \tag{8.34}$$

If the conversion achieved in time t_i is $X_{ALR}\big|_{t_i}$, the moles of limiting reactant in this fraction of outgoing material is

$$F_{ALR}\big|_{t_i} = (F_{ALR})_0\,dt \times fr\big|_{t_i}\left(1 - X_{ALR}\big|_{t_i}\right)$$

$$\Rightarrow \quad F_{ALR}\big|_{t_i} = (F_{ALR})_0\,dt\left(1 - X_{ALR}\big|_{t_i}\right)E\big|_{t_i}\,dt$$

The total moles of limiting reactant that would come out from the reactor would be

$$F_{ALR} = \sum_{i=0}^{\infty} F_{ALR}\big|_{t_i} = \sum_{i=0}^{\infty}(F_{ALR})_0\,dt\left(1 - X_{ALR}\big|_{t_i}\right)E\big|_{t_i}\,dt$$

$$\Rightarrow \quad \frac{F_{ALR}}{(F_{ALR})_0\,dt} = \sum_{i=0}^{\infty}\left(1 - X_{ALR}\big|_{t_i}\right)E\big|_{t_i}\,dt$$

$$\Rightarrow \quad \frac{F_{ALR}}{(F_{ALR})_0\,dt} = \int_0^{\infty}(1 - X_{ALR})E\,dt \qquad \text{as } dt \to 0$$

$$\Rightarrow \quad \frac{F_{ALR}}{(F_{ALR})_0\,dt} = \int_0^{\infty}E\,dt - \int_0^{\infty}X_{ALR}E\,dt$$

$$\Rightarrow \quad \frac{F_{ALR}}{(F_{ALR})_0\,dt} = 1 - \int_0^{\infty}X_{ALR}E\,dt$$

$$\Rightarrow \quad 1 - \frac{F_{ALR}}{(F_{ALR})_0\,dt} = \int_0^{\infty}X_{ALR}E\,dt$$

$$\Rightarrow \quad (X_{ALR})_{total} = \int_0^{\infty}X_{ALR}E\,dt \tag{8.35}$$

EXAMPLE 8.6

A second-order liquid-phase reaction

$$A \rightarrow \text{Products}$$

with reaction rate $-r_A = C_A^2$ mol/(L · min) is carried out in the reactor described in Example 8.1 under isothermal conditions with an initial concentration of A as 1 mol/L. What will be the conversion achieved if complete segregation of fluid elements is maintained in the reactor?

Solution

From the solution to Example 8.1, the RTD function for the reactor is represented in Table EXE8.6.

For a second-order reaction, from Equation 5.18,

$$X_A = \frac{kt(C_A)_0}{1+kt(C_A)_0} = \frac{t}{1+t} \tag{8.6.1}$$

From this equation we calculate X_A for different times and tabulate the values in the next column of Table EXE8.6. From Equation 8.35,

$$(X_A)_{total} = \int_0^\infty X_A E dt \tag{8.6.2}$$

In order to evaluate the integral, we calculate the quantity X_AE for each value of t and tabulate the values in the next column of Table EXE8.6. Next, we plot X_AE versus t as shown in Figure EXE8.6. From Equation EXE8.6.2,

$$(X_A)_{total} = \int_0^\infty X_A E dt = \text{Area ABCDEFGHI} = 0.884$$

TABLE EXE8.6
Calculation Table to Determine Conversion

t (min)	E	X_A	X_AE
0	0	0	0
2	0.004058	0.666667	0.002706
4	0.032468	0.8	0.025974
6	0.073052	0.857143	0.062616
8	0.243506	0.888889	0.21645
10	0.097403	0.909091	0.088548
12	0.040584	0.923077	0.037463
14	0.008117	0.933333	0.007576
16	0	0.941176	0

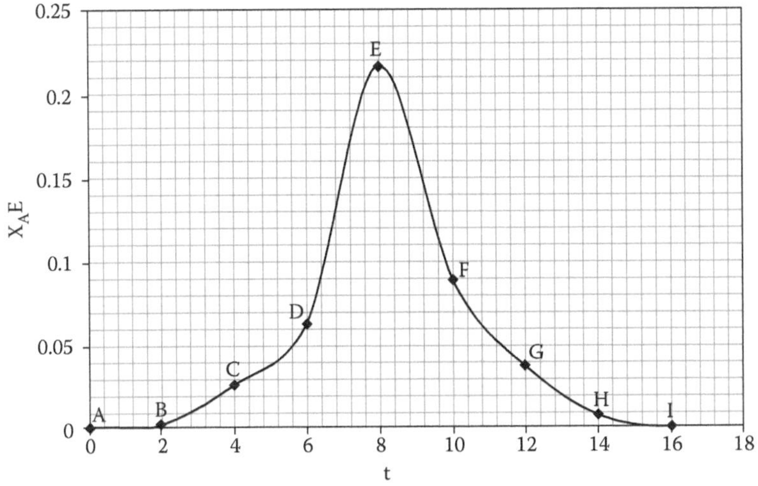

FIGURE EXE8.6 Plot for graphical integration.

8.4.2 Conversion under Complete Intermixing

Consider a reactor in which the fluid elements with different residence times, as they pass through the reactor, completely intermix with each other. The time spent by different fluid elements inside the reactor is according to the RTD function for the reactor. This is akin to a situation where we have a batch reactor into which the reactant is fed in installments. In the first installment, we will feed an amount of reactant that will stay in the reactor for an infinite time. In the second installment, we will feed an amount of reactant that will stay in the reactor for a time which is dt less than the first installment. When the second installment is added, the first installment would have already undergone a conversion dX_1, that is, the concentration now is less than the feed concentration $\left(C_{A_{LR}}\right)_0$, say $\left(C_{A_{LR}}\right)_1$. The second installment would mix with the partially converted first installment and the new concentration will increase from $\left(C_{A_{LR}}\right)_1$, say, $\left(C_{A_{LR}}\right)_1'$. In the third installment, we will feed an amount of reactant that will stay in the reactor for a time that is 2dt less than the first installment. When the third installment is added, the mixture of the first and second installment would have already undergone a conversion dX_2, that is, the concentration now is less than $\left(C_{A_{LR}}\right)_1'$, say $\left(C_{A_{LR}}\right)_2$. The third installment would mix with the partially converted mixture of the two earlier installments and the new concentration will increase from $\left(C_{A_{LR}}\right)_2$, say, $\left(C_{A_{LR}}\right)_2'$. This process will go on until we feed the last installment of reactant and immediately stop the reaction. This last installment would mix with the mixture obtained from all the earlier installments.

Let the total moles of the limiting reactant to be fed to the reactor be "$\left(N_{A_{LR}}\right)_{fed}$." Different fractions of this total amount will be fed to the reactor in different installments according to the RTD of the reactor. Consider a fraction of this feed that spends a time between t_i and $t_i + dt$ inside the reactor. From the definition of an RTD function, this fraction would be

$$fr\big|_{t_i} = E\big|_{t_i}\,dt$$

And the moles of limiting reactant within this fraction are

$$\left(N_{ALR}\right)_{fed} fr\big|_{t_i} = \left(N_{ALR}\right)_{fed} E\big|_{t_i} dt$$

If the total volume to be fed is V,

$$\left(N_{ALR}\right)_{fed} fr\big|_{t_i} = V\left(C_{ALR}\right)_0 E\big|_{t_i} dt$$

At this instant, the reactor will already contain all the fractions having residence times more than t_i. Therefore, the total amount of limiting reactant fed to the reactor before the ith fraction, that is, until the (i–1)th fraction, is

$$\left(N_{ALR}\right)_{fed}\big|_{i-1} = \int_{t_i}^{\infty} \left(N_{ALR}\right)_{fed} E\big|_{t_i} dt = \left(N_{ALR}\right)_{fed} \int_{t_i}^{\infty} E\big|_{t_i} dt$$

$$\Rightarrow \left(N_{ALR}\right)_{fed}\big|_{i-1} = \left(N_{ALR}\right)_{fed}\left(1-F\big|_{t_i}\right) = V\left(C_{ALR}\right)_0\left(1-F\big|_{t_i}\right) \tag{8.36}$$

Let the concentration of the limiting reactant in the reactor content at this instant be $\left(C_{ALR}\right)_i$. The amount of limiting reactant already existing in the reactor at this instant is

$$\text{LR already existing in the reactor} = V\left(C_{ALR}\right)_i\left(1-F\big|_{t_i}\right) \tag{8.37}$$

In the differential time interval dt after the ith installment is fed

$$\begin{aligned}&\text{LR already existing in the reactor} + \text{LR fed to the reactor}\\ &\qquad -\text{LR disappearing due to reaction} = \text{LR left in the reactor}\end{aligned}$$

$$\Rightarrow V\left(C_{ALR}\right)_i\left(1-F\big|_{t_i}\right) + V\left(C_{ALR}\right)_0 E\big|_{t_i} dt - \left(-r_{ALR}\right)V\left(1-F\big|_{t_i}\right)dt = V\left(C_{ALR}\right)_{i+1}\left(1-F\big|_{t_i+dt}\right)$$

$$\Rightarrow V\left[\left(C_{ALR}\right)_{i+1}\left(1-F\big|_{t_i+dt}\right) - \left(C_{ALR}\right)_i\left(1-F\big|_{t_i}\right)\right] = V\left(C_{ALR}\right)_0 E\big|_{t_i} dt - \left(-r_{ALR}\right)V\left(1-F\big|_{t_i}\right)dt$$

$$\Rightarrow \frac{\left(C_{ALR}\right)_{i+1}\left(1-F\big|_{t_i+dt}\right) - \left(C_{ALR}\right)_i\left(1-F\big|_{t_i}\right)}{dt} = \left(C_{ALR}\right)_0 E\big|_{t_i} - \left(-r_{ALR}\right)\left(1-F\big|_{t_i}\right)$$

$$\Rightarrow \frac{d\left(C_{ALR}\right)_i\left(1-F\big|_{t_i}\right)}{dt} = \left(C_{ALR}\right)_0 E\big|_{t_i} - \left(-r_{ALR}\right)\left(1-F\big|_{t_i}\right) \quad dt \to 0$$

Dropping the subscripts and generalizing, we have

$$\frac{d(C_{A_{LR}})(1-F)}{dt} = (C_{A_{LR}})_0 E - (-r_{A_{LR}})(1-F)$$

$$\Rightarrow \quad (1-F)\frac{d(C_{A_{LR}})}{dt} + (C_{A_{LR}})\frac{d(1-F)}{dt} = (C_{A_{LR}})_0 E - (-r_{A_{LR}})(1-F) \qquad (8.38)$$

Combining with Equation 8.22,

$$\Rightarrow \quad (1-F)\frac{d(C_{A_{LR}})}{dt} + (C_{A_{LR}})E = (C_{A_{LR}})_0 E - (-r_{A_{LR}})(1-F)$$

$$\Rightarrow \quad (1-F)\frac{d(C_{A_{LR}})}{dt} = \left[(C_{A_{LR}})_0 - (C_{A_{LR}})\right]E - (-r_{A_{LR}})(1-F) \qquad (8.39)$$

This equation is to be solved for the boundary conditions

$$\text{At } t = 0, \ (C_{A_{LR}}) = (C_{A_{LR}})_0$$

$$\text{At } t = \infty, \ (C_{A_{LR}}) = (C_{A_{LR}})_f$$

In terms of conversion, from Equation 5.8 (Chapter 5),

$$-(1-F)(C_{A_{LR}})_0 \frac{dX_{A_{LR}}}{dt} = \left[(C_{A_{LR}})_0 - (C_{A_{LR}})\right]E - (-r_{A_{LR}})(1-F)$$

$$\Rightarrow \quad (1-F)\frac{d(X_{A_{LR}})}{dt} = \frac{(-r_{A_{LR}})(1-F)}{(C_{A_{LR}})_0} - X_{A_{LR}}E \qquad (8.40)$$

This equation is to be solved for the boundary conditions

$$\text{At } t = 0, \ (X_{A_{LR}}) = 0$$

$$\text{At } t = \infty, \ (X_{A_{LR}}) = (X_{A_{LR}})_f$$

Equations 8.39 and 8.40 require numerical methods for their solutions.

8.5 MATHEMATICAL MODELS FOR REAL REACTORS

For real reactors, the discussion so far has primarily focused on the experimental determination of RTD. No attempt has been made to correlate the RTD so determined with the reactor particulars, or any physical phenomenon. In the absence of this, the RTD remains entirely empirical without any theoretical justification whatsoever. Over the years, some physical models with accompanying mathematical correlations have been proposed to explain the experimentally obtained RTD in terms of reactor particulars and/or physical characteristics of the system. In a real reactor, the flow does not necessarily follow any organized pattern. In order to mathematically analyze such a disorganized flow, each model proposes a physical picture of the flow through the reactor based on certain simplifying assumptions. The lesser the simplifying assumptions, the more complicated would be the accompanying mathematics. Of several such models, we discuss here two of them which are relatively simple to follow and provide a reasonably adequate explanation of the reactor RTD. Nevertheless, it should be kept in mind that no model provides a rigorous description of the flow through the reactor. All models are a trade-off between completeness of a physical model and the mathematical complexity.

8.5.1 Dispersed Plug Flow Model

Also known as the dispersion model, it relates a real reactor to an ideal PFR and explains its RTD in terms of the extent to which it deviates from the ideal one. In Chapter 6, we have seen that the ideal plug flow is one where nothing except a unidirectional uniform flow field in parallel trajectories causes the material to move. Even diffusion was excluded as a possible cause for the movement of material. The dispersed plug flow model visualizes the flow in a reactor as a unidirectional uniform flow field in parallel trajectories superimposed with diffusion of mass in the axial direction due to the concentration gradient existing. At the macroscopic scale, plugs of differential fluid volumes are moving along the reactor axis. At the microscopic scale, molecules or miniscule groups of molecules diffuse into and diffuse out of the differential fluid volume. Consider a differential volume of the reactor, through which tracer is moving, at a length L from the reactor inlet as shown in Figure 8.9. The cross-sectional area of the reactor is Ar_R.

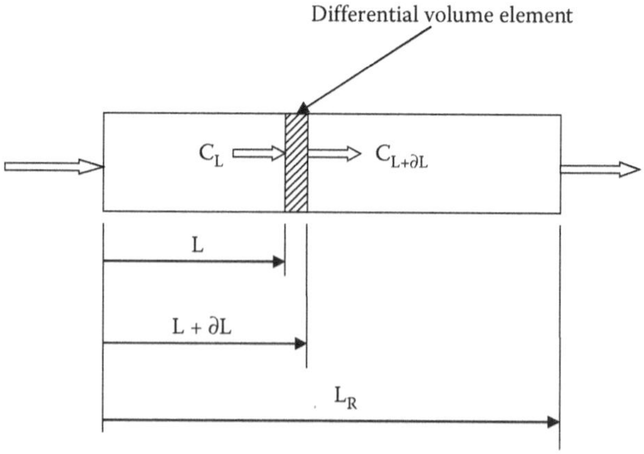

FIGURE 8.9 Schematic representation for dispersed plug flow model.

For the development of the model, we start with a material balance for the tracer over the differential volume. Since the tracer is a nonreacting material, we have

$$\text{Input} = \text{Output} + \text{accumulation}$$

$$\Rightarrow \quad \text{Input due to bulk flow} + \text{Input due to diffusion}$$
$$= \text{Output due to bulk flow} + \text{Output due to diffusion} + \text{accumulation}$$

$$\Rightarrow \quad C\big|_L v - DAr_R \left.\frac{\partial C}{\partial L}\right|_L = C\big|_{L+\partial L} v - DAr_R \left.\frac{\partial C}{\partial L}\right|_{L+\partial L} + \frac{\partial C}{\partial t} Ar_R \partial L$$

$$\Rightarrow \quad C\big|_L \frac{v}{Ar_R} - D \left.\frac{\partial C}{\partial L}\right|_L = C\big|_{L+\partial L} \frac{v}{Ar_R} - D \left.\frac{\partial C}{\partial L}\right|_{L+\partial L} + \frac{\partial C}{\partial t} \partial L$$

$$\Rightarrow \quad D\left[\frac{(\partial C/\partial L)\big|_{L+\partial L} - (\partial C/\partial L)\big|_L}{\partial L}\right] - u\left[\frac{C\big|_{L+\partial L} - C\big|_L}{\partial L}\right] = \frac{\partial C}{\partial t}$$

$$\Rightarrow \quad D\frac{\partial^2 C}{\partial L^2} - u\frac{\partial C}{\partial L} = \frac{\partial C}{\partial t} \tag{8.41}$$

For the ease of solution, this partial differential equation is transformed in terms of the following dimensionless parameters:

$$\lambda = \frac{L}{L_R} \tag{8.42}$$

$$\theta = \frac{t}{\tau} \tag{8.43}$$

$$\varphi = \frac{C}{C_0} \tag{8.44}$$

From Equation 8.42,

$$\frac{\partial \lambda}{\partial L} = \frac{1}{L_R} \tag{8.45}$$

From Equation 8.43,

$$\frac{\partial \theta}{\partial t} = \frac{1}{\tau} \tag{8.46}$$

From Equations 8.44 and 8.45,

$$C_0\frac{\partial\varphi}{\partial L}=\frac{\partial C}{\partial L}$$

$$\Rightarrow\quad C_0\frac{\partial\varphi}{\partial\lambda}\times\frac{\partial\lambda}{\partial L}=\frac{\partial C}{\partial L} \tag{8.47}$$

$$\Rightarrow\quad \frac{\partial C}{\partial L}=\frac{C_0}{L_R}\frac{\partial\varphi}{\partial\lambda}\quad\left(\text{From Equation 8.45, }\frac{\partial\lambda}{\partial L}=\frac{1}{L_R}\right)$$

$$\Rightarrow\quad \frac{\partial^2 C}{\partial L^2}=\frac{C_0}{L_R}\times\frac{\partial}{\partial L}\left(\frac{\partial\varphi}{\partial\lambda}\right)$$

$$\Rightarrow\quad \frac{\partial^2 C}{\partial L^2}=\frac{C_0}{L_R}\times\frac{\partial^2\varphi}{\partial L\partial\lambda}=\frac{C_0}{L_R^2}\frac{\partial^2\varphi}{\partial\lambda^2}\quad(\text{From Equation 8.45}, \partial L=L_R\partial\lambda) \tag{8.48}$$

From Equations 8.44 and 8.46,

$$C_0\frac{\partial\varphi}{\partial t}=\frac{\partial C}{\partial t}$$

$$\rightarrow\quad C_0\frac{\partial\varphi}{\partial\theta}\times\frac{\partial\theta}{\partial t}=\frac{\partial C}{\partial t} \tag{8.49}$$

$$\Rightarrow\quad \frac{\partial C}{\partial t}=\frac{C_0}{\tau}\frac{\partial\varphi}{\partial\theta}\left(\text{From Equation 8.46}\frac{\partial\theta}{\partial t}=\frac{1}{\tau}\right)$$

Substituting the values from Equations 8.47 through 8.49 into Equation 8.41,

$$\frac{DC_0}{L_R^2}\frac{\partial^2\varphi}{\partial\lambda^2}-\frac{uC_0}{L_R}\frac{\partial\varphi}{\partial\lambda}=\frac{C_0}{\tau}\frac{\partial\varphi}{\partial\theta}$$

$$\Rightarrow\quad \frac{D}{uL_R}\frac{\partial^2\varphi}{\partial\lambda^2}-\frac{\partial\varphi}{\partial\lambda}=\frac{\partial\varphi}{\partial\theta}\quad\left(\text{Since}\frac{L_R}{u\tau}=1\right) \tag{8.50}$$

The quantity (D/uL_R) is dimensionless and is known as the *dispersion number.* Its reciprocal, (uL_R/D), is known as the *reactor Peclet number.*

8.5.1.1 Pulse Tracer Input under Dispersed Plug Flow

Consider m_{tr} mass of tracer injected into the reactor instantaneously at $t=0$. The mass of tracer injected is such that

$$C_0=\frac{m_{tr}}{Ar_RL_R}$$

For a pulse tracer input, the solution to the differential Equation 8.41 or 8.50 depends on the boundary conditions. These in turn depend on the diffusivity "*D*" and the linear velocity "u." Since the model considers the flow as a plug flow superimposed with diffusion, the linear velocity is constant everywhere. The diffusivity, however, may vary with position, time, and concentration. Accordingly, we can have different sets of boundary conditions (Hulburt, 1944; Danckwerts, 1953; Wehner and Wilhelm, 1956) and different solutions for Equation 8.50. Here, we discuss two cases.

8.5.1.1.1 Closed Closed-Boundary Condition

Here, it is assumed that diffusion is present only inside the reactor vessel. At all locations preceding the entrance of the reactor and succeeding the exit of the reactor, the diffusivity D is zero. It abruptly acquires a finite value exactly at the reactor entrance, remains constant at this value throughout the length of the reactor, and then abruptly falls to zero exactly at the reactor exit. Mathematically it can be represented as

$$D = 0, \quad \text{at } \lambda < 0 \tag{8.51}$$

$$D = D, \quad \text{at } 0 \le \lambda \le 1 \tag{8.52}$$

$$D = 0, \quad \text{at } \lambda > 1 \tag{8.53}$$

Corresponding to these, the initial and boundary conditions would be as follows:
Initial condition; at $\theta = 0$

From Figure 8.9, the mass flow rate of tracer on the immediate left-hand side of the reactor entrance should be equal to the mass flow rate on the immediate RHS of the reactor entrance. In addition, the concentration of tracer on the immediate left-hand side of the reactor entrance is C_0. Mathematically, it can be represented as

$$C\Big|_{\lambda=0(-)} v = C\Big|_{\lambda=0(+)} v - DAr_R \frac{\partial C}{\partial L}\Big|_{\lambda=0(+)}$$

$$\Rightarrow \quad C\Big|_{\lambda=0(-)} v = C\Big|_{\lambda=0(+)} v - \frac{DAr_R}{L_R}\frac{\partial C}{\partial \lambda}\Big|_{\lambda=0(+)} \qquad \text{(From Equation, 8.45,} \quad \partial L = L_R \partial\lambda)$$

$$\Rightarrow \quad C_0 = C\Big|_{\lambda=0(+)} - \frac{D}{uL_R}\frac{\partial C}{\partial \lambda}\Big|_{\lambda=0(+)} \tag{8.54}$$

$$\Rightarrow \quad \varphi\Big|_{\lambda=0(+)} - \frac{D}{uL_R}\frac{\partial \varphi}{\partial \lambda}\Big|_{\lambda=0(+)} = 1 \tag{8.55}$$

Boundary conditions; at $\theta > 0$

Being a pulse input, the tracer concentration on the immediate left-hand side of the reactor entrance is zero, and so is the mass flow rate. Therefore, from the conservation of mass

$$0 = \mathrm{C}\big|_{\lambda=0(+)}\mathrm{v} - DAr_R \left.\frac{\partial \mathrm{C}}{\partial \mathrm{L}}\right|_{\lambda=0(+)}$$

$$\Rightarrow \quad 0 = \mathrm{C}\big|_{\lambda=0(+)}\mathrm{v} - \frac{DAr_R}{\mathrm{L_R}}\left.\frac{\partial \mathrm{C}}{\partial \lambda}\right|_{\lambda=0(+)} \quad (\text{From Equation 8.45,} \quad \partial\mathrm{L} = \mathrm{L_R}\partial\lambda)$$

$$\Rightarrow \quad \mathrm{C}\big|_{\lambda=0(+)} - \frac{D}{\mathrm{uL_R}}\left.\frac{\partial \mathrm{C}}{\partial \lambda}\right|_{\lambda=0(+)} = 0 \tag{8.56}$$

$$\Rightarrow \quad \varphi\big|_{\lambda=0(+)} - \frac{D}{\mathrm{uL_R}}\left.\frac{\partial \varphi}{\partial \lambda}\right|_{\lambda=0(+)} = 0 \tag{8.57}$$

Similarly, the mass flow rate of tracer on the immediate left-hand side of the reactor exit should be equal to the mass flow rate on the immediate RHS of the reactor exit. Mathematically, it can be represented as

$$\begin{aligned} &\mathrm{C}\big|_{\lambda=1(-)}\mathrm{v} - DAr_R\left.\frac{\partial \mathrm{C}}{\partial \mathrm{L}}\right|_{\lambda=1(-)} = \mathrm{C}\big|_{\lambda=1(+)}\mathrm{v} \\ \Rightarrow \quad &\left.\frac{\partial \mathrm{C}}{\partial \mathrm{L}}\right|_{\lambda=1(-)} = 0 \quad \left(\text{Since } \mathrm{C}\big|_{\lambda=1(-)} = \mathrm{C}\big|_{\lambda=1(+)}\right) \end{aligned} \tag{8.58}$$

$$\Rightarrow \quad \left.\frac{\partial \mathrm{C}}{\partial \lambda}\right|_{\lambda=1(-)} = 0 \tag{8.59}$$

$$\Rightarrow \quad \left.\frac{\partial \varphi}{\partial \lambda}\right|_{\lambda=1(-)} = 0 \tag{8.60}$$

Under these conditions, an analytical solution to Equation 8.41 or 8.50 is not possible. Numerical solutions are, however, available. These are in the form of plots between φ and θ at the reactor exit. From these plots, the RTD function can be determined. The RTD as a function of dimensionless time θ can be written as (Froment and Bischoff, 1979)

$$\begin{aligned} \mathrm{E}(\theta) = \tau\mathrm{E} = \exp\left(\frac{\mathrm{uL_R}}{2D}\right)\sum_{\mathrm{i}=1}^{\infty}\frac{(-1)^{\mathrm{i}+1}8\alpha_{\mathrm{i}}^2}{4\alpha_{\mathrm{i}}^2 + (4\mathrm{uL_R}/D) + (\mathrm{uL_R}/D)^2} \\ \times\exp\left[\frac{-\theta\left\{4\alpha_{\mathrm{i}}^2 + (\mathrm{uL_R}/D)^2\right\}}{(4\mathrm{uL_R}/D)}\right] \end{aligned} \tag{8.61}$$

where

$$\tan\alpha_i = \frac{(4uL_R/D)\times\alpha_i}{4\alpha_i - (uL_R/D)^2} \tag{8.62}$$

From a practical standpoint, we are interested in estimating the dispersion number (or the Peclet number) as it is a characteristic feature of the reactor. For this purpose, the following relation can be used:

$$t_{mean} = \tau \tag{8.63}$$

and

$$\frac{\sigma^2}{t_{mean}^2} = 2\left[\left(\frac{D}{uL_R}\right) - \left(\frac{D}{uL_R}\right)^2 \left\{1 - \exp\left(-\frac{uL_R}{D}\right)\right\}\right] \tag{8.64}$$

8.5.1.1.2 *Open Open-Boundary Condition*

In this case, diffusion is present at all locations, that is, preceding the entrance of the reactor, inside the reactor, as well as succeeding the exit of the reactor, with no discontinuity whatsoever. Mathematically it can be represented as

$$D = D, \quad \text{at} -\infty \le \lambda \le \infty \tag{8.65}$$

Corresponding to this, the initial and boundary conditions would be as follows:
Initial condition; at $\theta = 0$

$$C\big|_{\lambda=0(-)} = C\big|_{\lambda=0(+)} = C_0 \tag{8.66}$$

$$C = 0 \quad \text{at } \lambda \neq 0 \tag{8.67}$$

Boundary conditions; at $\theta > 0$

The mass flow rate on the immediate left-hand side of the reactor entrance is equal to the mass flow rate on the immediate RHS of the reactor entrance. In addition, the tracer concentration on the immediate left-hand side of the reactor entrance is equal to the tracer concentration on the immediate RHS of the reactor entrance. Mathematically,

$$C\big|_{\lambda=0(-)} = C\big|_{\lambda=0(+)} \tag{8.68}$$

and

$$C\Big|_{\lambda=0(-)} v - DAr_R \frac{\partial C}{\partial L}\Big|_{\lambda=0(-)} = C\Big|_{\lambda=0(+)} v - DAr_R \frac{\partial C}{\partial L}\Big|_{\lambda=0(+)}$$

$$DAr_R \frac{\partial C}{\partial L}\Big|_{\lambda=0(-)} = DAr_R \frac{\partial C}{\partial L}\Big|_{\lambda=0(+)} \quad \text{(From Equation 8.68)} \tag{8.69}$$

$$\Rightarrow \frac{\partial C}{\partial L}\Big|_{\lambda=0(-)} = \frac{\partial C}{\partial L}\Big|_{\lambda=0(+)}$$

$$\Rightarrow \frac{\partial C}{\partial \lambda}\Big|_{\lambda=0(-)} = \frac{\partial C}{\partial \lambda}\Big|_{\lambda=0(+)} \quad \text{(From Equation 8.45} \quad \partial L = L_R \partial \lambda) \tag{8.70}$$

Similarly, the mass flow rate on the immediate left-hand side of the reactor exit is equal to the mass flow rate on the immediate RHS of the reactor exit. In addition, the tracer concentration on the immediate left-hand side of the reactor exit is equal to the tracer concentration on the immediate RHS of the reactor exit. Mathematically,

$$C\Big|_{\lambda=1(-)} = C\Big|_{\lambda=1(+)} \tag{8.71}$$

and

$$C\Big|_{\lambda=1(-)} v - DAr_R \frac{\partial C}{\partial L}\Big|_{\lambda=1(-)} = C\Big|_{\lambda=1(+)} v - DAr_R \frac{\partial C}{\partial L}\Big|_{\lambda=1(+)}$$

$$DAr_R \frac{\partial C}{\partial L}\Big|_{\lambda=1(-)} = DAr_R \frac{\partial C}{\partial L}\Big|_{\lambda=1(+)} \quad \text{(From Equation 8.71)} \tag{8.72}$$

$$\Rightarrow \frac{\partial C}{\partial L}\Big|_{\lambda=1(-)} = \frac{\partial C}{\partial L}\Big|_{\lambda=1(+)}$$

$$\Rightarrow \frac{\partial C}{\partial \lambda}\Big|_{\lambda=1(-)} = \frac{\partial C}{\partial \lambda}\Big|_{\lambda=1(+)} \quad \text{(From Equation 8.45} \quad \partial L = L_R \partial \lambda) \tag{8.73}$$

With these boundary conditions, the solution to Equation 8.41 is (Levenspiel and Smith, 1957)

$$C = \frac{C_0 L_R}{2\sqrt{\pi D t}} \exp\left[-\frac{(L - ut)^2}{4Dt}\right] \tag{8.74}$$

From this equation, the tracer concentration at any instant at the reactor outlet would be

$$C\Big|_{\lambda=1} = \frac{C_0 L_R}{2\sqrt{\pi D t}} \exp\left[-\frac{(L_R - ut)^2}{4Dt}\right] \tag{8.75}$$

$$\Rightarrow \quad \varphi\big|_{\lambda=1} = \frac{1}{2\sqrt{\pi\theta(D/uL_R)}}\exp\left[-\frac{(1-\theta)^2}{4\theta(D/uL_R)}\right] \tag{8.76}$$

From Equation 8.75 in combination with Equation 8.7,

$$E = \frac{u}{2\sqrt{\pi Dt}}\exp\left[-\frac{(L_R - ut)^2}{4Dt}\right] \tag{8.77}$$

$$\Rightarrow \quad E(\theta) = \tau E = \frac{1}{2\sqrt{\pi(D/uL_R)}}\exp\left[-\frac{(1-\theta)^2}{4\theta(D/uL_R)}\right] \tag{8.78}$$

From Equation 8.77 in combination with Equations 8.28 and 8.31, we can estimate t_{mean} and σ^2 as

$$t_{mean} = \tau\left[1 + 2\left(\frac{D}{uL_R}\right)\right] \tag{8.79}$$

and

$$\frac{\sigma^2}{t_{mean}^2} = 2\left(\frac{D}{uL_R}\right) + 8\left(\frac{D}{uL_R}\right)^2 \tag{8.80}$$

From Equation 8.79, it is evident that *for open open-boundary conditions, the mean residence time is more than the space time.*

8.5.1.2 Dispersion Number as Characterizing Parameter under Dispersed Plug Flow

From Equation 8.76, the tracer concentration at the reactor exit is dependent on θ and the dispersion number. If we graphically represent Equation 8.76 as a plot between φ at reactor exit and θ for different values of the dispersion number we get a group of curves as shown in Figure 8.10.

From these curves and the preceding equations, the following general inferences can be drawn:

1. As the dispersion number approaches zero, the $\varphi - \theta$ curve approaches a vertical spike at $\theta = 1$—the behavior expected of an ideal PFR.
2. For small values of dispersion number, the reactor flow does not deviate much from the ideal plug flow. The $\varphi - \theta$ curve is symmetrical about $\theta = 1$.
3. As the dispersion number increases, the spread of the $\varphi - \theta$ curve, as measured by σ^2, increases. The spread is zero for an ideal PFR.
4. For small values of dispersion number, the maximum of φ occurs almost at $\theta = 1$. For an ideal PFR, this happens exactly at $\theta = 1$.

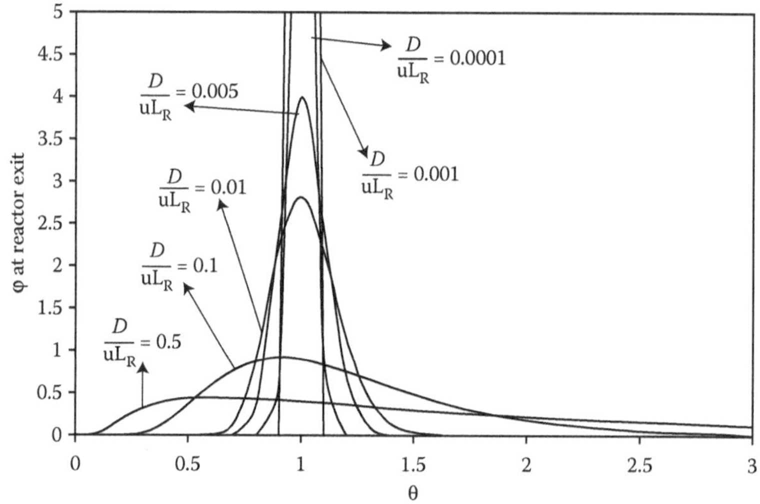

FIGURE 8.10 Effect of dispersion number on outlet tracer response; Equation 8.76.

5. For larger values of a dispersion number—more than about 0.01—the $\varphi - \theta$ curve is flatter.
6. For larger values of a dispersion number, the maximum of φ occurs appreciably to the left of $\theta = 1$.
7. For larger values of a dispersion number, the $\varphi - \theta$ curve is skewed about the maximum. The skewness increases as the dispersion number increases.

It is advisable to use the dispersed plug flow model for reactor analysis only if the $\varphi - \theta$ curve does not show appreciable skewness.

8.5.1.2.1 Small Dispersion Number

For small values of (D/uL_R) the term involving the square of (D/uL_R) can be safely neglected without an appreciable error in the calculated value of σ^2. In such a case, we can have a single equation for σ^2 for closed closed- and open open-boundary conditions as

$$\frac{\sigma^2}{t_{mean}^2} = 2\left(\frac{D}{uL_R}\right) \tag{8.81}$$

$$\Rightarrow \quad \sigma^2 = 2t_{mean}^2\left(\frac{D}{uL_R}\right)$$

$$\Rightarrow \quad \sigma^2 = 2\tau^2\left[1+4\left(\frac{D}{uL_R}\right)+4\left(\frac{D}{uL_R}\right)^2\right]\left(\frac{D}{uL_R}\right)$$

$$\Rightarrow \quad \sigma^2 = 2\tau^2\left(\frac{D}{uL_R}\right) = 2\left(\frac{DL_R}{u^3}\right) \tag{8.82}$$

For N reactors joined in series,

$$L_{total} = (L_R)_1 + (L_R)_2 + (L_R)_3 + \ldots + (L_R)_N$$

$$\Rightarrow \quad \frac{L_{total}}{u} = \frac{(L_R)_1}{u} + \frac{(L_R)_2}{u} + \frac{(L_R)_3}{u} + \ldots + \frac{(L_R)_N}{u}$$

$$\Rightarrow \quad (t_{mean})_{total} = (t_{mean})_1 + (t_{mean})_2 + (t_{mean})_3 + \ldots + (t_{mean})_N \tag{8.83}$$

Similarly,

$$(DL_R)_{total} = (DL_R)_1 + (DL_R)_2 + (DL_R)_3 + \ldots + (DL_R)_N$$

$$\Rightarrow \quad \frac{2(DL_R)_{total}}{u^3} = \frac{2(DL_R)_1}{u^3} + \frac{2(DL_R)_2}{u^3} + \frac{2(DL_R)_3}{u^3} + \ldots + \frac{2(DL_R)_N}{u^3}$$

$$\Rightarrow \quad \sigma_{total}^2 = \sigma_1^2 + \sigma_2^2 + \sigma_3^2 + \ldots + \sigma_N^2 \tag{8.84}$$

EXAMPLE 8.7

Estimate the dispersion number for the reactor system described in Example 8.5 for closed closed-reactor boundary conditions.

SOLUTION

From the solution to Example 8.5,

$$t_{mean} = 8.24 \text{ min}$$

$$\sigma^2 = 4.616 \text{ min}^2$$

For closed closed-boundary conditions, from Equation 8.63,

$$t_{mean} = \tau = 8.24 \text{ min}$$

From Equation 8.64,

$$2\left[\left(\frac{D}{uL_R}\right) - \left(\frac{D}{uL_R}\right)^2 \left\{1 - \exp\left(-\frac{uL_R}{D}\right)\right\}\right] = \frac{\sigma^2}{t_{mean}^2} = \frac{4.616}{8.24 \times 8.24} = 0.068$$

$$\Rightarrow \quad \left(\frac{D}{uL_R}\right) - \left(\frac{D}{uL_R}\right)^2 \left\{1 - \exp\left(-\frac{uL_R}{D}\right)\right\} = 0.034$$

This equation is solved by iteration to give

$$\frac{D}{uL_R} = 0.034$$

8.5.1.3 Conversion with Dispersed Plug Flow

For steady-state reactor operation, a material balance for the limiting reactant over the differential volume shown in Figure 8.9 gives (Danckwerts, 1953)

Input = Output + Disappearance due to chemical reaction

⇒ Input due to bulk flow + Input due to diffusion

= Output due to bulk flow + Output due to diffusion

+ Disappearance due to chemical reaction

$$\Rightarrow \quad C_{ALR}\Big|_{L} v - DAr_R \left.\frac{dC_{ALR}}{dL}\right|_{L} = C_{ALR}\Big|_{L+dL} v - DAr_R \left.\frac{dC_{ALR}}{dL}\right|_{L+dL} + \left(-r_{ALR}\right)Ar_R dL$$

$$\Rightarrow \quad D\left[\frac{(dC_{ALR}/dL)\big|_{L+dL} - (dC_{ALR}/dL)\big|_{L}}{dL}\right] - u\left[\frac{C_{ALR}\big|_{L+dL} - C_{ALR}\big|_{L}}{dL}\right] = \left(-r_{ALR}\right)$$

$$\Rightarrow \quad D\frac{d^2C_{ALR}}{dL^2} - u\frac{dC_{ALR}}{dL} = \left(-r_{ALR}\right) \tag{8.85}$$

In combination with Equation 8.45,

$$\frac{dC_{ALR}}{dL} = \frac{dC_{ALR}}{d\lambda} \times \frac{d\lambda}{dL} = \frac{1}{L_R} \times \frac{dC_{ALR}}{d\lambda} \tag{8.86}$$

$$\Rightarrow \quad \frac{d^2C_{ALR}}{dL^2} = \frac{d}{dL}\left(\frac{1}{L_R} \times \frac{dC_{ALR}}{d\lambda}\right)$$

$$\Rightarrow \quad \frac{d^2C_{ALR}}{dL^2} = \frac{1}{L_R} \times \frac{d^2C_{ALR}}{dL\,d\lambda}$$

$$\Rightarrow \quad \frac{d^2C_{ALR}}{dL^2} = \frac{1}{L_R^2}\frac{d^2C_{ALR}}{d\lambda^2} \tag{8.87}$$

Substituting these values in Equation 8.85,

$$\frac{D}{L_R^2}\frac{d^2C_{ALR}}{d\lambda^2} - \frac{u}{L_R} \times \frac{dC_{ALR}}{d\lambda} = \left(-r_{ALR}\right)$$

$$\Rightarrow \quad \frac{D}{uL_R}\frac{d^2C_{ALR}}{d\lambda^2} - \frac{dC_{ALR}}{d\lambda} = \tau\left(-r_{ALR}\right) \tag{8.88}$$

If the volume of the reactor system does not change with conversion, we have from Equation 2.25 (Chapter 2)

$$C_{ALR} = (C_{ALR})_0(1 - X_{ALR})$$

$$\Rightarrow \frac{dC_{ALR}}{d\lambda} = -(C_{ALR})_0 \frac{dX_{ALR}}{d\lambda} \quad (8.89)$$

$$\Rightarrow \frac{d^2C_{ALR}}{d\lambda^2} = -(C_{ALR})_0 \frac{d^2X_{ALR}}{d\lambda^2} \quad (8.90)$$

Substituting in Equation 8.88,

$$\frac{D}{uL_R}\frac{d^2X_{ALR}}{d\lambda^2} - \frac{dX_{ALR}}{d\lambda} = \frac{\tau}{(C_{ALR})_0}(-r_{ALR}) \quad (8.91)$$

For a first-order reaction, Equation 8.85 reduces to

$$D\frac{d^2C_{ALR}}{dL^2} - u\frac{dC_{ALR}}{dL} = kC_{ALR}$$

$$\Rightarrow \frac{d^2C_{ALR}}{dL^2} - \frac{u}{D}\frac{dC_{ALR}}{dL} = \frac{k}{D}C_{ALR} \quad (8.92)$$

Dankwerts has provided a solution for this equation under closed closed-boundary conditions. The mass flow rate of limiting reactant on the immediate left-hand side of the reactor entrance should be equal to the mass flow rate on the immediate RHS of the reactor entrance. Mathematically, it can be represented as

$$(C_{ALR})_0 v = C_{ALR}\big|_{L=0(+)} v - DAr_R \frac{dC_{ALR}}{dL}\bigg|_{L=0(+)} \quad (8.93)$$

Similarly, the mass flow rate of limiting reactant on the immediate left-hand side of the reactor exit should be equal to the mass flow rate on the immediate RHS of the reactor exit. Mathematically, it can be represented as

$$C_{ALR}\big|_{L=L_R(-)} v - DAr_R \frac{dC_{ALR}}{dL}\bigg|_{L=L_R(-)} = C_{ALR}\big|_{L=L_R(+)} v \quad (8.94)$$

$$\Rightarrow C_{ALR}\big|_{L=L_R(-)} v - DAr_R \frac{dC_{ALR}}{dL}\bigg|_{L=L_R(-)} = (C_{ALR})_0(1 - X_{ALR})v \quad (8.95)$$

Being a reactant, the concentration of the limiting reactant will be the lowest at the reactor exit compared to any other position within the reactor. It implies that

$$\left.\frac{dC_{A_{LR}}}{dL}\right|_{L=L_R(+)} = 0 \tag{8.96}$$

With these boundary conditions, the solution to Equation 8.92 gives (Danckwerts, 1953)

$$X_{A_{LR}} = 1 - \frac{4a}{(1+a)^2 \exp\{(-uL_R/2D)(1-a)\} - (1-a)^2 \exp\{(-uL_R/2D)(1+a)\}} \tag{8.97}$$

$$\Rightarrow \quad 1 - X_{A_{LR}} = \frac{C_{A_{LR}}}{(C_{A_{LR}})_0}$$

$$= \frac{4a}{(1+a)^2 \exp\{(-uL_R/2D)(1-a)\} - (1-a)^2 \exp\{(-uL_R/2D)(1+a)\}} \tag{8.98}$$

where

$$a = \sqrt{1 + \left(\frac{4kD}{u^2}\right)} = \sqrt{1 + 4k\tau\left(\frac{D}{uL_R}\right)} \tag{8.99}$$

At small values of dispersion number, the higher-order terms of the dispersion number in the expansion of the exponential function can be neglected. This reduces Equation 8.97 to

$$X_{A_{LR}} = 1 - \left(1 + \frac{k^2 D L_R}{u^3}\right) \exp(-k\tau) \tag{8.100}$$

$$\Rightarrow \quad X_{A_{LR}} = 1 - \left(1 + \frac{k^2\sigma^2}{2}\right) \exp(-k\tau) \quad \left(\text{From Equation 8.82} \quad \sigma^2 = 2\left(\frac{DL_R}{u^3}\right)\right) \tag{8.101}$$

Thus, for a first-order reaction, conversion can be predicted from knowledge of variance and space time alone. Rosas has extended this treatment for a two-step series reaction with irreversible first-order kinetics using numerical integration (Rosas, 1969).

Wehner and Wilhelm (1956) have shown that Equations 8.97 and 8.99 are valid even for open open-boundary conditions with τ replaced with t_{mean}.

EXAMPLE 8.8

The first-order reaction

$$A \rightarrow \text{Products}$$

with $-r_A = 0.05C_A$ mol/(L · min) is carried out in the reactor described in Example 8.5 for closed closed-reactor boundary conditions. What is the expected conversion?

By how much percentage is the conversion reduced compared to an ideal PFR of the same size?

Solution

From the solution to Example 8.7, for closed closed-boundary conditions

$$t_{mean} = \tau = 8.24 \text{ min}$$

$$\sigma^2 = 4.616 \text{ min}^2$$

$$\frac{D}{uL_R} = 0.034$$

From Equation 8.99,

$$a = \sqrt{1 + 4k\tau\left(\frac{D}{uL_R}\right)} = \sqrt{1 + 4 \times 0.05 \times 8.24 \times 0.034} = 1.028$$

From Equation 8.97,

$$X_A = 1 - \frac{4a}{(1+a)^2 \exp\{(-uL_R/2D)(1-a)\} - (1-a)^2 \exp\{(-uL_R/2D)(1+a)\}}$$

$$\Rightarrow X_A = 1 - \frac{4 \times 1.028}{(1+1.028)^2 \exp\{-(1-1.028)/2 \times 0.034\} - (1-1.028)^2 \exp\{-(1+1.028)/2 \times 0.034\}}$$

$$\Rightarrow X_A = 1 - \frac{4.112}{4.113 \times \exp\{-(1-1.028)/2 \times 0.034\} - 0.0008 \times \exp\{-(1+1.028)/2 \times 0.034\}} = 0.337$$

For the conversion achieved in an ideal PFR, from Equation 6.24 (Chapter 6),

$$k\tau = -\ln\left(1 - (X_A)_{PFR}\right)$$

$$\Rightarrow \ln\left(1 - (X_A)_{PFR}\right) = -0.05 \times 8.24$$

$$\Rightarrow 1 - (X_A)_{PFR} = 0.662$$

$$\Rightarrow (X_A)_{PFR} = 0.338$$

$$\%\text{reduction in conversion} = \frac{0.338 - 0.337}{0.338} \times 100 = 0.3\%$$

8.5.2 Tanks-in-Series Model

Developed by MacMullin and Weber (1935), this model visualizes the real reactor as several equal-sized ideal BFRs joined in series such that the RTD function for the series of reactors of the model matches that obtained experimentally for the real reactor. Consider a series of N equal-sized ideal BFRs as shown in Figure 8.11.

A pulse of m_{tr} mass of tracer is injected to the inlet of the first reactor at time $t = 0$. Since each reactor is the same size, the space time for each reactor will be the same, so that

$$\tau_1 = \tau_2 = \tau_3 = \ldots = \tau_i = \ldots = \tau_{N-1} = \tau_N = \tau(\text{say}) \tag{8.102}$$

In the time interval between t and t + dt after the injection of the tracer pulse, where dt → 0, a material balance for the tracer over the first BFR gives

$$\text{Input} = \text{Output} + \text{Accumulation}$$

$$\Rightarrow \quad 0 = (C_{out})_1 v + \frac{d(C_{out})_1}{dt} V_{BFR}$$

$$\Rightarrow \quad \frac{d(C_{out})_1}{dt} = -\frac{(C_{out})_1}{\tau}$$

$$\Rightarrow \quad \int_{C_0}^{(C_{out})_1} \frac{d(C_{out})_1}{(C_{out})_1} = -\frac{1}{\tau}\int_0^t dt \quad \left(C_0 = \frac{m_{tr}}{V_{BFR}}\right)$$

$$\Rightarrow \quad \ln\frac{(C_{out})_1}{C_0} = -\frac{t}{\tau}$$

$$\Rightarrow \quad (C_{out})_1 = C_0 e^{-(t/\tau)} \tag{8.103}$$

Outlet from the first reactor enters the second reactor. Therefore, in the same time interval a material balance for the tracer over the second BFR gives

$$\text{Input} = \text{Output} + \text{Accumulation}$$

$$\Rightarrow \quad (C_{out})_1 v = (C_{out})_2 v + \frac{d(C_{out})_2}{dt} V_{BFR}$$

$$\Rightarrow \quad (C_{out})_1 = (C_{out})_2 + \frac{d(C_{out})_2}{dt}\tau \tag{8.104}$$

$$\Rightarrow \quad \frac{d(C_{out})_2}{dt} + \frac{1}{\tau}(C_{out})_2 = \frac{C_0}{\tau} e^{-(t/\tau)}$$

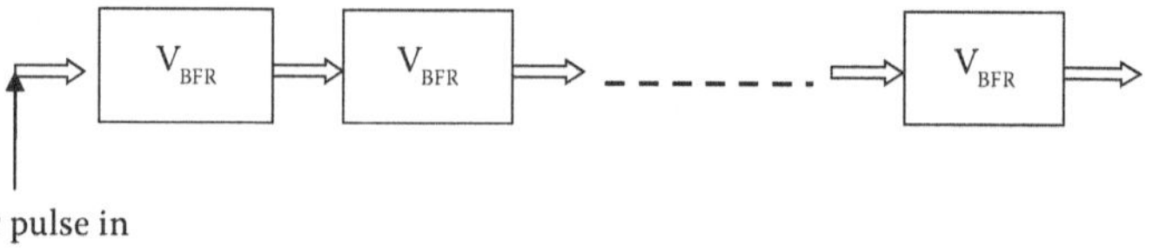

FIGURE 8.11 Tracer pulse injected to N ideal equal-sized BFRs in series.

The solution to this first-order differential equation is

$$(C_{out})_2 e^{\int (dt/\tau)} = \int \frac{C_0}{\tau} e^{-\frac{t}{\tau}} e^{\int (dt/\tau)} dt + \text{constant}$$

$$\Rightarrow \quad (C_{out})_2 e^{(t/\tau)} = \frac{C_0}{\tau} \int dt + \text{constant}$$

$$\Rightarrow \quad (C_{out})_2 e^{(t/\tau)} = \frac{C_0 t}{\tau} + \text{constant} \tag{8.105}$$

Using the initial condition of $(C_{out})_2 = 0$ at $t = 0$, the value of the constant in Equation 8.105 is estimated to be zero. Therefore, from Equation 8.105,

$$(C_{out})_2 = \frac{C_0 t}{\tau} e^{-(t/\tau)} \tag{8.106}$$

A material balance for the tracer over the third BFR gives

$$(C_{out})_2 v = (C_{out})_3 v + \frac{d(C_{out})_3}{dt} V_{BFR}$$

$$\Rightarrow \quad (C_{out})_2 = (C_{out})_3 + \frac{d(C_{out})_3}{dt} \tau \tag{8.107}$$

$$\Rightarrow \quad \frac{d(C_{out})_3}{dt} + \frac{1}{\tau}(C_{out})_3 = \frac{C_0 t}{\tau^2} e^{-(t/\tau)}$$

The solution to this first-order differential equation is

$$(C_{out})_3 e^{\int (dt/\tau)} = \int \frac{C_0 t}{\tau^2} e^{-\frac{t}{\tau}} e^{\int (dt/\tau)} dt + \text{constant}$$

$$\Rightarrow \quad (C_{out})_3 e^{(t/\tau)} = \frac{C_0}{\tau^2} \int t\,dt + \text{constant}$$

$$\Rightarrow \quad (C_{out})_3 e^{(t/\tau)} = \frac{C_0 t^2}{2\tau^2} + \text{constant} \tag{8.108}$$

Using the initial condition of $(C_{out})_3 = 0$ at $t = 0$, the value of the constant in Equation 8.108 is estimated to be zero. Therefore, from Equation 8.108,

$$(C_{out})_3 = \frac{C_0 t^2}{2\tau^2} e^{-(t/\tau)} \tag{8.109}$$

Similarly, for the fourth reactor,

$$\frac{d(C_{out})_4}{dt}+\frac{1}{\tau}(C_{out})_4=\frac{C_0t^2}{2\tau^3}e^{-(t/\tau)}$$

$$\Rightarrow\quad (C_{out})_4 e^{\int(dt/\tau)}=\int\frac{C_0t^2}{2\tau^3}e^{-(t/\tau)}e^{\int(dt/\tau)}dt+\text{constant}$$

$$\Rightarrow\quad (C_{out})_4 e^{(t/\tau)}=\frac{C_0}{2\tau^3}\int t^2dt+\text{constant}$$

$$\Rightarrow\quad (C_{out})_4 e^{(t/\tau)}=\frac{C_0t^3}{(2\times 3)\tau^3}+\text{constant}$$

$$\Rightarrow\quad (C_{out})_4 e^{(t/\tau)}=\frac{C_0t^3}{(3!)\tau^3}+\text{constant} \tag{8.110}$$

From the initial condition the constant is zero, so

$$(C_{out})_4=\frac{C_0t^3}{(3!)\tau^3}e^{-(t/\tau)} \tag{8.111}$$

By similar treatment, the outlet concentration of tracer from the Nth BFR at time t is

$$(C_{out})_N=\frac{C_0t^{N-1}}{(N-1)!\times\tau^{N-1}}e^{-\frac{t}{\tau}}=\frac{C_0\theta^{N-1}}{(N-1)!}e^{-\theta}\quad\left(\theta=\frac{t}{\tau}\right) \tag{8.112}$$

$$\Rightarrow\quad \varphi_{out}=\frac{\theta^{N-1}}{(N-1)!}e^{-\theta} \tag{8.113}$$

$$\Rightarrow\quad \varphi_{out}=\frac{(N\theta_N)^{N-1}}{(N-1)!}e^{-N\theta_N}\quad(\theta=N\theta_N) \tag{8.114}$$

$$\Rightarrow\quad (\varphi_{out})_N=\frac{N^N(\theta_N)^{N-1}}{(N-1)!}e^{-N\theta_N}\quad\left((\varphi_{out})_N=\frac{(C_{out})_N\times NV_{BFR}}{m_{tr}}\right) \tag{8.115}$$

Let the RTD function for the series of BFRs at time t be E_N. From Equation 8.7 in combination with Equation 8.112,

$$E_N = \frac{(C_{out})_N v}{\int_0^\infty (C_{out})_N v dt} = \frac{C_0 t^{N-1} v e^{-(t/\tau)}}{(N-1)! \times \tau^{N-1} \int_0^\infty (C_{out})_N v dt}$$

$$\Rightarrow \quad E_N = \frac{C_0 t^{N-1} v e^{-(t/\tau)}}{(N-1)! \times \tau^{N-1} \times (\text{amount of tracer leaving the reactor between 0 and } \infty \text{ time})}$$

$$\Rightarrow \quad E_N = \frac{C_0 t^{N-1} v e^{-(t/\tau)}}{(N-1)! \times \tau^{N-1} \times (\text{total amount of tracer fed to the reactor})}$$

$$\Rightarrow \quad E_N = \frac{C_0 t^{N-1} v e^{-(t/\tau)}}{(N-1)! \times \tau^{N-1} \times m_{tr}} = \frac{C_0 t^{N-1} v e^{-(t/\tau)}}{(N-1)! \times \tau^{N-1} \times C_0 V}$$

$$\Rightarrow \quad E_N = \frac{t^{N-1}}{\tau^N (N-1)!} e^{-(t/\tau)} \tag{8.116}$$

$$\Rightarrow \quad \tau E_N = \frac{\theta^{N-1}}{(N-1)!} e^{-\theta} \tag{8.117}$$

$$\Rightarrow \quad E_N = \frac{N^N t^{N-1}}{\tau_N^N (N-1)!} e^{-(Nt/\tau_N)} \tag{8.118}$$

where $\tau_N = N\tau$ is the space time for the entire series of BFRs.

In deducing Equation 8.116, we have used the material balance consideration to evaluate the integral $\int_0^\infty (C_{out})_N v dt$ as equal to $C_0 V$. We will now use this result to evaluate another integral, which will be useful in our subsequent deduction of variance. We have, in combination with Equation 8.112,

$$\int_0^\infty (C_{out})_N v dt = C_0 V$$

$$\Rightarrow \quad \int_0^\infty (C_{out})_N dt = C_0 \tau$$

$$\Rightarrow \quad \frac{C_0}{(N-1)! \times \tau^{N-1}} \int_0^\infty t^{N-1} e^{-(t/\tau)} dt = C_0 \tau$$

$$\Rightarrow \quad \int_0^\infty t^{N-1} e^{-(t/\tau)} dt = (N-1)! \times \tau^N$$

$$\Rightarrow \quad \int_0^\infty t^{N+1} e^{-(t/\tau)} dt = (N+1)! \times \tau^{N+2} \tag{8.119}$$

From Equations 8.116, 8.119, and 8.31,

$$\sigma_N^2 = \int_0^\infty [t - \tau_N]^2 E_N dt$$

$$\Rightarrow \quad \sigma_N^2 = \int_0^\infty [t - N\tau]^2 E_N dt$$

$$\Rightarrow \quad \sigma_N^2 = \int_0^\infty [t^2 + N^2\tau^2 - 2N\tau t] E_N dt$$

$$\Rightarrow \quad \sigma_N^2 = \int_0^\infty t^2 E_N dt + N^2\tau^2 \int_0^\infty E_N dt - 2N\tau \int_0^\infty t E_N dt$$

$$\Rightarrow \quad \sigma_N^2 = \frac{1}{\tau^N (N-1)!} \int_0^\infty (t^{N-1} \times t^2) e^{-(t/\tau)} dt + N^2\tau^2$$

$$-2N\tau \times \tau_N \quad \text{(From Equations 8.25 and 8.28)}$$

$$\Rightarrow \quad \sigma_N^2 = \frac{1}{\tau^N (N-1)!} \int_0^\infty t^{N+1} e^{-(t/\tau)} dt - N^2\tau^2$$

$$\Rightarrow \quad \sigma_N^2 = \frac{(N+1)! \times \tau^{N+2}}{\tau^N (N-1)!} - N^2\tau^2$$

$$\Rightarrow \quad \sigma_N^2 = N(N+1)\tau^2 - N^2\tau^2$$

$$\Rightarrow \quad \sigma_N^2 = N\tau^2 \tag{8.120}$$

$$\Rightarrow \quad \sigma_N^2 = N \times \frac{\tau_N^2}{N^2} = \frac{\tau_N^2}{N} \tag{8.121}$$

EXAMPLE 8.9

Estimate the number of tanks if the reactor described in Example 8.5 is modeled as tanks in series.

Solution

From the solution to Example 8.5,

$$t_{mean} = \tau_N = 8.24 \text{ min}$$

$$\sigma_N^2 = 4.616 \text{ min}^2$$

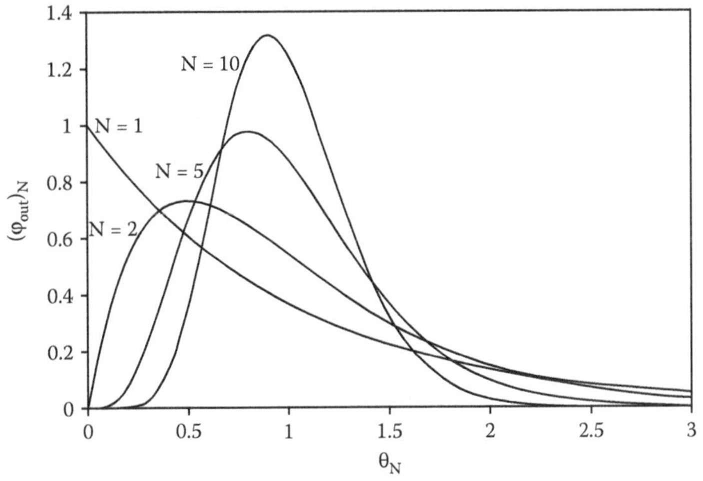

FIGURE 8.12 Dimensionless output concentration response for tracer pulse in tanks-in-series model; Equation 8.115.

From Equation 8.121,

$$\sigma_N^2 = \frac{\tau_N^2}{N}$$

$$\Rightarrow \quad N = \frac{\tau_N^2}{\sigma_N^2} = \frac{8.24 \times 8.24}{4.616} = 14.71$$

Figure 8.12 is a graphical representation of Equation 8.115. From this figure and the preceding equations we can summarize the following ones:

1. For N = 1, the output concentration response is the one for an ideal BFR.
2. As N increases, the position of maximum outlet concentration shifts closer and closer to $\theta_N = 1$.
3. As N increases, the maximum outlet concentration also increases. In the limit $N \rightarrow \infty$, $(\varphi_{out})_N \rightarrow \infty$ at $\theta_N = 1$.
4. As N increases, σ_N^2 decreases being inversely proportional to N. The output concentration response and E curve become narrower and narrower with the increase in N.
5. With an increase in N, the output concentration response and the E curve decrease in skewness.

8.5.2.1 Conversion from Tanks-in-Series Model

We have modeled the real reactor as an assembly of N equal-sized BFRs joined in series such that the RTD function of a real reactor resembles that for the tanks in series. Therefore,

$$(\sigma^2)_{real} = \sigma_N^2, \quad \tau_N = \tau_{real}$$

For a first-order reaction from Equation 6.148 (Chapter 6),

$$(X_A)_N = 1 - \frac{1}{[1+k\tau]^N}$$

$$\Rightarrow \ (X_A)_N = 1 - \frac{1}{\left[1+k\left(\tau_N^2/N^2\right)\right]^N} \tag{8.122}$$

$$\Rightarrow \ (X_A)_N = 1 - \frac{1}{\left[1+k\left(\sigma_N^2/N\right)\right]^N} \quad \left(\text{From Equation 8.121} \quad \sigma_N^2 = \frac{\tau_N^2}{N}\right) \tag{8.123}$$

We can predict the conversion for a first-order reaction from the real reactor as follows:

Step 1

From the pulse tracer experiment on the real reactor, compute the RTD function, and hence the variance $(\sigma^2)_{real}$, which should be equal to the variance expected from the modeled sequence of N tanks σ_N^2.

Step 2

Using Equation 8.121, compute N—this can be a fraction as well—the characteristic parameter of the real reactor under the tanks-in-series model.

Step 3

From Equation 8.123, compute the conversion expected.

EXERCISES

8.1 Plot the E–t graph for the reactor for which the output concentration response from a pulse tracer experiment is as follows:

a.

t (s)	**0**	**30**	**60**	**90**	**120**	**150**	**180**	**210**	**240**
C_{out} (mmol/L)	0	0	6	24	90	22	6.5	0	0

b.

t (s)	**0**	**5**	**10**	**20**	**30**	**40**	**50**	**60**	**70**
C_{out} (mmol/L)	0	1.5	3	5.5	4.5	1.5	0.6	0.2	0

c.

t (s)	**0**	**5**	**10**	**20**	**30**	**40**	**50**	**60**	**70**
C_{out} (mmol/L)	0	3.8	7	5	3	1.5	0.6	0.2	0

d.

t (s)	0	5	10	20	30	40	50	60	70
C_{out} (mmol/L)	0	3.5	3	5.5	4.5	1.5	0.6	0.2	0

e.

t (s)	0	5	10	20	30	40	50	60	70
C_{out} (mmol/L)	0	1.5	3	5.5	4.2	2.5	3.8	0.4	0

8.2 Plot the F–t graph for the reactors described in Exercise 8.1.

8.3 Plot the E–t graph for the reactor for which the output concentration response from a step tracer experiment is as follows:

a.

t (s)	0	30	60	90	120	150	180	210	240
C_{out} (mmol/L)	0	0.3	6	24	75	92	95	95.2	95.2

b.

t (s)	0	30	60	90	120	150	180	210	240	270	300
C_{out} (mmol/L)	0	5	19	40	60	75	85	90	92	93	93

c.

t (s)	0	30	60	90	120	150	180	210	240	270	300
C_{out} (mmol/L)	0	2	5	8	14	40	62	78	88	90	90

d.

t (s)	0	30	60	90	120	150	180	210	240	270	300
C_{out} (mmol/L)	0	2	5	8	17	40	50	78	88	90	90

e.

t (s)	0	30	60	90	120	150	180	210	240	270	300
C_{out} (mmol/L)	0	0.2	0.5	0.8	3	60	70	78	88	90	90

8.4 Plot the F–t graph for the reactors described in Exercise 8.3.

8.5 If the volumetric flow rate through the reactors in the pulse experiment of Exercise 8.1 was 10 L/min, how much tracer was injected into the reactor in each case?

8.6 The reactors of Exercise 8.1 are injected with a step tracer input of 50 mmol/L. Predict the outlet concentration response in each case.

8.7 The reactors of Exercise 8.3 are injected with a pulse tracer input of 20 mmol. The volumetric flow rate through the reactors is 10 L/min. Predict the outlet concentration response in each case.

8.8 Compute the mean residence time and variance of residence times for reactors of Exercise 8.1 in each case.

8.9 Compute the mean residence time and variance of residence times for reactors of Exercise 8.3 in each case.

8.10 In the tracer experiment of Exercise 8.1 part (a), the outlet from the reactor is fed into an ideal BFR of space time 100 s. Compute the outlet concentration after 120 s for the reactor combination.

8.11 In the tracer experiment of Exercise 8.1 part (a), the outlet from the reactor is fed into an ideal PFR of space time 100 s. Compute the outlet concentration response for the reactor combination.

8.12 From qualitative observations, which reactor in Exercise 8.1 is closest to an ideal PFR?

8.13 The first-order liquid-phase reaction

$$A \rightarrow \text{Products}$$

with reaction rate $-r_A = 0.2\ C_A$ mol/(L · min) is carried out in the reactor of Exercise 8.3, part (a). What will be the conversion achieved if complete segregation of fluid elements is maintained in the reactor?

8.14 The second-order liquid-phase reaction

$$A \rightarrow \text{Products}$$

with reaction rate $-r_A = 0.2C_A^2$ mol/(L · min) is carried out in the reactor of Exercise 8.3, part (a) with an initial concentration of 0.5 mol/L. What will be the conversion achieved if complete segregation of fluid elements is maintained in the reactor?

8.15 In order to analyze the flow through a reactor of 100 L volume, an ideal pulse of tracer is injected. The tracer concentration at the reactor outlet is recorded as follows:

t (s)	**0**	**5**	**10**	**15**	**20**	**25**	**30**	**35**	**40**	**45**	**50**
C_{out} (mmol/L)	0	0.2	0.5	0.8	3	50	230	55	4	0.8	0

Calculate the dispersion number for the reactor if the boundaries are closed closed. What is the volumetric flow rate of fluid through the reactor?

8.16 In Exercise 8.15, what will be the dispersion number and volumetric flow rate if the reactor boundaries are open open?

8.17 Liquid is flowing through a reactor of 500 L volume at a flow rate of 30 L/min. An ideal pulse of tracer is injected into the reactor. The tracer concentration at the reactor outlet is recorded as follows:

t (min)	**0**	**5**	**10**	**15**	**20**	**25**	**30**	**35**	**40**	**45**	**50**
C_{out} (mg/L)	0	1	5	7.5	6.3	5	3.7	2.3	1.1	0.5	0

Are the reactor boundaries closed closed or open open? How much tracer was injected into the reactor?

8.18 Once you have answered Exercise 8.17, calculate the dispersion number for the reactor. What additional inference can you draw about the reactor?

8.19 In Exercise 8.17, how much tracer will be remaining in the reactor after 30 min?

8.20 The first-order reaction of Exercise 8.13 is carried out in the reactor of Exercise 8.15. What will be the conversion achieved? What is the size of an ideal PFR required to achieve the same conversion?

8.21 Qualitatively discuss the effect of dispersion number on conversion of reactions of higher order.

8.22 Develop an algorithm to determine dispersion number of a reactor by step tracer experiment.

8.23 Fit the output concentration responses of the reactors in Exercise 8.1 into tanks-in- series model.

8.24 The output concentration response for a step tracer experiment on a reactor is given below.

t (min)	0	2	5	15	25	35	50	70
C_{out}/C_0	0	0.01	0.02	0.08	0.2	0.42	0.85	1

Fit the data into tanks-in-series model.

8.25 The first-order liquid-phase reaction

$$A \rightarrow \text{Products}$$

with reaction rate $-r_A = 0.05\ C_A$ mol/(L · min) is carried out in the reactor of Exercise 8.23. What will be the conversion achieved?

REFERENCES

Danckwerts, P.V., Continuous flow systems. Distribution of residence times. *Chem. Eng. Sci.*, 2, 1, 1953.

Danckwerts, P.V., The effect of incomplete mixing on homogeneous reactions. *Chem. Eng. Sci.*, 8, 93, 1958.

Froment, G.F. and Bischoff, K.B., *Chemical Reactor Analysis and Design*, John Wiley & Sons, New York, 618, 1979.

Hulburt, H.M., Chemical processes in continuous flow systems. *Ind. Eng. Chem.*, 36, 1012, 1944.

Levenspiel, O., *Chemical Reaction Engineering*, 3rd ed., John Wiley & Sons, New York, 270, 1999.

Levenspiel, O. and Smith, W.K., Notes on the diffusion-type model for the longitudinal mixing of fluid in flow. *Chem. Eng. Sci.*, 6, 227, 1957.

MacMullin, R.B. and Weber, M. Jr., The theory of short-circuiting in continuous-flow mixing vessels in series and kinetics of chemical reactions in such systems. *Trans. AIChE*, 31, 409, 1935.

Rosas, C.B., Axial mixing and selectivity. Two consecutive first-order reactions. *Ind. Eng. Chem. Fundamen.*, 8, 361, 1969.

Wehner, J.F. and Wilhelm, R.H., Boundary conditions of flow reactor. *Chem. Eng. Sci.*, 6, 89, 1956.

Zwietering, T.N., The degree of mixing in continuous flow systems. *Chem. Eng. Sci.*, 11, 1, 1959.

9 Reactors of Special Types

9.1 ELECTROCHEMICAL REACTORS

Electrochemical reactions are those where the energy shortfall to drive a chemical reaction is supplied in the form of electrical energy. Alternatively, the energy released in the course of the reaction appears in the form of electrical energy. They constitute a very important class of reactions with immense industrial significance. They find applications in chlor alkali industries, metallurgical and metal-processing industries, wastewater treatment, specialty chemicals, energy storage, etc., to name a few. Their importance has led to the development of electrochemical engineering as an independent engineering discipline. Here, the reader is given a brief overview of the fundamentals of electrochemical reactions and the reactors for such applications.

9.1.1 Electrochemical Reaction Fundamentals

Electrochemical reactions are heterogeneous reactions taking place at the interface of an electrically charged solid—the electrode—and a liquid—the electrolyte. The basic steps involved are

- Transport of reactant moiety from the bulk of the electrolyte to the electrode–electrolyte interface.
- Adsorption of reactant moiety on the surface of the electrode.
- Exchange of electron—charge transfer—between the adsorbed reactant moiety and the electrode. This charge transfer can be of two types.
  - The electrode is at a positive potential—called the anode—so that the reactant moiety loses an electron that crosses over to the solid matrix of the electrode and travels through the external circuit. In other words, the reactant moiety is oxidized.
  - The electrode is at a negative potential—called the cathode—so that the reactant moiety gains an electron. This, in turn, crosses over from the solid matrix of the electrode into the liquid electrolyte and travels through the electrolyte to the opposite electrode. In other words, the reactant moiety is reduced.
- Desorption of product moiety from the surface of an electrode.
- Transport product moiety from the electrode–electrolyte interface to the bulk of the electrolyte.

9.1.1.1 Equilibrium in Electrochemical Reaction

Like all other reacting systems, the electrochemical reacting systems are also governed by the laws of thermodynamics. Referring to Chapter 4, the reaction proceeds in the direction of decreasing free energy and attains equilibrium at the minimum of free energy when it has no more inclination to do external work. In electrochemical reactions, the direction of charge transfer—oxidation or reduction—is dependent on the potential of the electrode. For oxidative charge transfer, the electrode potential must be above a threshold positive potential. Opposite to this reductive charge transfer requires the electrode potential to be below a threshold negative potential. Just at the threshold potential the system is at equilibrium. For an electrochemical reaction

$$\mathrm{A} \rightleftharpoons \mathrm{B} + ne^- \tag{9.1}$$

at standard conditions, the potential of the electrode is related to the free energy change by the relation

$$\Delta G_\mathrm{r}^0 = (\Delta G_\mathrm{f}^0)_\mathrm{B} - (\Delta G_\mathrm{f}^0)_\mathrm{A} = -nFE^0 \tag{9.2}$$

E^0 is known as the standard electrode potential. At other than standard conditions, the electrode potential is related to the standard electrode potential by the Nernst equation (Atkins and de Paula, 2006)

$$E = E^0 + \frac{\mathrm{RT}}{nF} \ln \frac{a_\mathrm{A}}{a_\mathrm{B}} \tag{9.3}$$

For dilute solutions, activities can be approximated with concentrations. Electrode potentials are always measured in comparison to some standard reference electrode. Standard hydrogen electrode and standard calomel electrode are examples of reference electrodes.

An electrochemical reactor consists of two sets of electrodes—a cathode and an anode. The cell voltage for the electrochemical reactor is defined as

$$E_\mathrm{cell} = E_\mathrm{an} - E_\mathrm{cat} \tag{9.4}$$

EXAMPLE 9.1

The reactions in a water electrolyzer may be written as follows:

At anode

$$2\mathrm{H_2O} \rightleftharpoons \mathrm{O_2(g)} + 4\mathrm{H^+} + 4\mathrm{e^-} \quad E^0 = 1.23\,\mathrm{V}\ \text{versus SHE}$$

At cathode

$$2\mathrm{H^+} + 2\mathrm{e^-} \rightleftharpoons \mathrm{H_2(g)} \quad E^0 = 0.00\,\mathrm{V}\ \text{versus SHE}$$

Calculate the standard free energy change for the electrolysis of water.

Solution

The reaction for electrolysis of water may be written as

$$2H_2O \rightleftharpoons 2H_2 + O_2$$

which is the sum of the two electrode reactions written above.
For the reaction

$$2H_2O \rightleftharpoons O_2(g) + 4H^+ + 4e^-$$

$$n_1 = 4$$

From Equation 9.2,

$$(\Delta G_r^0)_1 = -nF(E^0)_1 = -4 \times 96,487 \times 1.23 = 474716\,\text{J/mol} = 474.716\,\text{kJ/mol}$$

For the reaction

$$2H^+ + 2e^- \rightleftharpoons H_2(g)$$

$$(E^0)_2 = 0.00\,\text{V}$$

So, from Equation 9.2,

$$(\Delta G_r^0)_2 = -nF(E^0)_2 = 0$$

For the water electrolysis reaction,

$$\Delta G_r^0 = (\Delta G_r^0)_1 + (\Delta G_r^0)_2 = 474.716\,\text{kJ/mol}$$

9.1.1.2 Kinetics of Electrochemical Reaction

The extent of chemical transformation in an electrochemical reaction is related to the number of electrons transferred in the charge transfer step by the well-known Faraday's law. Referring to reaction 9.1 above, in differential time dt, where $dt \rightarrow 0$

$$\text{Number of moles of A reacted} = \frac{\text{Net moles of electron transferred}}{n}$$

$$\Rightarrow \quad -dN_A = \frac{\text{Moles of electron lost by A} - \text{moles of electrons gained by B}}{n}$$

$$\Rightarrow \quad -dN_A = \frac{\text{Net charge transferred}}{nF} = \frac{dq_{an} - dq_{cat}}{nF} = \frac{dq}{nF}$$

$$\Rightarrow \quad \frac{-dN_A}{dt} = \frac{1}{nF}\left[\frac{dq_{an}}{dt} - \frac{dq_{cat}}{dt}\right] = \frac{1}{nF}\frac{dq}{dt}$$

$$\Rightarrow \quad \frac{-dN_A}{dt} = \frac{I_{an} - I_{cat}}{nF} = \frac{I}{nF} \tag{9.5}$$

From Equation 9.5, we can express the rate of an electrochemical reaction in the following ways:

In terms of electrode surface area,

$$(-r_A)' = \frac{I}{Ar_{el}nF} = \frac{i}{nF} \tag{9.6}$$

In terms of the volume of the reacting system,

$$-r_A = \frac{I}{VnF} = \frac{Ar_{el}i}{VnF} \tag{9.7}$$

At equilibrium, the net rate of reaction of species A at the electrode is zero; therefore, from Equation 9.5,

$$(I_{cat})_{eq} = (I_{an})_{eq} = I_{ex} \tag{9.8}$$

$$=> (i_{cat})_{eq} = (i_{an})_{eq} = i_{ex} \tag{9.9}$$

This is another statement for the kinetic interpretation of equilibrium, which states that equilibrium is not the cessation of all chemical activity. Rather, it is the state when the rate of forward reaction is equal to the rate of backward reaction. In the context of reaction 9.1, the forward reaction is the loss of electron by A and the backward reaction is the gain of electron by B.

For reaction 9.1 to have a net rate of conversion of A to B, I_{an} must be more than I_{cat}. For this to happen, the electrode must be at a more positive potential than that prescribed by Equation 9.3. *The difference between the actual potential of the electrode and its potential at the equilibrium is known as the overpotential.* Thus

$$\eta = E_{act} - E \tag{9.10}$$

The net current flowing through the electrode, hence the rate of electrolytic reaction, is dependent on the overpotential. The current overpotential relationship depends on the controlling mechanism of the overall electrode process. In the low overpotential regime, when the net current is low and so is the reaction rate, the current overpotential relationship is given by the Butler–Volmer kinetic equation (Butler, 1924) as given below

$$I = I_{ex}\left[\exp\left\{\frac{(1-\alpha)nF\eta}{RT}\right\} - \exp\left\{-\frac{\alpha nF\eta}{RT}\right\}\right] \tag{9.11}$$

$$=> i = i_{ex}\left[\exp\left\{\frac{(1-\alpha)nF\eta}{RT}\right\} - \exp\left\{-\frac{\alpha nF\eta}{RT}\right\}\right] \tag{9.12}$$

Here, α is a transfer coefficient having value between 0 and 1.
In combination with Equation 9.6,

$$(-r_A)' = \frac{i_{ex}}{nF}\left[\exp\left\{\frac{(1-\alpha)nF\eta}{RT}\right\} - \exp\left\{-\frac{\alpha nF\eta}{RT}\right\}\right] \tag{9.13}$$

And, from Equation 9.7,

$$-r_A = \frac{Ar_{el}i_{ex}}{VnF}\left[\exp\left\{\frac{(1-\alpha)nF\eta}{RT}\right\} - \exp\left\{-\frac{\alpha nF\eta}{RT}\right\}\right] \tag{9.14}$$

In the Butler–Volmer equation, the first exponential term represents the rate of forward reaction and the second exponential term represents the rate of the backward reaction. Practically, it is always desirable to have the rate of backward reaction as low as possible. To this end, the industrial processes are often carried out at significant overpotential. Under this condition, the second exponential term in the Butler–Volmer equation can be neglected and the rate of reaction is approximated as

$$(-r_A)' = \frac{i_{ex}}{nF}\exp\left\{\frac{(1-\alpha)nF\eta}{RT}\right\} \tag{9.15}$$

And

$$-r_A = \frac{Ar_{el}i_{ex}}{VnF}\exp\left\{\frac{(1-\alpha)nF\eta}{RT}\right\} \tag{9.16}$$

These equations are known as Tafel equations (Tafel, 1905).

As the overpotential is increased, the current, and hence the rate of electrolytic reaction, acquires a limiting value and becomes independent of overpotential. This happens when the overall electrode process becomes mass transfer limited. Now, the rate is not governed by the availability of sufficiently energetic electrons at the electrode surface. Rather, it is governed by the availability of reactant moiety at the electrode surface, which is limited by the rate at which these moieties are transferred from the bulk of the electrolyte to the electrode surface. Assuming Whitman's film theory (Whitman, 1923) to be valid, the limiting current under mass transfer control is given by

$$I_{lim} = \frac{nFDAr_{el}(C_A)_{bulk}}{\delta} \tag{9.17}$$

$$\Rightarrow i_{lim} = \frac{nFD(C_A)_{bulk}}{\delta} \tag{9.18}$$

From Equations 9.6 and 9.7,

$$(-r_A)' = \frac{i_{\lim}}{nF} = \frac{D(C_A)_{bulk}}{\delta} \tag{9.19}$$

And

$$-r_A = \frac{Ar_{el}i_{\lim}}{VnF} = \frac{Ar_{el}D(C_A)_{bulk}}{V\delta} \tag{9.20}$$

9.1.2 Constant Volume Batch Electrochemical Reactors

For the design and performance evaluation of batch electrochemical reactors, we can use Equation 5.6 (Chapter 5) in combination with any of the rate equations developed above. When the reaction is carried out under conditions of negligible mass transfer resistance, Equation 9.14 is to be used. In this case,

$$(C_A)_0 \frac{dX_A}{dt} = -r_A = \frac{Ar_{el}i_{ex}}{VnF}\left[\exp\left\{\frac{(1-\alpha)nF\eta}{RT}\right\} - \exp\left\{-\frac{\alpha nF\eta}{RT}\right\}\right] \tag{9.21}$$

When the potential is kept constant during electrolysis, Equation 9.21 can be integrated by separating the variables, so that

$$(C_A)_0 \int_0^{X_A} dX_A = \frac{Ar_{el}i_{ex}}{VnF}\left[\exp\left\{\frac{(1-\alpha)nF\eta}{RT}\right\} - \exp\left\{-\frac{\alpha nF\eta}{RT}\right\}\right]\int_0^t dt$$

$$=> \quad X_A = \frac{Ar_{el}i_{ex}t}{(C_A)_0 VnF}\left[\exp\left\{\frac{(1-\alpha)nF\eta}{RT}\right\} - \exp\left\{-\frac{\alpha nF\eta}{RT}\right\}\right] \tag{9.22}$$

$$=> \quad X_A = \frac{I_{ex}t}{(C_A)_0 VnF}\left[\exp\left\{\frac{(1-\alpha)nF\eta}{RT}\right\} - \exp\left\{-\frac{\alpha nF\eta}{RT}\right\}\right] \tag{9.23}$$

In mass transfer control regime, combining Equation 5.6 (Chapter 5) with Equation 9.20 gives

$$(C_A)_0 \frac{dX_A}{dt} = -r_A = \frac{Ar_{el}D(C_A)_{bulk}}{V\delta}$$

$$=> \quad (C_A)_0 \frac{dX_A}{dt} = \frac{Ar_{el}D(C_A)_0(1-X_A)}{V\delta}$$

$$=> \quad \int_0^{X_A} \frac{dX_A}{(1-X_A)} = \frac{Ar_{el}D}{V\delta}\int_0^t dt$$

$$=> \quad \ln(1-X_A) = -\frac{Ar_{el}Dt}{V\delta} \tag{9.24}$$

$$\Rightarrow X_A = 1 - \exp\left(-\frac{Ar_{el}Dt}{V\delta}\right) \tag{9.25}$$

It is important to note that conversion under mass transfer control is independent of initial reactant concentration provided that the diffusivity is not appreciably affected by concentration.

For operational reasons, in most of the practical electrochemical reactors, the current is kept constant and the cell voltage is controlled through a suitable power-conditioning device. In such a case, Equation 5.6 (Chapter 5) can be directly combined with Equation 9.7 to give

$$(C_A)_0 \frac{dX_A}{dt} = -r_A = \frac{Ar_{el}i}{VnF}$$

$$\Rightarrow \quad (C_A)_0 \int_0^{X_A} dX_A = \frac{Ar_{el}i}{VnF} \int_0^t dt$$

$$\Rightarrow \quad X_A = \frac{Ar_{el}it}{(C_A)_0 VnF} \tag{9.26}$$

$$\Rightarrow X_A = \frac{It}{(C_A)_0 VnF} \tag{9.27}$$

Equation 9.27 is nothing but a reassertion of Faraday's law.

EXAMPLE 9.2

Nickel is to be electrodeposited on a stainless- steel plate. For this we prepare a 50-L solution of nickel salt so as to have 2 M concentration of Ni^{2+} and H^+ concentration of 1 M. In the batch electrolytic cell, the cathode is made up of the stainless-steel plate while the anode is made of a platinum-coated titanium sheet. The electrode reactions are

At cathode

$$Ni^{2+}(aq) + 2e^- \rightleftharpoons Ni(s) \quad E^0 = -0.25\,V \text{ versus SHE}$$

At anode

$$H_2O(l) \rightleftharpoons 1/2O_2(g) + 2H^+(aq) + 2e^- \quad E^0 = 1.23\,V \text{ versus SHE}$$

O_2 is released at the anode at 0.5 atm pressure.

The ohmic resistance of the cell is 0.5 Ω. The cathode and anode overpotentials are 0.5 V. The cell is operated at 25°C. Calculate

a. Cell voltage required for an initial current of 1 A.
b. Time required to achieve 80% conversion assuming that the current is kept constant by continuously increasing the cell voltage.

Solution

a. For the cathode reaction, from Equation 9.3

$$E_{cathode} = E^0_{cathode} + \frac{RT}{nF}\ln\frac{a_{Ni(s)}}{a_{Ni^{2+}(aq)}}$$

$$\Rightarrow \quad E_{cathode} = E^0_{cathode} + \frac{RT}{nF}\ln\frac{1}{a_{Ni^{2+}(aq)}} \quad \text{(activity of pure solid is 1)} \qquad (9.2.1)$$

Similarly, for the anode

$$E_{anode} = E^0_{anode} + \frac{RT}{nF}\ln\frac{a_{H_2O(l)}}{a^{0.5}_{O_2}a^2_{H^+}}$$

$$\Rightarrow \quad E_{anode} = E^0_{anode} + \frac{RT}{nF}\ln\frac{1}{a^{0.5}_{O_2}a^2_{H^+}} \quad \text{(Water in excess is treated as pure water)}$$

(9.2.2)

From Equations 9.2.1, 9.2.2, and 9.4,

$$E_{cell} = E_{anode} - E_{cathode} = E^0_{anode} - E^0_{cathode} + \frac{RT}{nF}\ln\frac{a_{Ni^{2+}(aq)}}{a^{0.5}_{O_2}a^2_{H^+}}$$

$$\Rightarrow \quad E_{cell} = 1.23-(-0.25)+\frac{2.303RT}{nF}\log\frac{C_{Ni^{2+}(aq)}}{p^{0.5}_{O_2}C^2_{H^+}}$$

$$\Rightarrow \quad E_{cell} = 1.48+\frac{0.059}{2}\log\frac{2}{0.5^{0.5}\times 1} = 1.493\,\text{V}$$

This is the minimum cell voltage required in the absence of overpotentials and ohmic resistance. Therefore

$$(E_{cell})_{total} = E_{cell} + IR + \eta_{cathode} + \eta_{anode}$$

$$\Rightarrow (E_{cell})_{total} = 1.493 + (1\times 0.5) + 0.5 + 0.5 = 2.993\,\text{V}$$

b. From Equation 9.27,

$$X_A = \frac{It}{(C_A)_0 VnF}$$

$$\Rightarrow \quad t = \frac{(C_A)_0 VnFX_A}{I} = \frac{2\times 50\times 2\times 96487\times 0.8}{1} = 15437920\,\text{s} = 4288.3\,\text{h}$$

9.1.3 Electrochemical Reactor with Ideal Plug Flow of Electrolyte

For the design and performance evaluation of plug flow electrochemical reactors we can use Equation 6.1 (Chapter 6) in combination with any of the rate equations for electrochemical reaction developed above. Under conditions of negligible mass transfer resistance and constant potential we have, from Equations 9.14 and 6.1 (Chapter 6),

$$V_{PFR} = (F_A)_0 \int_0^{X_A} \frac{dX_A}{-r_A}$$

$$\Rightarrow \quad V_{PFR} = (F_A)_0 \int_0^{X_A} \frac{dX_A}{(Ar_{el}i_{ex}/V_{PFR}nF)[\exp\{(1-\alpha)nF\eta/RT\} - \exp\{-\alpha nF\eta/RT\}]}$$

$$\Rightarrow \quad X_A = \frac{Ar_{el}i_{ex}}{(F_A)_0 nF}\left[\exp\left\{\frac{(1-\alpha)nF\eta}{RT}\right\} - \exp\left\{-\frac{\alpha nF\eta}{RT}\right\}\right] \tag{9.28}$$

The equation can be used to predict conversion for a given electrode area and molar flow rate of reactant. For design purposes, it can be used to estimate the electrode area required to achieve a target conversion for a given molar feed rate.

Under mass transfer control regime, Equation 9.20 is combined with Equation 6.1 (Chapter 6). Thus,

$$V_{PFR} = (F_A)_0 \int_0^{X_A} \frac{dX_A}{Ar_{el}D(C_A)_{bulk}/V_{PFR}\delta}$$

$$\Rightarrow \quad \frac{Ar_{el}D}{(F_A)_0\delta} = \int_0^{X_A} \frac{dX_A}{(C_A)_0(1-X_A)}$$

$$\Rightarrow \quad \ln(1-X_A) = -\frac{Ar_{el}D(C_A)_0}{(F_A)_0\delta} = -\frac{Ar_{el}D}{v\delta} \tag{9.29}$$

$$\Rightarrow X_A = 1 - \exp\left(-\frac{Ar_{el}D}{v\delta}\right) \tag{9.30}$$

For electrochemical PFR also the conversion is independent of the initial reactant concentration and depends only on the electrode area and volumetric flow rate.

For electrolysis under constant current conditions,

$$V_{PFR} = (F_A)_0 \int_0^{X_A} \frac{dX_A}{Ar_{el}i/V_{PFR}nF}$$

$$\Rightarrow \quad X_A = \frac{Ar_{el}i}{(F_A)_0 nF} = \frac{I}{(F_A)_0 nF} \tag{9.31}$$

9.1.4 Electrochemical Reactor with Ideal Backmix Flow of Electrolyte

Here we can use Equation 6.86 (Chapter 6) with the rate equations for an electrochemical reaction. For negligible mass transfer resistance, this gives

$$V_{BFR} = \frac{(F_A)_0 X_A}{-r_A} = \frac{(F_A)_0 X_A}{(Ar_{el}i_{ex}/V_{BFR}nF)[\exp\{(1-\alpha)nF\eta/RT\} - \exp\{-\alpha nF\eta/RT\}]}$$

$$\Rightarrow \quad X_A = \frac{Ar_{el}i_{ex}}{(F_A)_0 nF}\left[\exp\left\{\frac{(1-\alpha)nF\eta}{RT}\right\} - \exp\left\{-\frac{\alpha nF\eta}{RT}\right\}\right] \tag{9.32}$$

This equation can be used to predict conversion for a given electrode area and molar flow rate of reactant. For design purposes, it can be used to estimate the electrode area required to achieve a target conversion for a given molar feed rate.

For conditions of mass transfer control, we have

$$V_{BFR} = \frac{(F_A)_0 X_A}{-r_A} = \frac{(F_A)_0 X_A}{Ar_{el}D(C_A)_{bulk}/V_{BFR}\delta}$$

$$\Rightarrow \quad X_A = \frac{Ar_{el}D(C_A)_0(1-X_A)}{(F_A)_0\delta} = \frac{Ar_{el}D(1-X_A)}{v\delta}$$

$$\Rightarrow \quad X_A = \frac{Ar_{el}D}{v\delta + Ar_{el}D} \tag{9.33}$$

Under conditions of constant current, we have

$$V_{BFR} = \frac{(F_A)_0 X_A}{-r_A} = \frac{(F_A)_0 X_A}{Ar_{el}i/V_{BFR}nF}$$

$$\Rightarrow \quad X_A = \frac{Ar_{el}i}{(F_A)_0 nF} = \frac{I}{(F_A)_0 nF} \tag{9.34}$$

It is interesting to note that Equation 9.31 for plug flow electrochemical reaction and Equation 9.34 for backmix flow electrochemical reactor are identical. This is not unexpected as under constant current conditions the rate is independent of reactant concentration. However, a complete backmix flow reactor will have intense agitation of the electrolyte so that chances of mass transfer control with a limiting constant current are very remote. For these reactors, Equation 9.32 is most likely to be applicable.

9.2 BIOCHEMICAL REACTORS

Biochemical reactions are enzyme-catalyzed chemical reactions. All life processes are based on these reactions. Like all catalyzed reactions, enzymes, which are biocatalysts, provide a low activation energy pathway for the reactant molecules to change into product molecules. For this to happen, the reactant molecule—known as a substrate in biochemical engineering terminology—binds itself to one of the active sites present on the enzyme. This excites the substrate molecule and facilitates its transformation into the product molecule. Subsequently, the product molecule detaches from the active site and makes it available for another substrate molecule to bind. Although it is only recently that we have made some headway into understanding the basics of these reactions, they have been used by humankind for ages. Beer, wine, vinegar, curd, etc., are just a few examples. In industrial application of biochemical reactions, the metabolic processes of microorganisms are channelized to useful ends. The reactors used for these purposes are known as biochemical reactors.

9.2.1 Biochemical Reaction Fundamentals

Essentially, all biochemical reactions are enzyme-mediated chemical transformations of substrate molecules. Besides catalytic activity, the most striking aspect of enzymes are

1. Specificity. Enzymes are highly specific in their activity. Specificity can be of the following types:
   a. Absolute specificity. Enzyme catalyzes only a particular substrate—product transformation.
   b. Group specificity. Enzyme acts on molecules with a specific functional group.
   c. Linkage specificity. Enzyme acts on a particular type of chemical bond; say, a double bond, for example.
   d. Stereochemical specificity. Enzyme activity is specific to a particular steric or optical isomer.
2. Regulation. The action of enzymes is regulated by a complex array of positive and negative feedbacks, which allows precise control over the rate of reaction. These are of various kinds.
   a. Inhibition. Catalytic action of enzymes is suppressed by substances called enzyme inhibitors. They are of different types. Irreversible inhibitors or poisons completely deactivate the enzyme. A competitive inhibitor has a shape similar to that of the substrate molecule so that it competes with substrate molecules for binding with the active sites on the enzyme. Besides the active sites, there are other sites on the enzyme—known as the allosteric sites—where some molecule other than the substrate molecule can bind. Even though the active sites of the enzyme are free, the catalytic activity of the enzyme is modified by such bindings on the allosteric sites. The entities that bind on the allosteric sites are called effectors or modifiers. If the effect is that of increasing the enzyme activity, it is known as allosteric activation. The reverse is known as allosteric inhibition.
   b. Cooperation. If the binding of one substrate molecule positively affects the binding of subsequent substrate molecules, the phenomenon is known as cooperation.

9.2.1.1 Kinetics of Biochemical Reaction

Experiments suggest that biochemical reactions do not follow the law of mass action. So, these are nonelementary reactions. Michaelis and Menten (Johnson and Goody, 2011; Michaelis and Menten, 1913), in the beginning of twentieth century, were the first to propose a mechanism for the biochemical reaction in terms of the enzyme substrate interaction. Their proposed mechanism is as follows:

$$S + E \underset{k_2}{\overset{k_1}{\rightleftharpoons}} S.E \tag{9.35}$$

$$S.E \underset{k_4}{\overset{k_3}{\rightleftharpoons}} P + E \tag{9.36}$$

The overall reaction being

$$S + E \rightarrow P + E \tag{9.37}$$

For each elementary reaction step in Equations 9.35 and 9.36, we can apply the law of mass action to have

$$-\frac{dC_S}{dt} = k_1C_SC_E - k_2C_{S.E} \tag{9.38}$$

$$\frac{dC_{S.E}}{dt} = k_1C_SC_E - k_2C_{S.E} - k_3C_{S.E} + k_4C_PC_E \tag{9.39}$$

$$\frac{dC_P}{dt} = k_3C_{S.E} - k_4C_PC_E \tag{9.40}$$

Michaelis and Menten assumed that the substrate is in instantaneous equilibrium with the enzyme substrate complex, so that

$$k_1C_SC_E = k_2C_{S.E}$$

$$\Rightarrow \quad C_{S.E} = \frac{k_1}{k_2}C_SC_E = C_SC_E/K_1 \tag{9.41}$$

Since the total concentration of the enzyme remains constant, we have

$$C_E = (C_E)_0 - C_{S.E} \tag{9.42}$$

Combining Equations 9.41 and 9.42,

$$C_{S.E} = \frac{(C_E)_0 C_S}{K_1 + C_S} \tag{9.43}$$

where $K_1 = k_2/k_1$. If $k_4 \ll k_3$, that is to say that reaction 9.36 is practically irreversible, we have from Equation 9.40

$$r_P = \frac{dC_P}{dt} = \frac{k_3(C_E)_0 C_S}{K_1 + C_S} = \frac{V_{max} C_S}{K_1 + C_S} \tag{9.44}$$

Here, $V_{max} = k_3(C_E)_0$ is the maximum rate of formation of P when the entire enzyme is attached to the substrate. From the stoichiometry of reaction 9.37

$$-r_S = r_P$$

$$=> \quad -r_S = -\frac{dC_S}{dt} = \frac{dC_P}{dt} = \frac{V_{max} C_S}{K_1 + C_S} \tag{9.45}$$

This rate equation can also be deduced using the quasi-steady-state approximation, which gives

$$-r_S = -\frac{dC_S}{dt} = \frac{dC_P}{dt} = \frac{V_{max} C_S}{K_m + C_S} \tag{9.46}$$

where $K_m = (k_2 + k_3)/k_1$. When $k_2 \gg k_3$, $K_m \simeq K_1$ and Equations 9.45 and 9.46 are equivalent.

From Equation 9.46,

$$\frac{1}{-r_S} = \frac{1}{V_{max}} + \frac{K_m}{V_{max}} \frac{1}{C_S} \tag{9.47}$$

This equation predicts that a plot between $1/-r_S$ and $1/C_S$ is a straight line with slope K_m/V_{max} and the y-axis intercept $1/V_{max}$. These plots are known as Lineweaver–Burk plots (Lineweaver and Burk, 1934) and are useful in experimental determination of the kinetic parameters of biochemical reactions.

We can see from Equation 9.46 that when $K_m = C_S$, $-r_S = V_{max}/2$. Therefore, K_m is numerically equal to that substrate concentration when the reaction rate is half the maximum reaction rate.

EXAMPLE 9.3

From the following data given for a biochemical reaction, calculate the kinetic parameters V_{max} and K_m.

Substrate concentration (mg/L)	500	400	300	200	100	50	20	10
Reaction rate (mg/(L·h))	0.6	0.55	0.5	0.46	0.3	0.2	0.1	0.05

TABLE EXE9.3
Calculation Table Lineweaver Burk Plot

$1/C_S$	0.002	0.0025	0.003333	0.005	0.01	0.02	0.05	0.1
$1/-r_S$	1.666667	1.818182	2	2.173913	3.333333	5	10	20

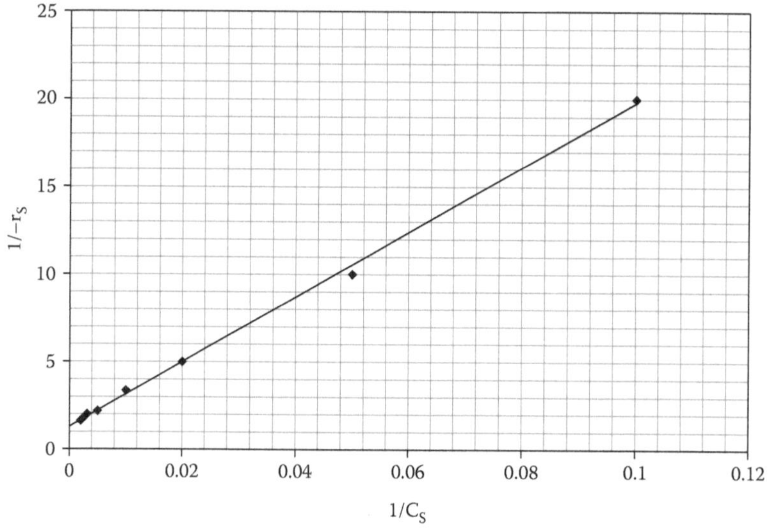

FIGURE EXE9.3 Lineweaver Burk plot.

Solution

We are given the data for substrate concentration, C_S, and reaction rate, $-r_S$. First, we calculate the quantities $1/-r_S$, and $1/C_S$. These are tabulated in Table EXE9.3.

Next, we plot $1/-r_S$ against $1/C_S$ and fit a straight line through the plotted data points. This is shown in Figure EXE9.3.

From Equation 9.47,

$$\frac{1}{V_{max}} = 1.3$$

$$=> \quad V_{max} = 0.77\,h^{-1}$$

$$\frac{K_m}{V_{max}} = \frac{20-5}{0.1-.02} = 187.5$$

$$=> \quad K_m = 187.5 \times 0.77 = 144.4\,mg/L$$

9.2.1.2 Kinetics of Microbial Growth

When a small number of microorganisms (inoculum) are placed in a comparatively large amount of nutrients (substrate), the following distinct stages of growth are observed:

1. The lag phase. An initial short period when the microorganisms acquaint themselves with the new environment.

2. The log phase. The substrate is in abundance and the rate of growth is independent of substrate concentration and only depends on the microbial ability to process their food, the substrate. Microorganisms are young and death is minimal. In essence, the rate of growth is directly proportional to the microbial concentration (mass per unit volume) so that

$$\frac{dC_M}{dt} = \mu C_M \tag{9.48}$$

$$=> C_M = (C_M)_0 e^{\mu t} \tag{9.49}$$

3. The stationary phase. Here the microbial population remains stationary as the production of new microorganisms is balanced by the death of old ones.
4. Endogenous phase. Substrate is depleted and the rate of death of old microorganisms far exceeds the appearance of new ones.

9.2.1.2.1 *Substrate-Limited Growth*

When the availability of substrate is limited, the specific growth rate μ in Equation 9.48 is no longer independent of substrate concentration. The dependence of μ on substrate concentration is given by Monod's equation (Monod, 1949)

$$\mu = \frac{\mu_{max} C_S}{K_m + C_S} \tag{9.50}$$

$$=> \quad \frac{dC_M}{dt} = \frac{\mu_{max} C_S C_M}{K_m + C_S} \tag{9.51}$$

9.2.1.2.2 *Net Microbial Growth Rate*

Equation 9.51 represents the appearance of new cells as a result of consumption of substrate. At the same time, some microorganisms also die so that the net rate of microbial growth is less than that expressed by Equation 9.51. The death rate (endogenous decay) is found to be proportional to the concentration of microbial mass present at a time. Hence,

$$\left(\frac{dC_M}{dt}\right)_{dth} = -k_{dth} C_M \tag{9.52}$$

Combining Equations 9.51 and 9.52,

$$\left(\frac{dC_M}{dt}\right)_{net} = \frac{dC_M}{dt} + \left(\frac{dC_M}{dt}\right)_{dth}$$

$$=> \quad \left(\frac{dC_M}{dt}\right)_{net} = \frac{\mu_{max} C_S C_M}{K_m + C_S} - k_{dth} C_M \tag{9.53}$$

$$=> \left(\frac{dC_M}{dt}\right)_{net} = \left[\frac{\mu_{max}C_S}{K_m + C_S} - k_{dth}\right]C_M \tag{9.54}$$

9.2.1.2.3 Substrate Utilization

All the substrate that is consumed by the microorganisms does not appear as new microbial mass. A part of it is metabolized by the microorganisms to sustain their life processes. The rate of disappearance of substrate and the rate of growth of microorganisms are related as

$$\frac{dC_M}{dt} = Y\left(-\frac{dC_S}{dt}\right) \tag{9.55}$$

$$=> -\frac{dC_S}{dt} = \frac{1}{Y}\frac{dC_M}{dt} = \frac{\mu_{max}C_SC_M}{Y(K_m + C_S)} \tag{9.56}$$

where Y is the yield coefficient. From Equation 9.53

$$\left(\frac{dC_M}{dt}\right)_{net} = Y\left(-\frac{dC_S}{dt}\right) - k_{dth}C_M \tag{9.57}$$

9.2.2 Batch Biochemical Reactor

For a singular enzyme substrate reaction—a purified extracellular enzyme acting on a specific substrate—Equation 9.46 is used in conjunction with Equation 5.6 (Chapter 5). Thus,

$$(C_S)_0\frac{dX_S}{dt} = -r_S = \frac{V_{max}C_S}{K_m + C_S}$$

$$=> -\frac{dC_S}{dt} = \frac{V_{max}C_S}{K_m + C_S}$$

$$=> -\int_{(C_S)_0}^{C_S}\left[\frac{K_m}{V_{max}C_S} + \frac{1}{V_{max}}\right]dC_S = \int_0^t dt = t$$

$$=> K_m \ln\frac{(C_S)_0}{C_S} + (C_S)_0 - C_S = V_{max}t \tag{9.58}$$

For substrate-limited growth in batch biochemical reactors from Equation 9.55

$$\frac{dC_M}{dt} = Y\left(-\frac{dC_S}{dt}\right)$$

$$=> C_M - (C_M)_0 = Y[(C_S)_0 - C_S]$$

$$=> C_S = (C_S)_0 - \frac{C_M - (C_M)_0}{Y} = \frac{Y(C_S)_0 - C_M + (C_M)_0}{Y} \tag{9.59}$$

If the initial microorganism concentration is negligible, Equation 9.59 reduces to

$$C_S = (C_S)_0 - \frac{C_M}{Y} \tag{9.60}$$

Combining Equations 9.54 and 9.60,

$$\frac{dC_M}{dt} = \left[\frac{\mu_{max} C_S}{K_m + C_S} - k_{dth}\right] C_M$$

$$=> \int_{(C_M)_0}^{C_M} \frac{dC_M}{(\mu_{max} C_S C_M/(K_m + C_S)) - k_{dth} C_M} = \int_0^t dt = t$$

$$=> \int_{(C_M)_0}^{C_M} \frac{YK_M + Y(C_S)_0 - C_M}{\mu_{max} Y(C_S)_0 C_M - \mu_{max} C_M^2 - Yk_{dth} K_M C_M - k_{dth} Y(C_S)_0 C_M + k_{dth} C_M^2} dC_M = t$$

$$=> \int_{(C_M)_0}^{C_M} \frac{Y\{K_M + (C_S)_0\} - C_M}{\{\mu_{max} Y(C_S)_0 - Yk_{dth} K_M - k_{dth} Y(C_S)_0\} C_M + \{k_{dth} - \mu_{max}\} C_M^2} dC_M = t$$

$$=> \int_{(C_M)_0}^{C_M} \frac{K_1 - C_M}{K_2 C_M + K_3 C_M^2} dC_M = t$$

$$=> \int_{(C_M)_0}^{C_M} \frac{K_1}{K_2 C_M + K_3 C_M^2} dC_M - \int_{(C_M)_0}^{C_M} \frac{C_M}{K_2 C_M + K_3 C_M^2} dC_M = t$$

$$=> \int_{(C_M)_0}^{C_M} \frac{K_1}{K_2 C_M} dC_M - \int_{(C_M)_0}^{C_M} \frac{K_1 K_3}{K_2(K_2 + K_3 C_M)} dC_M - \int_{(C_M)_0}^{C_M} \frac{1}{(K_2 + K_3 C_M)} dC_M = t$$

$$=> \frac{K_1}{K_2} \ln \frac{C_M}{(C_M)_0} - \left(\frac{K_1}{K_2} + \frac{1}{K_3}\right) \ln \frac{K_2 + K_3 C_M}{K_2 + K_3 (C_M)_0} = t \tag{9.61}$$

where

$$K_1 = Y\{K_M + (C_S)_0\} \tag{9.62}$$

$$K_2 = \mu_{max} Y(C_S)_0 - Yk_{dth} K_M - k_{dth} Y(C_S)_0 \tag{9.63}$$

$$K_3 = k_{dth} - \mu_{max} \tag{9.64}$$

9.2.3 Plug Flow Biochemical Reactor

For a singular enzyme, substrate reaction Equation 9.46 is used in conjunction with Equation 6.6 (Chapter 6). Thus,

$$\tau_{PFR} = (C_S)_0 \int_0^{(X_S)} \frac{dX_S}{-r_S}$$

$$=> \quad \tau_{PFR} = -\int_{(C_S)_0}^{C_S} \frac{dC_S}{-r_S} = \int_{C_S}^{(C_S)_0} \frac{(K_m + C_S)dC_S}{V_{max}C_S}$$

$$=> \quad \tau_{PFR} = \int_{C_S}^{(C_S)_0} \left[\frac{K_m}{V_{max}C_S} + \frac{1}{V_{max}}\right] dC_S$$

$$=> \quad V_{max}\tau_{PFR} = K_m \ln\frac{(C_S)_0}{C_S} + (C_S)_0 - C_S \tag{9.65}$$

For substrate-limited growth in plug flow biochemical reactors, from Equations 9.54, 9.60, and 6.6 (Chapter 6),

$$\tau_{PFR} = \int_{(C_M)_0}^{C_M} \frac{dC_M}{r_M} = \int_{(C_M)_0}^{C_M} \frac{dC_M}{(\mu_{max}C_SC_M/(K_m + C_S)) - k_{dth}C_M}$$

$$=> \quad \tau_{PFR} = \int_{(C_M)_0}^{C_M} \frac{YK_M + Y(C_S)_0 - C_M}{\mu_{max}Y(C_S)_0C_M - \mu_{max}C_M^2 - Yk_{dth}K_MC_M - k_{dth}Y(C_S)_0C_M + k_{dth}C_M^2} dC_M$$

$$=> \quad \tau_{PFR} = \frac{K_1}{K_2}\ln\frac{C_M}{(C_M)_0} - \left(\frac{K_1}{K_2} + \frac{1}{K_3}\right)\ln\frac{K_2 + K_3C_M}{K_2 + K_3(C_M)_0} \tag{9.66}$$

For substrate consumption, from Equation 6.14,

$$\tau_{PFR} = -\int_{(C_S)_0}^{C_S} \frac{1}{(-r_S)} dC_S = \int_{C_S}^{(C_S)_0} \frac{1}{(-r_S)} dC_S \tag{9.67}$$

EXAMPLE 9.4

A microorganism culture has the following kinetic parameters:

$$\mu_{max} = 1.8\,d^{-1},\ K_m = 100\,mg/L,\ Y = 0.5\,mg/mg,\ k_{dth} = 0.05\,d^{-1}$$

This culture is to be grown in a biochemical PFR so as to have an outlet microorganism concentration of 2500 mg/L. The initial concentration of the microorganisms is 10 mg/L. The concentration of limiting substrate in the feed is 7000 mg/L. How much space time of the PFR would be required? What is the outgoing substrate concentration?

SOLUTION

From Equation 9.62,

$$K_1 = Y\{K_M + (C_S)_0\} = 0.5 \times (100 + 7000) = 3550$$

From Equation 9.63,

$$K_2 = \mu_{max} Y (C_S)_0 - Y k_{dth} K_M - k_{dth} Y (C_S)_0$$

$$=> \quad K_2 = 1.8 \times 0.5 \times 7000 - 0.5 \times 0.05 \times 100 - 0.05 \times 0.5 \times 7000 = 6122.5$$

From Equation 9.64,

$$K_3 = k_{dth} - \mu_{max} = 0.05 - 1.8 = -1.75$$

From Equation 9.66,

$$\tau_{PFR} = \frac{K_1}{K_2} \ln \frac{C_M}{(C_M)_0} - \left(\frac{K_1}{K_2} + \frac{1}{K_3} \right) \ln \frac{K_2 + K_3 C_M}{K_2 + K_3 (C_M)_0}$$

$$=> \quad \tau_{PFR} = \frac{3550}{6122.5} \ln \frac{2500}{10} - \left(\frac{3550}{6122.5} - \frac{1}{1.75} \right) \ln \frac{6122.5 - 1.75 \times 2500}{6122.5 - 1.75 \times 10}$$

$$=> \quad \tau_{PFR} = 3.2015 + 0.0105 = 3.2\,d$$

From Equation 9.60,

$$C_S = (C_S)_0 - \frac{C_M - (C_M)_0}{Y}$$

$$=> \quad C_S = 7000 - \frac{2500 - 10}{0.5} = 2020\,mg/L$$

9.2.4 BACKMIX FLOW BIOCHEMICAL REACTOR

For a singular enzyme substrate reaction Equation 9.46 is used in conjunction with Equation 6.88 (Chapter 6). Thus,

$$\tau_{BFR} = \frac{(C_S)_0 X_S}{-r_s} = \frac{\{(C_S)_0 - C_S\}(K_m + C_S)}{V_{max} C_S} \tag{9.68}$$

$$=> \quad C_S^2 + \{V_{max} + K_m - (C_S)_0\}C_S - K_m(C_S)_0 = 0$$

$$=> \quad C_S = \frac{1}{2}\left[\{(C_S)_0 - V_{max}\tau_{BFR} - K_m\} + \sqrt{\{(C_S)_0 - V_{max}\tau_{BFR} - K_m\}^2 + 4K_m(C_S)_0}\right] \tag{9.69}$$

For substrate-limited growth in backmix flow biochemical reactors, from Equations 9.54, 9.60, and 6.130 (Chapter 6),

$$\tau_{BFR} = \frac{\{C_M - (C_M)_0\}}{r_M} = \frac{C_M}{\mu_{max}C_S C_M/(K_m + C_S) - k_{dth}C_M} \quad \text{(when } (C_M)_0 \text{ is negligible)}$$

$$=> \quad \frac{1}{\tau_{BFR}} = \frac{\mu_{max}C_S}{K_m + C_S} - k_{dth}$$

$$=> \quad \frac{1}{\tau_{BFR}} = \frac{\mu_{max}C_S}{K_m + C_S} - k_{dth} \tag{9.70}$$

$$=> \quad \frac{1}{\tau_{BFR}} = \frac{\mu_{max}\{Y(C_S)_0 - C_M\}}{YK_m + Y(C_S)_0 - C_M} - k_{dth} \tag{9.71}$$

Rearranging and solving for C_M we get

$$C_M = \frac{\tau_{BFR}\mu_{max}Y(C_S)_0 - \tau_{BFR}YK_m k_{dth} - \tau_{BFR}Yk_{dth}(C_S)_0 - YK_m - Y(C_S)_0}{\tau_{BFR}\mu_{max} - \tau_{BFR}k_{dth} - 1}$$

$$=> \quad C_M = \frac{Y(C_S)_0(\tau_{BFR}\mu_{max} - \tau_{BFR}k_{dth} - 1) - \tau_{BFR}YK_m k_{dth} - YK_m}{\tau_{BFR}\mu_{max} - \tau_{BFR}k_{dth} - 1}$$

$$=> \quad C_M = Y(C_S)_0 - \frac{\tau_{BFR}YK_m k_{dth} + YK_m}{\tau_{BFR}\mu_{max} - \tau_{BFR}k_{dth} - 1} \tag{9.72}$$

Equation 9.72 suggests that a plot between the outgoing microbial concentration from the BFR and the inlet substrate concentration is a straight line with slope Y and y-axis intercept as $(-(\tau_{BFR}YK_m k_{dth} + YK_m/\tau_{BFR}\mu_{max} - \tau_{BFR}k_{dth} - 1))$. This allows for the estimation of the yield coefficient Y and death coefficient k_{dth} from BFR runs if the parameters μ_{max} and K_m are known.

For substrate consumption, from Equation 6.90 (Chapter 6), for biochemical reactions being liquid-phase reaction

$$\tau_{BFR} = \frac{(C_S)_0 - C_S}{-r_S} \tag{9.73}$$

EXAMPLE 9.5

What will be the space time required if the reaction of Example 9.4 is carried out in a biochemical BFR?

Solution

From the solution to Example 9.4,

$$C_S = 2020\,\text{mg}/\text{L}$$

From Equation 9.70,

$$\frac{1}{\tau_{BFR}} = \frac{\mu_{max} C_S}{K_m + C_S} - k_{dth} = \frac{1.8 \times 2020}{100 + 2020} - 0.05$$

$$\frac{1}{\tau_{BFR}} = 1.66\,\text{d}^{-1}$$

$$\Rightarrow \quad \tau_{BFR} = 0.6\,\text{d}$$

Comparing the results from Examples 9.4 and 9.5, we can see that unlike the normal nth-order reaction, a biochemical reaction may require less space time, hence a smaller reactor, if carried out in a BFR. This anomaly may be explained by the form of rate expression for microbial growth and substrate utilization. Biochemical reactions fall under the category of reactions known as autocatalytic reactions where the product itself acts as the catalyst—microorganisms are the product as well as the source of enzyme, the catalyst. Unlike an nth- order reaction, the rate of substrate utilization first increases, passes through a maximum, and then decreases as the substrate concentration is increased. Consequently, the $(1/-r_S)$ versus C_S plot passes through a minimum. An illustration of such behavior is shown in Figure 9.1. From Equation 9.67, the space time required for a biochemical

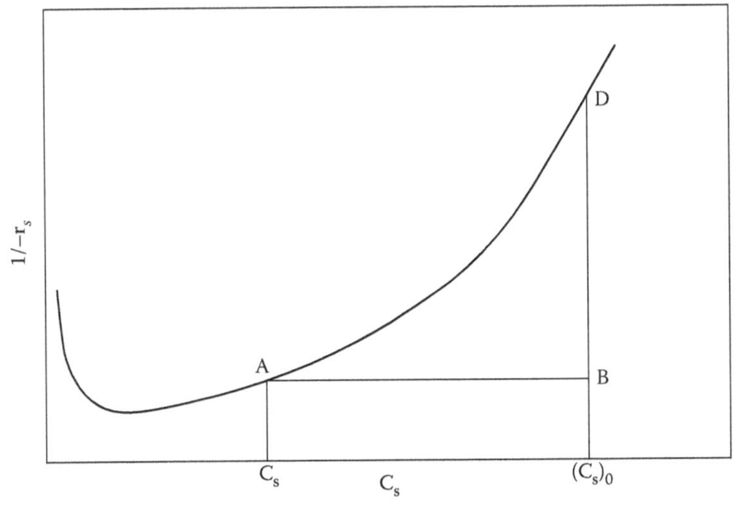

FIGURE 9.1 Substrate utilization rate as a function of substrate concentration.

PFR to reduce the substrate concentration from $(C_S)_0$ to C_S is the area $AC_S(C_S)_0D$. From Equation 9.73, the space time required for a biochemical BFR to reduce the substrate concentration from $(C_S)_0$ to C_S is the area $AC_S(C_S)_0B$. Clearly, less space time is required for the BFR as compared to the PFR. If a microbial BFR is operated at the substrate concentration corresponding to the maximum substrate utilization rate (minimum of $1/-r_S$), the volume of the reactor will be minimum. If the outgoing substrate concentration is more than that corresponding to the maximum rate, the required volume of BFR would be less than the volume required for PFR. Such behavior is expected for other autocatalytic reactions as well.

9.3 CATALYTIC PACKED BED REACTORS

Extensively used for fluid–solid catalytic reactions, a packed bed reactor is a fixed bed of catalyst particles—mostly granular in nature—through which the reactant fluid is passing. Their industrial uses are many; the following being a select list of industrial processes where packed bed reactors find application:

- Ammonia synthesis. Promoted iron catalysts
- Sulfuric acid production. The step converting SO_2 into SO_3. V_2O_5 or Pt catalyst
- Steam reforming of methane. Ni catalyst
- Catalytic reforming of naphtha. Supported Pt or Ru catalysts
- Fischer–Tropsch synthesis. Co, Fe, and Ru catalysts
- NO_x removal from flue gases
- Reactors for immobilized enzyme technology

The functioning of a packed bed reactor depends not only on the kinetic aspects of the process but also to a large extent on the hydrodynamic characteristic of the bed. An understanding on these is essential before we can discuss the performance and design of packed bed reactors. While the kinetic aspects of solid–fluid catalytic reactions have been discussed in Chapter 3, a brief description of the hydrodynamics of packed beds follows.

9.3.1 Pressure Drop in Packed Beds

Compared to an unrestricted open-flow path of an equivalent cross-section area, a fixed bed of particles offers significantly more resistance to the flow of fluid. As a result, a higher pressure drop must be maintained across the two sides of the bed to maintain the same volumetric flow rate. The higher the requirement of pressure drop, the more is the energy that must be spent by the fluid motive device (pump, fan, blower, etc.). Pressure drop estimation for the fixed bed, therefore, is important from the design and operational standpoint. For gaseous reactions, a drop in pressure also results in a drop in the local reactant concentration.

When a fluid flows past a solid, it experiences a drag force arising due to the friction between the solid and the fluid. This drag force always resists the flow of the fluid. Its magnitude depends on the shape of the solid, the relative velocity of the fluid with respect to the solid, the nature of the solid surface, and the physical properties of the fluid. When a fluid is flowing through a bed of granular solid particles, the total frictional resistance offered by the bed is the resultant of all the drag forces of individual particles constituting the bed. This is the most fundamental approach to calculating the frictional resistance and hence the pressure drop for flow of fluids through packed beds. However, such an approach is too rigorous for practical utility. Rather, the bed is visualized as a network of intricately interconnected narrow channels and the principles of fluid flow through conduits are applied. The estimation of friction factor, as done for flow through conduits, then becomes the guiding criterion for predicting the pressure drop through packed beds. Friction factor is defined as

$$f = \frac{\text{Wall shear stress}}{(\text{density}) \times (\text{velocity head})} \tag{9.74}$$

The most widely used correlation for the estimation of friction factor for packed beds is the Ergun equation (Ergun, 1952). In brief,

$$f_{PB} = \frac{\Delta p}{L_{PB}} \frac{d_p}{\rho_{fl} u_s^2} \left(\frac{\epsilon^3}{1-\epsilon} \right) = \frac{150}{Re_{PB}} + 1.75 \tag{9.75}$$

where

$$Re_{PB} = \frac{d_p u_s \rho_{fl}}{(1-\epsilon)\mu} \tag{9.76}$$

and

$$d_p = \frac{6V_p}{Ar_p} \tag{9.77}$$

From Equations 9.75 and 9.76,

$$\frac{\Delta p}{L_{PB}} = \frac{150 \rho_{fl} u_s^2}{Re_{PB} d_p} \left(\frac{1-\epsilon}{\epsilon^3} \right) + \frac{1.75 \rho_{fl} u_s^2}{d_p} \left(\frac{1-\epsilon}{\epsilon^3} \right)$$

$$\Rightarrow \frac{\Delta p}{L_{PB}} = \frac{150 \mu u_s}{d_p^2} \frac{(1-\epsilon)^2}{\epsilon^3} + \frac{1.75 \rho_{fl} u_s^2}{d_p} \left(\frac{1-\epsilon}{\epsilon^3} \right) \tag{9.78}$$

$$\Rightarrow \Delta p = \frac{150 \mu u_s L_{PB}}{d_p^2} \frac{(1-\epsilon)^2}{\epsilon^3} + \frac{1.75 \rho_{fl} u_s^2 L_{PB}}{d_p} \left(\frac{1-\epsilon}{\epsilon^3} \right) \tag{9.79}$$

The equation can be used to calculate the expected pressure drop for a given height of the packed bed.

9.3.2 Packed Bed Reactor Models

Just as in the case of modeling of any other process equipment, a model for a packed bed reactor must be

- Based on well-explained physical phenomenon
- Easy to use
- In reasonable conformity to experimental findings
- Based on not too complicated mathematics

There are numerous models available that claim to represent the packed bed reactor but no model is perfect. Traditionally, packed bed reactors have been explained based on "continuum models." Continuum models can be further classified into "pseudohomogeneous" and "heterogeneous" models. Pseudohomogeneous models neglect the fluid film mass and heat-transfer resistances. On the other hand, heterogeneous models take into account all the transport resistances. More recently, a more sophisticated approach known as the "wave model" has been proposed for the packed bed reactors. We present here two simple packed bed reactor models for isothermal operations.

9.3.2.1 Plug Flow Packed Bed Reactor Model

This is the simplest of the packed bed reactor models, which may also be termed a one-dimensional pseudohomogeneous model. Consider the schematic of a packed bed reactor shown in Figure 9.2.

For isothermal operation, writing a material balance for component A over the differential length element

$$C_A\Big|_L u_s A r_R = C_A\Big|_{L+dL} u_s A r_R + (-r_A)' dA r_{cat}$$

$$\Rightarrow \quad C_A\Big|_L u_s = C_A\Big|_{L+dL} u_s + (-r_A)' \frac{a A r_R dL}{A r_R}$$

$$\Rightarrow \quad u_s \frac{C_A\Big|_L - C_A\Big|_{L+dL}}{dL} = (-r_A)' a$$

$$\Rightarrow \quad -u_s \frac{dC_A}{dL} = (-r_A)' a \tag{9.80}$$

$$\Rightarrow \quad L_{PB} = -\frac{u_s}{a} \int_{(C_A)_0}^{C_A} \frac{dC_A}{(-r_A)'} \tag{9.81}$$

$$\Rightarrow \quad a L_{PB} A r_R = -u_s A r_R \int_{(C_A)_0}^{C_A} \frac{dC_A}{(-r_A)'}$$

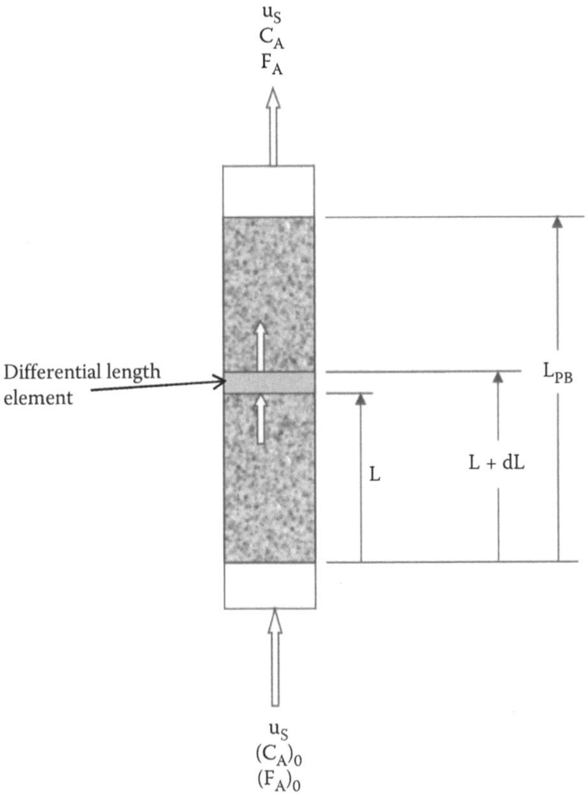

FIGURE 9.2 Schematic representation of a packed bed reactor.

$$=> \quad Ar_{cat} = -v \int_{(C_A)_0}^{C_A} \frac{dC_A}{(-r_A)'} \tag{9.82}$$

$$=> V_{cat} = -\frac{v}{a} \int_{(C_A)_0}^{C_A} \frac{dC_A}{(-r_A)'} = \frac{v}{a} \int_{C_A}^{(C_A)_0} \frac{dC_A}{(-r_A)'} \tag{9.83}$$

where a is the active area per unit volume of the catalyst bed. These equations along with Equation 9.78 can be used for the design or performance evaluation of a packed bed reactor. The suitable rate expression for $(-r_A)'$ as described in Chapter 3 is to be used.

Sometimes the catalyst active area is expressed per unit mass of the catalyst. In such a case, we can use the relation

$$Ar_{cat} = aV_{cat} = a'm_{cat}$$

$$=> \quad a = a'\frac{m_{cat}}{V_{cat}} = a'\rho_{cat} \tag{9.84}$$

where a′ is active area per unit mass of the catalyst and ρ_{cat} is the density of the catalyst. Equation 9.83 then changes to

$$m_{cat} = -\frac{v}{a'}\int_{(C_A)_0}^{C_A}\frac{dC_A}{(-r_A)'} = \frac{v}{a'}\int_{C_A}^{(C_A)_0}\frac{dC_A}{(-r_A)'} \tag{9.85}$$

$$\Rightarrow \quad a'\frac{m_{cat}}{v} = \int_{C_A}^{(C_A)_0}\frac{dC_A}{(-r_A)'}$$

$$\Rightarrow \quad a'\frac{d}{dC_A}\left[\frac{m_{cat}}{v}\right] = \frac{-1}{(-r_A)'} \tag{9.86}$$

If the mass of the catalyst in the bed is constant,

$$a'm_{cat}\frac{d}{dC_A}\left[\frac{1}{v}\right] = \frac{-1}{(-r_A)'} \tag{9.87}$$

And, if the volumetric flow rate of the gas through the bed is constant,

$$\frac{a'}{v}\frac{dm_{cat}}{dC_A} = \frac{-1}{(-r_A)'} \tag{9.88}$$

Equations 9.86 through 9.88 are very useful in the experimental study of solid-catalyzed reactions with unknown kinetics. The reaction for which the kinetics is to be studied is carried out in a packed bed reactor with known catalyst mass. Different experimental runs are carried out with different volumetric flow rates of the fluid with constant initial concentration of the reactant and the resulting outlet concentration is noted. From this a relation—often in the form of a graph—is obtained between (1/v) and C_A. From this relation the quantity $(d/dC_A)[1/v]$ is determined for different values of C_A. Multiplying the different values of $(d/dC_A)[1/v]$ so obtained by $a'm_{cat}$ one can determine $1/(-r_A)'$ for different values of C_A. This provides the rate of reaction at different concentrations. In an alternate procedure, the volumetric flow rate through the reactor is kept constant and different experimental runs are carried out with different mass of catalyst in the bed. From this, a relation is obtained between m_{cat} and C_A. From this relation, the quantity (dm_{cat}/dC_A) is determined for different values of C_A. Multiplying the different values of (dm_{cat}/dC_A) so obtained by (a′/v) one can determine $1/(-r_A)'$ for different values of C_A. This provides the rate of reaction at different concentration.

EXAMPLE 9.6

For a liquid–solid catalytic reaction, the following rate data are obtained in an experimental packed bed reactor using various amounts of catalyst and a fixed feed rate of $(F_A)_0 = 10$ kmol/h. The packing density of the catalyst bed is 750 kg/m³.

Catalyst mass (kg)	1	2	3	4	5	6	7
X_A	0.12	0.2	0.27	0.33	0.38	0.41	0.43

Plot the reaction rate (expressed per unit volume of the catalyst) versus conversion.

Solution

From Equations 9.84 and 9.88,

$$\frac{1}{v}\frac{dm_{cat}}{dC_A} = \frac{-1}{a'(-r_A)'}$$

$$\Rightarrow \quad \frac{1}{v}\left[\frac{dm_{cat}}{dX_A} \times \frac{dX_A}{dC_A}\right] = \frac{-\rho_{PB}}{a(-r_A)'}$$

$$\Rightarrow \quad \frac{1}{v(C_A)_0}\frac{dm_{cat}}{dX_A} = \frac{\rho_{PB}}{-r_A} \quad \left(X_A = 1 - \frac{C_A}{(C_A)_0}\right)$$

$$\Rightarrow \quad \frac{1}{(F_A)_0\rho_{PB}}\frac{dm_{cat}}{dX_A} = \frac{1}{-r_A}$$

$$\Rightarrow \quad -r_A = (F_A)_0\rho_{PB}\frac{dX_A}{dm_{cat}} \tag{9.6.1}$$

For estimating the quantity (dX_A/dm_{cat}), we plot X_A versus m_{cat} and measure the slope of the curve at different conversions. This is shown in Figure EXE9.6a.

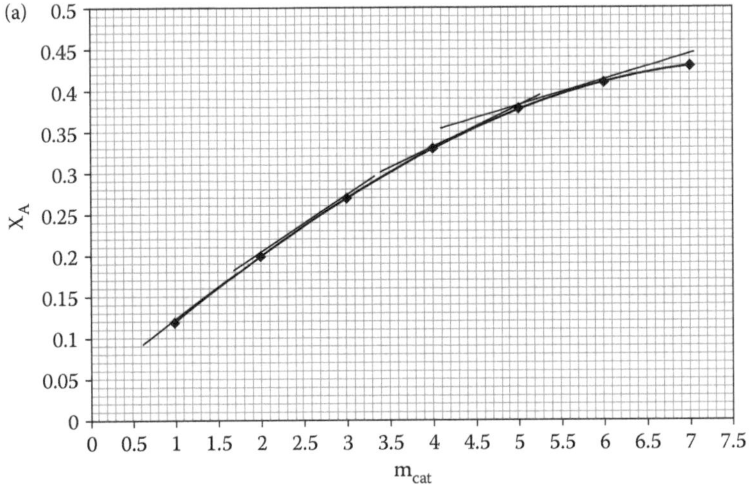

FIGURE EXE9.6A Graphical procedure to determine dX_A/dm_{cat}.

TABLE EXE9.6
Rate at Different Conversion

X_A	0.15	0.2	0.25	0.3	0.35	0.4
dX_A/dm_{cat} (kg^{-1})	0.08	0.073	0.063	0.056	0.045	0.033
$-r_A$ ($kmol/m^3 \cdot h$)	600	547.5	472.5	420	337.5	247.5

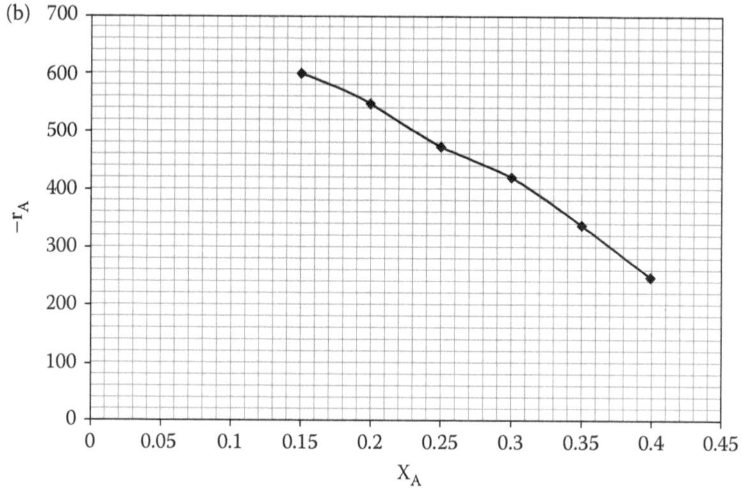

FIGURE EXE9.6B Rate–conversion plot.

From this figure, we prepare Table EXE9.6.
Next, we plot rate versus conversion shown in Figure EXE9.6.b

9.3.2.2 One-Dimensional Heterogeneous Packed Bed Reactor Model

Unlike in the pseudohomogeneous model, transport processes in the solid and the fluid phases are separately accounted for in the heterogeneous model. We again consider the differential section of the reactor shown in Figure 9.2. Writing the material balance for component A in the fluid phase

$$u_s \frac{dC_A}{dL} = k_{fl} a_{ES} \{(C_A)_{fl} - (C_A)_{ES}\} \tag{9.89}$$

Similarly, material balance for solid-phase (catalyst particles) gives

$$k_{fl} a_{ES} \{(C_A)_{fl} - (C_A)_{ES}\} = -(-r_A)'a \tag{9.90}$$

From EquationS 9.89 and 9.90,

$$-u_s \frac{dC_A}{dL} = (-r_A)'a \tag{9.91}$$

This equation is the same as Equation 9.80. So, the pseudohomogeneous and heterogeneous models essentially lead to the same results under isothermal conditions. The results predicted by the two models will be different for nonisothermal conditions. However, this difference is normally not very large to suggest the use of more complex mathematics involved in the heterogeneous model.

9.4 CATALYTIC FLUIDIZED BED REACTORS

Despite the simplicity of operation, the packed bed reactors for solid–fluid catalytic reactions have their own shortcomings. Notable among them is the difficulty of temperature control—nonuniform intraparticle temperature gradients at different parts of the bed. When the fluid velocity is sufficiently high, it results in the solid particles being suspended in the fluid so that the bed acquires the characteristics of a fluid. In a fluidized bed reactor, the catalyst particles are not fixed in their position and the individual particles are not in contact with each other. Some of the benefits of fluidized bed reactors for fluid–solid catalytic reactions are

- Uniformity of fluid–solid contact
- High intensity of mixing
- Uniform temperature gradient

Fluidized bed reactors find their largest applications in gas–solid catalytic reactions and these will be discussed in the subsequent sections. A schematic of a fluidized bed reactor is shown in Figure 9.3.

The gas distributor spreads the gas evenly across the reactor cross section allowing uniform fluidization. A plenum chamber below the distributor provides the function of predistributing the gas. A distinct upper bed surface may or may not be

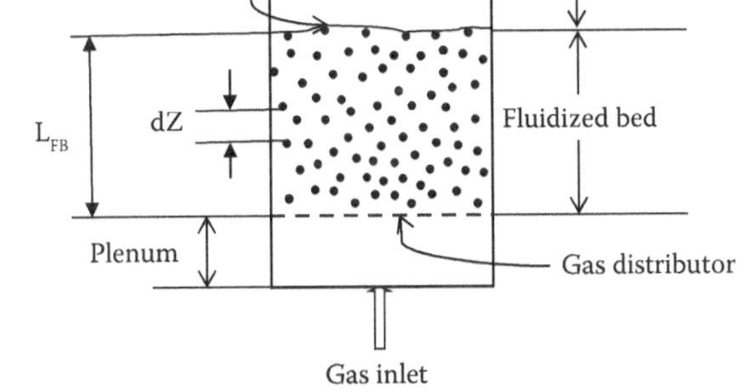

FIGURE 9.3 Schematic of a fluidized bed reactor.

present depending on the bed characteristics. Above the upper bed surface, sufficient vertical height is provided in the reactor for the entrained particles to be separated from the gas stream and returned to the bed.

9.4.1 The Fluidization Phenomenon

Starting with a packed bed of solids, the pressure drop across the bed increases as the gas velocity through the bed increases as obtained from Equation 9.78. At a gas velocity known as the minimum fluidizing velocity, u_{mf}, the resulting pressure drop exerts a force sufficient to balance the weight of the bed so that the particles are lifted and suspended. The contact between individual particles is lost and they begin to move around randomly. The minimum fluidizing velocity can be estimated as (Kunii and Levenspiel, 1997)

$$\frac{150(1-\epsilon_{mf})}{\epsilon_{mf}^3}u_{mf}^* + \frac{1.75(u_{mf}^*)^2 d_p^*}{\epsilon_{mf}^3} = (d_p^*)^2 \tag{9.92}$$

For most fluidized beds, the second term on the left-hand side of Equation 9.92 is much less compared to the first term and can be neglected. So, Equation 9.92 can be approximated as

$$\frac{150(1-\epsilon_{mf})}{\epsilon_{mf}^3}u_{mf}^* = (d_p^*)^2$$

$$=> \quad u_{mf}^* = \frac{(d_p^*)^2\epsilon_{mf}^3}{150(1-\epsilon_{mf})} \tag{9.93}$$

where

$$u_{mf}^* = u_{mf}\left[\frac{\rho_g^2}{\mu(\rho_S-\rho_g)g}\right]^{1/3} \tag{9.94}$$

and

$$d_p^* = d_p\left[\frac{\rho_g(\rho_S-\rho_g)g}{\mu^2}\right]^{1/3} \tag{9.95}$$

The bed porosity at minimum fluidization is either obtained experimentally or by the use of empirical correlations. One such correlation that can be used for small particle sizes is (Broadhurst and Becker, 1975)

$$\epsilon_{mf} = 0.586\Theta^{-0.72}\left[\frac{\mu^2}{\rho_g(\rho_S-\rho_g)gd_p^3}\right]^{0.029}\left(\frac{\rho_g}{\rho_S}\right)^{0.021} \tag{9.96}$$

Typical values of ϵ_{mf} are around 0.5.

In the Stokes regime (Reynolds number < 0.4) the terminal settling velocity of the particles can be estimated as

$$u_t = \frac{(\rho_S - \rho_g)gd_p^2}{18\mu} \tag{9.97}$$

At higher gas velocities (Reynolds number between 0.4 and 500) the following equation is proposed for estimating the terminal settling velocity (Kunii and Levenspiel, 1969)

$$u_t = \left[\frac{1.78 \times 10^{-2}\{(\rho_S - \rho_g)g\}^2}{\rho_g \mu}\right]^{1/3} d_p \text{ (quantities in CGS unit)} \tag{9.98}$$

Beyond minimum fluidization, the bed behavior depends on the gas velocity, the physical properties of the gas—its density and viscosity—and the physical properties of the particles—their particle size and density. For a given solid and gas characteristics, the fluidized bed exhibits different flow regimes with increasing velocity of the gas. Though the transition is not abrupt, the fluidized beds are generally categorized as follows:

Bubbling fluidized beds are obtained with gas velocities above minimum fluidizing velocity but below the terminal settling velocity of the particles. In fact, this flow regime is observed when the superficial gas velocity reaches a minimum bubbling velocity u_{mb}. At this point, a part of the gas flow starts bypassing the rest of the bed. As a result, distinct gas bubbles, with very few particles inside, appear within the bed surrounded by the solid rich emulsion phase. There also exists an intermediate phase between the bubble and the emulsion known as the cloud phase. The solids in the emulsion phase rapidly move around and the entire bed appears as a boiling liquid. The bubbles rise through the bed, increasing in size, and burst on reaching the upper end of the bed, ejecting particles above the surface of the bed. These ejected solids eventually fall back into the bed. With increasing gas velocity, the bubbles appear more frequently and the particles in the emulsion phase move faster. Such beds are also known as *aggregative fluidized beds.* In the regime of bubbling fluidization, the pressure drop almost remains constant and independent of gas velocity.

Turbulent fluidized beds are obtained with the gas velocities exceeding the terminal settling velocity of the particles. When the bubbles burst at the top of the bed and the particles are ejected, they are entrained by the high velocity gas stream. These particles will move out of the bed unless some gas solid separation device is present at the top of the column to collect these particles and return them to the bed.

Fast fluidized beds are obtained at still higher gas velocities so that the bubbles coalesce and become so large that they occupy the entire central core of the bed. The cloud and the emulsion phases merge and are restricted to only the regions near the wall of the column. The upper bed level almost disappears.

Pneumatic conveying beds are obtained at still higher velocity when even the wall region of cloud and emulsion phases disappear and discrete particles are carried away by the flow of gas. Fluidized bed reactors operating under pneumatic conveying regime are also known as *straight- through-transport reactors.*

Beyond the bubbling fluidization regime, a sharp decrease in pressure drop is observed with increasing gas velocity.

9.4.2 Kunii–Levenspiel Model for Fluidized Bed Reactors

9.4.2.1 Bubbling Fluidized Bed

In a bubbling fluidized bed, the bubbles form very close to the distributor and the entire bed must be considered as a two-phase system. The bubbles are approximately hemispherical from the top but the bottom often has an upward dent. The gas within a bubble remains almost isolated from the gas in the emulsion. At the bubble boundary, there is a thin buffer region where the gas from the bubble intermixes with the gas in the emulsion. This buffer region is known as the cloud. On the bottom side of each gas bubble there is a wake containing a concentration of solids. As the gas bubbles rise, so does the wake below it with its solids. This gives a net downward flow of solids in the emulsion phase. A schematic of the bubble, cloud, wake, and emulsion is shown in Figure 9.4.

Davidson and Harrison, based on experiments on single bubbles, deduced the following equation for the velocity of a single rising bubble (Davidson and Harrison, 1963):

$$u_{br} = 0.71(gd_b)^{1/2} \quad \text{(quantities in CGS units)} \tag{9.99}$$

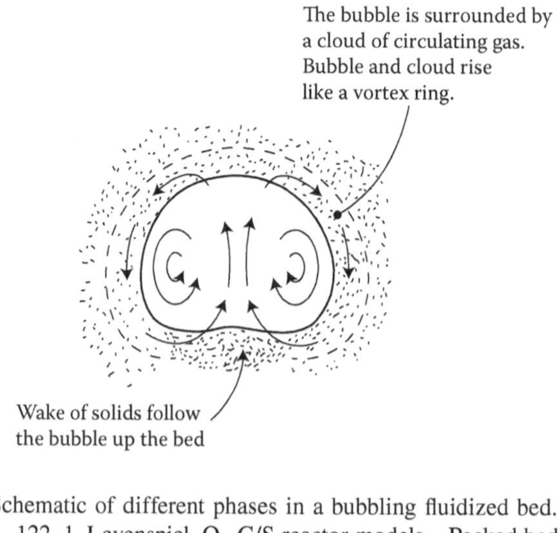

FIGURE 9.4 Schematic of different phases in a bubbling fluidized bed. (Reprinted from *Powder Technol.*, 122, 1, Levenspiel, O., G/S reactor models—Packed beds, bubbling fluidized beds, turbulent fluidized beds and circulating (fast) fluidized beds, Copyright 2002, with permission from Elsevier.)

Thereafter, from the analogy with liquid–gas systems they proposed an estimation of the rising bubble velocity in a bubbling fluidized bed as

$$u_b = u_{br} + (u_S - u_{mf}) \tag{9.100}$$

$$\Rightarrow \quad u_b = u_S - u_{mf} + 0.71(gd_b)^{1/2} \quad \text{(quantities in CGS units)} \tag{9.101}$$

The bubble diameter has been found to depend on the bed diameter, height from the distributor plate, superficial gas velocity, the physical properties of the gas, and the physical properties of the solids. By correlating the experimental data from several studies, Mori and Wen proposed an empirical relation to predict the bubble diameter as (Mori and Wen, 1975)

$$\frac{d_{bm} - d_b}{d_{bm} - d_{b0}} = \exp\left(-\frac{0.3L_{FB}}{d_{FB}}\right) \tag{9.102}$$

The maximum bubble diameter, d_{bm}, can be estimated as

$$d_{bm} = 0.652[Ar_R(u_S - u_{mf})]^{0.4} \quad \text{(Quantities in CGS units)} \tag{9.103}$$

For porous plate distributors the initial bubble diameter, d_{b0}, can be estimated as

$$d_{b0} = 0.00376[u_S - u_{mf}]^2 \quad \text{(Quantities in CGS units)} \tag{9.104}$$

And, for perforated plate distributors, it is

$$d_{b0} = 0.652\left[\frac{Ar_R(u_S - u_{mf})}{n_{per}}\right]^{0.4} \quad \text{(Quantities in CGS units)} \tag{9.105}$$

The Kunii–Levenspiel model for bubbling fluidized beds is based on the following assumptions:

1. All bubbles are of the same size.
2. Solids in the emulsion phase move down the bed in plug flow.
3. The emulsion phase exists at minimum fluidization condition. The porosity in the emulsion phase is equal to ϵ_{mf}.
4. The porosity in the wake is equal to ϵ_{mf}.
5. The gas and the solids in the wake move upward with the same average velocity as the velocity of the bubbles.

Since the total mass of solids in the bed remains constant, a material balance on the solids in the bed gives

Downward mass flow rate of solids in the emulsion

= Upward mass flow rate of solids in wake

$$=> \quad A r_R \rho_S (1 - \nu_b - \nu_w) u_{ems} = A r_R \rho_S \nu_w u_b$$

$$=> \quad u_{ems} = \frac{\nu_w u_b}{1 - \nu_b - \nu_w} \tag{9.106}$$

Similarly, a material balance on the gas gives

Total gas flowrate = Gas flowrate in bubbles + Gas flowrate in wake

+ Gas flowrate in emulsion

$$=> \quad A r_R u_S = A r_R \nu_b u_b + A r_R \nu_w \epsilon_{mf} u_b + A r_R (1 - \nu_b - \nu_w) \epsilon_{mf} u_{emg}$$

$$=> \quad u_S = \nu_b u_b + \nu_w \epsilon_{mf} u_b + (1 - \nu_b - \nu_w) \epsilon_{mf} u_{emg} \tag{9.107}$$

The emulsion phase is at minimum fluidization condition with the solids moving downward. The minimum fluidizing velocity will now refer to gas velocity relative to the moving solids. Since the minimum fluidizing velocity is measured based on the cross-section area of the empty reactor, we have

$$u_{emg} = \frac{u_{mf}}{\epsilon_{mf}} - u_{ems} \tag{9.108}$$

$$=> \quad u_{emg} = \frac{u_{mf}}{\epsilon_{mf}} - \frac{\nu_w u_b}{1 - \nu_b - \nu_w} \quad \text{(substituting } u_{ems} \text{ from Equation 9.106)}$$

$$=> \quad u_{emg} = \frac{u_{mf}(1 - \nu_b - \nu_w) - \nu_w u_b \epsilon_{mf}}{\epsilon_{mf}(1 - \nu_b - \nu_w)} \tag{9.109}$$

Substituting this value of u_{emg} in Equation 9.107, we have

$$u_S = \nu_b u_b + \nu_w \epsilon_{mf} u_b + u_{mf}(1 - \nu_b - \nu_w) - \nu_w u_b \epsilon_{mf}$$

$$=> u_S = \nu_b u_b + u_{mf}(1 - \nu_b - \nu_w)$$

$$=> u_S = \nu_b u_b + u_{mf}(1 - \nu_w) - u_{mf} \nu_b$$

$$=> (u_b - u_{mf}) \nu_b = u_S - u_{mf}(1 - \nu_w)$$

$$=> \nu_b = \frac{u_S - u_{mf}(1 - \nu_w)}{u_b - u_{mf}} \tag{9.110}$$

In most bubbling fluidized beds, $u_b \gg u_{mf}$, and $\nu_w \ll 1$ so Equation 9.110 can be further simplified as

$$\nu_b = \frac{u_S - u_{mf}}{u_b} \tag{9.111}$$

Like any gas–solid-reacting system, the transport of reactant species from the gas to the external solid surface constitutes one of the steps in the fluidized bed reactor. At high superficial velocities, the agitation in the bed is sufficiently high so that the mass transfer from gas to solid may not be the rate-limiting step. The transport processes typical of the fluidized bed reactors are the exchange of material between the bubble and the cloud and that between the cloud and the emulsion.

Consider an nth-order gas–solid catalytic reaction

$$A \rightarrow B \tag{9.112}$$

taking place in a fluidized bed reactor. For exchange between the bubble and the cloud per unit volume of the bed

$$\left(-\frac{dN_A}{dt}\right)_{b\rightarrow c} = (k_v)_{b\rightarrow c}[(C_A)_b - (C_A)_c] \tag{9.113}$$

and

$$\left(-\frac{dN_B}{dt}\right)_{c\rightarrow b} = (k_v)_{c\rightarrow b}[(C_B)_c - (C_B)_b] \tag{9.114}$$

Kunii and Levenspiel have proposed the following equation to estimate the mass transfer coefficient (Kunii and Levenspiel, 1969):

$$(k_v)_{b\rightarrow c} = (k_v)_{c\rightarrow b} = 4.5\left(\frac{u_{mf}}{d_b}\right) + 5.85\left(\frac{D_{AB}^{0.5} g^{0.25}}{d_b^{1.25}}\right) \tag{9.115}$$

Similarly, for exchange between cloud and emulsion per unit volume of the bed,

$$\left(-\frac{dN_A}{dt}\right)_{c\rightarrow em} = (k_v)_{c\rightarrow em}[(C_A)_c - (C_A)_{em}] \tag{9.116}$$

and

$$\left(-\frac{dN_B}{dt}\right)_{em\rightarrow c} = (k_v)_{em\rightarrow c}[(C_B)_{em} - (C_B)_c] \tag{9.117}$$

where the following correlation, based on Higbie's penetration theory, is used to estimate the mass transfer coefficient:

$$(k_v)_{c\rightarrow em} = (k_v)_{em\rightarrow c} = 6.77\left[\frac{\epsilon_{mf} D_{AB} u_b}{d_b^3}\right]^{0.5} \tag{9.118}$$

The rate of reaction in the bubble, emulsion, and the cloud can be expressed, respectively, as

$$(-r_A)_b = k_b (C_A)_b^n \tag{9.119}$$

$$(-r_A)_c = k_c (C_A)_c^n \tag{9.120}$$

and

$$(-r_A)_{em} = k_{em} (C_A)_{em}^n \tag{9.121}$$

Here, all the specific reaction rate constants are based on the unit volume of the bed.

Consider the differential section of the fluidized bed of height dZ shown schematically in Figure 9.3. Writing a material balance on species A over the bubble phase

$$\text{Bulk flow input} = \text{Bulk flow output} + \text{Mass transport output}$$

$$+ \text{Converted by chemical reaction}$$

$$=> \quad A r_R v_b u_b (C_A)_b\big|_Z = A r_R v_b u_b (C_A)_b\big|_{Z+dZ} + (k_v)_{b\to c} A r_R v_b dZ[(C_A)_b$$

$$-(C_A)_c] + k_b (C_A)_b^n A r_R v_b dZ$$

$$=> \quad -u_b \frac{d(C_A)_b}{dZ} = (k_v)_{b\to c}[(C_A)_b - (C_A)_c] + k_b (C_A)_b^n$$

$$-u_b \frac{d(C_A)_b}{dZ} = (k_v)_{b\to c}[(C_A)_b - (C_A)_c] + k_b (C_A)_b^n \tag{9.122}$$

A material balance on species A over the cloud phase (including wakes) gives

$$\text{Bulk flow input} + \text{Mass transport input} = \text{Bulk flow output} + \text{Mass transport output}$$

$$+ \text{Converted by chemical reaction}$$

$$=> \quad A r_R v_w u_b (C_A)_c\big|_Z + (k_v)_{b\to c} A r_R v_b dZ[(C_A)_b - (C_A)_c] = A r_R v_w u_b (C_A)_c\big|_{Z+dZ}$$

$$+ (k_v)_{c\to em} A r_R v_b dZ[(C_A)_c - (C_A)_{em}] + k_c (C_A)_c^n A r_R v_b dZ$$

$$=> \quad -u_b \frac{v_w}{v_b} \frac{d(C_A)_c}{dZ} = (k_v)_{c\to em}[(C_A)_c - (C_A)_{em}] - (k_v)_{b\to c}[(C_A)_b - (C_A)_c] + k_c (C_A)_c^n \tag{9.123}$$

The model assumes that the quantity on the left-hand side of Equation 9.123 is negligible in comparison to the quantity on the RHS; so, the equation further simplifies to

$$(k_v)_{b\to c}[(C_A)_b-(C_A)_c]=(k_v)_{c\to em}[(C_A)_c-(C_A)_{em}]+k_c(C_A)_c^n \tag{9.124}$$

Material balance on species A over the emulsion phase gives

Bulk flow input + Mass transport input = Bulk flow output

+ Converted by chemical reaction

$$=> \quad Ar_R(1-\nu_b-\nu_w)u_{emg}(C_A)_{em}\big|_Z+(k_v)_{c\to em}Ar_R\nu_b dZ[(C_A)_c-(C_A)_{em}]$$

$$=Ar_R(1-\nu_b-\nu_w)u_{emg}(C_A)_{em}\big|_{Z+dZ}+k_{em}(C_A)_{em}^n Ar_R\nu_b dZ$$

$$=> \quad -u_{emg}\frac{(1-\nu_b-\nu_w)}{\nu_b}\frac{d(C_A)_{em}}{dZ}=k_{em}(C_A)_{em}^n-(k_v)_{c\to em}[(C_A)_c-(C_A)_{em}] \tag{9.125}$$

Neglecting the term on the left-hand side of Equation 9.125 in comparison to the terms on the RHS, the equation simplifies to

$$(k_v)_{c\to em}[(C_A)_c-(C_A)_{em}]=k_{em}(C_A)_{em}^n \tag{9.126}$$

Equations 9.122, 9.124, and 9.126 are the governing equations for the Kunii–Levenspiel model. For the solution to these equations, four new parameters are defined as

$$\gamma_b=\frac{\text{volume of catalyst in the bubbles}}{\text{volume of bed}} \tag{9.127}$$

$$\gamma_c=\frac{\text{volume of catalyst in the clouds and wakes}}{\text{volume of bed}} \tag{9.128}$$

$$\gamma_{em}=\frac{\text{volume of catalyst in the emulsion}}{\text{volume of bed}} \tag{9.129}$$

The parameter γ_b is to be determined experimentally. The other two fractions may be estimated as

$$\gamma_c=(1-\epsilon_{mf})\nu_b\left[\frac{3u_{mf}}{(u_{br}\epsilon_{mf}-u_{mf})}+\frac{\nu_w}{\nu_b}\right] \tag{9.130}$$

And, for parameter γ_{em}

$$\gamma_{em}=(1-\epsilon_{mf})(1-\nu_b)-\gamma_c-\gamma_b \tag{9.131}$$

$$k_{cat} = \text{Specific reaction rate constant based on unit volume of catalyst} \quad (9.132)$$

Now, we have

$$k_b = \gamma_b k_{cat} \quad (9.133)$$

$$k_c = \gamma_c k_{cat} \quad (9.134)$$

$$k_{em} = \gamma_{em} k_{cat} \quad (9.135)$$

For a first-order reaction, the model provides the performance equation for the bubbling fluidized bed reactor as

$$\ln\frac{(C_A)_0}{C_A} = \left[\gamma_b k_{cat} + \cfrac{1}{\cfrac{1}{\nu_b (k_v)_{b\to c}} + \cfrac{1}{\gamma_c k_{cat} + \cfrac{1}{\cfrac{1}{\nu_b (k_v)_{c\to em}} + \cfrac{1}{\gamma_{em} k_{cat}}}}}\right]\left(\frac{L_{FB}}{u_s}\right) \quad (9.136)$$

Consider a hypothetical reactor where the gases are flowing under ideal plug flow with the same amount of solids in the reactor as in the case of the bubbling fluidized bed reactor. For this hypothetical PFR using the performance equation of an ideal PFR,

$$\ln\left(\frac{(C_A)_0}{(C_A)_{ex}}\right)_{hyp} = \frac{(\gamma_b + \gamma_c + \gamma_{em})L_{FBd}}{u_s} k_{cat} \quad (9.137)$$

Here, the model defines a new parameter known as the contacting efficiency of the bed, given by

$$\eta_{FB} = \frac{\ln((C_A)_0/(C_A)_{ex})}{\ln((C_A)_0/(C_A)_{ex})_{hyp}}$$

$$\Rightarrow \quad \eta_{FB} = \frac{\left[\gamma_b + \cfrac{1}{\cfrac{k_{cat}}{\nu_{FBd} (k_v)_{b\to c}} + \cfrac{1}{\gamma_c + \cfrac{1}{\cfrac{k_{cat}}{\nu_{FBd} (k_v)_{c\to em}} + \cfrac{1}{\gamma_{em}}}}}\right]}{(\gamma_b + \gamma_c + \gamma_{em})} \quad (9.138)$$

9.4.2.2 Circulating Fluidized Bed

In a fluidized bed reactor, not all the solid particles are of the same size; rather, they have a particle size distribution. Even at the minimum fluidization condition, the gas velocity may exceed the terminal settling velocity of the finest fraction of the particles. Normally, the reactor is to be operated at a superficial velocity much higher than the minimum fluidization velocity for satisfactory fluidization and turbulence in the bed. This leads to a sizable fraction of the smaller-sized particles to leave the bed, which needs to be replenished with makeup catalyst particles to maintain a steady mass of catalyst in the reactor. Such fluidized beds are known as circulating fluidized beds.

In a circulating fluidized bed, there is a lower dense region that may exist in one of the previously discussed fluidization regimes—bubbling, turbulent, fast fluidized, or pneumatic transport. In the dense region, the average solids fraction in the bed is reasonably constant and may be expressed as fr_d. Above the dense lower region there exists a lean upper region with lower fraction of solids. In their model, Kunii and Levenspiel (1997, 2000) describe the upper lean region as consisting of three different phases. One of the phases is made up of a lean mixture of solids dispersed in the gas having a mean density ρ_1. These solids rise with a velocity u_1. Another phase is made up of clumps of solids with mean density ρ_2. These clumps are rising with a velocity u_2. There is a third phase made up of clumps of solids as well as solids near the wall with mean density ρ_3. These solids are flowing down with velocity u_3. Further, the model proposes exchange of solids between phases—with a rate coefficient k_1 between clumps and lean mixture, and a rate coefficient k_2 between upflowing and downflowing solids. In the lean region, the average solids fraction in the bed decreases exponentially from fr_d. For an infinitely tall column, the solids fraction reaches a limiting value fr*, referred to as the saturation carrying capacity of the gas, which is independent of u_S. For columns of finite height, the exiting solids fraction is more than fr*.

The solid fraction in the exit gas is represented by

$$fr_{ex} = fr^* + (fr_d - fr^*)\exp(-aL_{FBl}) \tag{9.139}$$

$$\Rightarrow L_{FBl} = \frac{1}{a}\ln\left(\frac{fr_d - fr^*}{fr_{ex} - fr^*}\right) \tag{9.140}$$

And, at any point in the lean region of the bed

$$fr_l = fr^* + (fr_d - fr^*)\exp(-az_{FBl}) \tag{9.141}$$

a is known as the decay constant for solids in the lean part of the column. The mean fraction of solids in the lean region would be

$$(fr_{mean})_l = fr^* + \left(\frac{fr_d - fr_{ex}}{aL_{FBl}}\right) \tag{9.142}$$

For the entire column, the mean solids fraction can be calculated as

$$(fr_{mean})_{tot} = \frac{(fr_{mean})_l L_{FBl} + fr_d L_{FBd}}{L_{FB}} \tag{9.143}$$

Case 1. Lower bed under fast fluidization

Under these conditions, the model predicts different performance equations for the dense and lean regions of the reactor. For the lower dense region, instead of discrete bubbles there is a central core of upflowing gas and a region of dispersed solids near the wall. The performance equation for the dense region is given by

$$\ln\frac{(C_A)_0}{(C_A)_d} = \left[fr_{cr}k_{cat} + \frac{1}{(1/\nu_{FBd}(k_v)_{cr\to wl}) + (1/fr_{wl}k_{cat})}\right]\left(\frac{L_{FBd}}{u_s}\right) \tag{9.144}$$

For the lean region of the reactor, a material balance over a differential section at height Z_{FBl} from the start of the lean region gives

$$-u_s\frac{dC_A}{dz_{FBl}} = \eta_l fr_l k_{cat} C_A \tag{9.145}$$

where (Kunii and Levenspiel, 1997)

$$\eta_l = 1-(1-\eta_d)\exp(-bz_{FBl}) \tag{9.146}$$

with $b = 6.62\ m^{-1}$ is determined experimentally. Combining Equation 9.145 with Equations 9.141 and 9.146,

$$-u_s\frac{dC_A}{dz_{FBl}} = \{1-(1-\eta_d)\exp(-bz_{FBl})\}\{fr^* + (fr_d - fr^*)\exp(-az_{FBl})\}k_{cat}C_A$$

$$=> \int_{(C_A)_{ex}}^{(C_A)_d}\frac{dC_A}{C_A} = \frac{k_{cat}}{u_s}$$

$$\int_0^{L_{FBl}}[fr^* + (fr_d - fr^*)\exp(-az_{FBl}) - fr^*(1-\eta_d)\exp(-bz_{FBl})$$

$$-(1-\eta_d)(fr_d - fr^*)\exp\{-(a+b)z_{FBl}]dz_{FBl}$$

$$=> \ln\frac{(C_A)_d}{(C_A)_{ex}} = \frac{k_{cat}fr^*}{u_s}L_{FBl} + \frac{k_{cat}(fr_d - fr^*)}{u_s a}\{1-\exp(-aL_{FBl})\}$$

$$-\frac{k_{cat}fr^*(1-\eta_d)}{u_s b}\{1-\exp(-bL_{FBl})\} - \frac{k_{cat}(1-\eta_d)(fr_d - fr^*)}{u_s(a+b)}\{1-\exp(-(a+b)L_{FBl})\}$$

$$\Rightarrow \ln\frac{(C_A)_d}{(C_A)_{ex}} = \frac{k_{cat}fr^*}{u_s}\left[L_{FBl} - \frac{(1-\eta_d)}{b}\{1-\exp(-bL_{FBl})\}\right]$$
$$+\frac{k_{cat}(fr_d - fr^*)}{u_s}\left[\frac{\{1-\exp(-aL_{FBl})\}}{a} - \frac{(1-\eta_d)\{1-\exp(-(a+b)L_{FBl})\}}{(a+b)}\right] \tag{9.147}$$

For the entire reactor,

$$\frac{(C_A)_0}{(C_A)_{ex}} = \frac{(C_A)_0}{(C_A)_d} \times \frac{(C_A)_d}{(C_A)_{ex}}$$
$$\Rightarrow \quad \ln\frac{(C_A)_0}{(C_A)_{ex}} = \ln\frac{(C_A)_0}{(C_A)_d} + \ln\frac{(C_A)_d}{(C_A)_{ex}} \tag{9.148}$$

Case 2. Lower bed under pneumatic conveying

In pneumatic conveying, all the particles are evenly dispersed in the gas and the gas flow can be reasonably approximated as plug flow. From the performance equation of a PFR, under this regime

$$\ln\frac{(C_A)_0}{(C_A)_{ex}} = \frac{(fr_{mean})_l L_{FBl} + fr_d L_{FBd}}{u_s} k_{cat} \tag{9.149}$$

The quantity $\frac{(fr_{mean})_l L_{FBl} + fr_d L_{FBd}}{u_s}$ can be visualized as a modified space time for the reactor. Clearly, in this case, the contacting efficiency is 1. In all likelihood, there will not be a distinct dense region and lean region in the reactor and a mean solids fraction for the entire reactor may be used in Equation 9.149.

As evident from the above paragraphs, the Kunii–Levenspiel model has several parameters characterizing the bed. The success of the model in realistically predicting the performance of fluidized bed reactors depends on the realistic determination of these parameters. Table 9.1 summarizes some of the values for these parameters as available in the technical literature.

TABLE 9.1
Values of Parameters in Kunii–Levenspiel Model

Parameter	Range	Reference
I_{mf}	0.5–0.6	Kunii and Levenspiel (1997)
fr_d	Fast-fluidized bed 0.06–0.2	Kunii and Levenspiel (1997)
	Pneumatic conveying 0.01–0.06	
au_s	2–7 s^{-1}	Kunii and Levenspiel (1969, 1995)
fr_{cr}	0.01	Kunii and Levenspiel (1997)
fr_{wl}	0.4–0.5	Kunii and Levenspiel (1997)
$(k_v)_{cr\to wl}$	5–20 s^{-1}	Kunii and Levenspiel (1997)
v_{FBd}	0.6–0.9	Kunii and Levenspiel (1997)
γ_b	0.001–0.01	Levenspiel (1999)

EXERCISES

9.1 What cell voltage will be required and the amount of nickel that can be deposited per hour in an electrochemical PFR if the reaction of Example 9.2 is carried out with an electrolyte solution with 1.5 M Ni^{2+} and 1 M H^+ concentration? Other conditions remain the same.

9.2 For a one-electron reduction reaction, the cathodic exchange current density is 300 μA/cm² and the transfer coefficient is 0.5. How much cathode area is needed in a PFR for 80% conversion of the reducing electrochemical species for a molar flow rate of 100 mol/h if the cathode is maintained at an overpotential of –150 mV? Under the given conditions, the reaction follows the Butler–Volmer kinetics and is carried out at ambient temperature.

9.3 For the reaction of Exercise 9.2, it is found that the current density reaches a limiting value of 9.5 mA/cm² at an overpotential of –600 mV. What will be the cathode area required for the same conversion if the cathode is maintained at –600 mV overpotential?

9.4 Repeat Exercises 9.2 and 9.3 for a backmix flow reactor. Comment on your results.

9.5 The exchange current density for hydrogen evolution in acidic water electrolysis over Pt catalyst is found to be 0.2 mA/cm². The transfer coefficient can be assumed to be 0.5. How much cathode area will be needed for the production of 100 mol hydrogen per second in a backmix flow electrolyzer working at ambient temperature? The cathodic overpotential is –400 mV. The reaction on the cathode is

$$2H^+ + 2e^- \rightleftharpoons H_2$$

The overall reaction is

$$2H_2O \rightarrow 2H_2 + O_2$$

9.6 In an experimental study for the anaerobic digestion of effluents, the following kinetic parameters were determined:

$$\mu_{max} = 0.7\,d^{-1},\ K_m = 0.9\,g\ COD/L,\ Y = 0.25\,g/g\ COD,\ k_{dth} = 0.05\,d^{-1}$$

The effluent has 8000 mg/L COD and is to be treated to reduce its COD to 3000 mg/L. What will be the volume of a PFR required if 100 m³/day of this effluent is to be treated?

9.7 For Exercise 9.6, check if we can use a smaller BFR for the same treatment.

9.8 In Exercise 9.6, if the COD is to be reduced to 500 mg/L suggest the best combination of BFR and PFR to have the minimum total volume of the reactor system.

9.9 The bacteria *Escherichia coli* was found to grow in a laboratory batch culture containing galactose as the substrate and the following kinetic parameters were obtained:

$$V_{max} = 0.9\,h^{-1}, K_m = 90\,\mu g/L$$

How much time will be required to reduce the galactose concentration from 1 to 0.2 mg/L? (Data adapted from Lendenmann et al., 2000.)

9.10 1-Butene is selectively oxidized to 1,3-butadiene by passing through a fixed bed of iron molybdate catalyst. The catalyst has specific surface area of 50 m^2/g. The feed is 6.5% butene, 7% oxygen, and rest helium, flowing at 120 mL/min, at atmospheric pressure and 400°C temperature. The mass of catalyst in the fixed bed is 0.2 g. 50% conversion of butene is obtained. The reaction is first order in butene. Find out the reaction rate constant for the reaction based on unit area of the catalyst. (Data adapted from Straguzzi et al., 1987a, b.)

9.11 n-Hexane, at a feed rate of 100 m^3/min, is cracked to produce light olefins in a packed bed reactor using nano crystalline ZSM-5 catalyst. The kinetic information for the reaction may be taken from Exercise 3.23. How much catalyst will be needed in the bed for 70% conversion of n-hexane? Assume ε to be zero.

9.12 In the reforming of normal hexane over $Pt–Al_2O_3$ catalyst in a fixed bed, the following conversion results were obtained:

$\frac{m_{cat}}{(F_A)_0}$ (g · h/mol)	16	24	30	46	62	78
Conversion	0.08	0.15	0.2	0.3	0.38	0.44

Plot the rate versus conversion for the reaction. (Data adapted from Marin and Froment, 1982.)

9.13 A pilot fluidized bed reactor of 1 m diameter and a porous disk distributor plate is used for a gas–solid catalytic reaction with a gas flow rate of 2100 L/min. The solid catalyst particles may be assumed to be spherical with an average diameter of 250 μm. The density of the gas at the reactor- operating conditions is 1.03 kg/m^3 and the viscosity of the gas is 1.2×10^{-2} cp. The catalyst particles have a density of 1250 kg/m^3. Under these conditions, the average bubble diameter is 4 cm. How much fraction of the bed is occupied by bubbles?

9.14 A fluidized bed reactor of 1.5 m diameter is filled with solid catalyst particles of mean size 100 μm and 0.7 sphericity with a density of 2000 kg/m^3. The gas flow rate through the bed is 5000 L/min. The density of the gas is 0.8 kg/m^3 and its viscosity is 3×10^{-5} kg/(m · s). Check if the bed is within the bubbling fluidization regime.

9.15 In Exercise 9.13, if the bed height is 1 m, estimate the minimum and the maximum bubble diameter. Also, estimate the bubble diameter at a height of 60 cm from the distributor plate.

9.16 In Exercise 9.13, what is the mass of catalyst required if the bed height is 1 m and the bubbles are free from catalyst particles?

9.17 The solid-catalyzed reaction

$$A \rightarrow B \quad -r_A = 0.1C_A \text{mol/(L catalyst)} \cdot s$$

is carried out in the fluidized bed reactor of Exercise 9.13 with a bed height of 1 m. The diffusivity of A under given conditions is 6×10^{-5} m^2/s. Under the given conditions of fluidization, the volume fraction of catalyst in the bubble is found to be 0.01. Estimate the conversion according to the Kunii-Levenspiel model.

9.18 A laboratory bubbling fluidized bed reactor of 16 mm internal diameter is filled with 1.5 g catalyst particles of 150 μm particle size with sphericity 0.8. The terminal settling velocity and minimum fluidization velocity were determined to be 38 and 0.6 cm/s, respectively. The bed is fluidized with a gas mixture at 750 K and 1 atm pressure, consisting of H_2, CO, and N_2 in the ratio 3:1:1 flowing at the rate of 60 L/(h · g) catalyst. The viscosity of the gas mixture is 2×10^{-5} kg/(m · s). The average bubble diameter is 0.5 cm. Calculate the porosity at minimum fluidization, and bed height. (Data adapted from Liu et al., 2015.)

9.19 From the definition of contacting efficiency, derive an equation for the contacting efficiency for the lower dense region of a circulating fluidized bed where the dense region is under fast fluidization.

REFERENCES

Atkins, P. and de Paula, J., *Physical Chemistry*, 8th ed., Freeman, New York, 221, 2006.

Broadhurst, T.E. and Becker, H.A., Onset of fluidization and slugging in beds of uniform particles. *AIChE J.*, 21, 238, 1975.

Butler, J.A.V., Studies in heterogeneous equilibria. Part III. A kinetic theory of reversible oxidation potentials at inert electrodes. *Trans. Faraday Soc.*, 19, 734, 1924.

Davidson, J.F. and Harrison, D., *Fluidized Particles*, Cambridge University Press, New York, 1963.

Ergun, S., Fluid flow through packed columns. *Chem. Eng. Prog.*, 48, 89, 1952.

Johnson, K.A. and Goody, R.S., The original Michaelis constant: Translation of the 1913 Michaelis-Menetn paper. *Biochemistry*, 50, 8264, 2011.

Kunii, D. and Levenspiel, O., *Fluidization Engineering*, Wiley, New York, 1969.

Kunii, D. and Levenspiel, O., Circulating fluidized-bed reactors. *Chem. Eng. Sci.*, 52, 2471, 1997.

Kunii, D. and Levenspiel, O., The K-L reactor model for circulating fluidized beds. *Chem. Eng. Sci.*, 55, 4563, 2000.

Lendenmann, U., Snozzi, M., and Egli, T., Growth kinetics of *Escherichia coli* with galactose and several other sugars in carbon limited chemostat culture. *Can. J. Microbiol.*, 46, 72, 2000.

Levenspiel, O., *Chemical Reaction Engineering*, 3rd ed., John Wiley & Sons, New York, 270, 1999.

Levenspiel, O., G/S reactor models—Packed beds, bubbling fluidized beds, turbulent fluidized beds and circulating (fast) fluidized beds. *Powder Technol.*, 122, 1, 2002.

Lineweaver, H. and Burk, D., The determination of enzyme dissociation constants. *J. Am. Chem. Soc.*, 56, 658, 1934.

Liu, J., Cui, D., Yu, J., Su, F., and Xu, G., Performance characteristics of fluidized bed syngas methanation over Ni-Mg/Al_2O_3 catalyst. *Chin. J. Chem. Eng.*, 23, 86, 2015.

Marin, G.B. and Froment, G.F., Reforming of C_6 hydrocarbons on a Pt Al_2O_3 catalyst. *Chem. Eng. Sci.*, 37, 759, 1982.

Michaelis, L. and Menten, M.L., Die kinetik der invertinwirkung. *Biochem. Z.*, 49, 333, 1913.

Monod, J., The growth of bacterial cultures. *Ann. Rev. Microbiol.*, 3, 371, 1949.

Mori, S. and Wen, C.Y., Estimation of bubble diameter in gaseous fluidized beds. *AIChE J.*, 21, 109, 1975.

Straguzzi, G.I., Bischoff, K.B., Koch, T.A., and Schuit, G.C.A., Selective oxidation catalysts containing antimony for the conversion of 1-butene to butadiene I. Preparation and characterization. *J. Catal.*, 103, 357, 1987a.

Straguzzi, G.I., Bischoff, K.B., Koch, T.A., and Schuit, G.C.A., Selective oxidation catalysts containing antimony for the conversion of 1-butene to butadiene II. Selective oxidation of 1-butene and CO. *J. Catal.*, 104, 47, 1987b.

Tafel, J., Uber die polarisation bei kathodischer wasserstoffentwicklung. *Z. Physik. Chem.*, 50A, 641, 1905.

Whitman, W.G., The two film theory of absorption. *Chem. Metallurg. Eng.*, 29, 147, 1923.

Appendix A: The International System of Units

Physical Quantity	Unit	Symbol
Mass	Kilogram	kg
Length	Meter	m
Time	Second	s
Temperature	Kelvin	K
Electric current	Ampere	A
Amount of substance	Mole	mol

Appendix B: Commonly Used Units for Some Physical Quantities with Conversion Factors

Physical Quantity	SI Unit	Other Common Units with Symbols in Brackets
Mass	Kilogram	Gram (g) = 10^{-3} kg
		Milligram (mg) = 10^{-6} kg
		Metric ton (MT) = 10^{3} kg
Length	Meter	Centimeter (cm) = 10^{-2} m
		Millimeter (mm) = 10^{-3} m
		Micrometer (μm) = 10^{-6} m
		Nanometer (nm) = 10^{-9} m
Time	Second	Minute (min) = 60 s
		Hour (h) = 3600 s
		Day (d) = 86,400 s
Temperature	Kelvin	Celsius (°C) = K – 273
Electric current	Ampere	Milliampere (mA) = 10^{-3} A
		Microampere (μA) = 10^{-6} A
Electric potential	Volt (V)	Millivolt (mV) = 10^{-3} V
Electrical resistance	Ohm	Milliohm (mohm) = 10^{-3} Ω
		Kiloohm (kohm) = 10^{3} Ω
Amount of substance	Mole	Kilomole (kmol) = 10^{3} mol
		Millimole (mmol) = 10^{-3} mol
Area	Square meter (m^2)	Square centimeter (cm^2) = 10^{-4} m^2
		Square millimeter (mm^2) = 10^{-6} m^2
Volume	Cubic meter (m^3)	Cubic centimeter (cm^3) = 10^{-6} m^3
		Cubic millimeter (mm^3) = 10^{-9} m^3
		Liter (L) = 10^{-3} m^3
		Milliliter (mL) = 10^{-3} L = cm^3
Density	Kilogram per cubic meter	$g/cm^3 = 10^{-3}$ kg/m^3
Viscosity	Poiseuille (Pl)	kg/(m · s) = Pl
		Poise = 10^{-1} Pl
		g/(cm · s) = poise = 10^{-1} Pl
		Centipoise (cp) = 10^{-3} Pl

(Continued)

Physical Quantity	SI Unit	Other Common Units with Symbols in Brackets
Pressure	Pascal (Pa)	N/m^2 = Pa
		Kilopascal (kPa) = 10^3 Pa
		atm = 101,325 Pa
		bar = 10^5 Pa
		mm Hg = 133.32 Pa
		Dyne/cm^2 = 10^{-1} Pa
Energy, heat	Joule (J)	Kilojoule (kJ) = 10^3 J
		Calorie (cal) = 4.18 J
		Kilocalorie (kcal) = 10^3 cal = 4180 J

Appendix C: Some Important Physical Constants

Constant	Value
Universal gas constant	8.314 J/(mol · K)
	8.314 (Pa · m^3)/(mol · K)
	0.082 (L · atm)/(mol · K)
	1.987 cal/(mol · K)
Faraday	96,487 C/gequiv
Avogadro's number	6.022×10^{23} mol^{-1}
Molar volume of an ideal gas at 273.15 K and 1 atm pressure	2.241×10^{-2} m^3/mol
Standard atm	101,325 Pa

Index

D

E

L

M

N

O

P

T

U

V

W